Green Empowerment and High Quality Development

China Council for International Cooperation on
Environment and Development (CCICED)
Secretariat

Green Empowerment and High Quality Development

CCICED Annual Policy Report 2023

CEPG CCICED Springer

China Council for International
Cooperation on Environment and
Development (CCICED) Secretariat
Beijing, China

ISBN 978-981-96-4220-5 ISBN 978-981-96-4218-2 (eBook)
https://doi.org/10.1007/978-981-96-4218-2

Jointly published with China Environment Publishing Group Co., Ltd.
The print edition is not for sale in China (Mainland). Customers from China (Mainland) please order the print book from: China Environment Publishing Group Co., Ltd.

This work was supported by China Environment Publishing Group Co., Ltd.

This Springer imprint is published by the registered company Springer Nature Singapore Pte Ltd.
The registered company address is: 152 Beach Road, #21-01/04 Gateway East, Singapore 189721, Singapore

Preface

Promoting Global Cooperation on Environment and Development

Modernization for the Harmonious Coexistence of Human Beings and Nature[1]

Man and nature are the community of life, and adhering to the harmonious coexistence of man and nature is an inherent requirement for sustainable development. The 20th National Congress of the Communist Party of China held in October 2022 outlined and elaborated on the theory of Chinese path to modernization and clarified the Chinese characteristics of the five aspects of the Chinese path to modernization, of which modernization of man and nature in harmony with each other is one of them. On July 17–18, 2023, the Chinese government once again convened the National Conference on Ecological Environmental Protection in Beijing after a lapse of five years, at which President Xi Jinping attended and delivered an important speech, making strategic arrangements for comprehensively promoting the Beautiful China initiative and accelerating the modernization of human beings in harmony with nature, which provided directional guidance and fundamental guidelines for further strengthening ecological environmental protection and promoting the construction of an ecological civilization.

Since the 18th National Congress of the Communist Party of China, China has placed the construction of ecological civilization in a prominent position in the governance of the country, and ecological environmental protection has undergone historic, transformative, and overall changes from theory to practice, and the Beautiful China initiative has taken significant steps forward.

[1] This preface is an excerpt from the speech of Huang Runqiu, Minister of Ecology and Environment, at the opening ceremony of the 2023 Annual Meeting of the China Council for International Cooperation on Environment and Development (CCICED) on August 28, 2023.

We use the concept of ecological civilization to guide development. The concept of ecological civilization and the construction of ecological civilization have been written into the *Constitution of the People's Republic of China* and incorporated into the overall approach of socialism with Chinese characteristics. Guided by Xi Jinping's thought on ecological civilization, we have implemented the new development philosophy, implemented the requirements of green development into the entire process of economic and social development, promoted the formation of spatial patterns, industrial structures, modes of production and lifestyles that conserve resources and protect the environment, and unswervingly taken the path of civilized development that is characterized by increasing production, higher living standards, and healthy ecosystems. We will adhere to the concept that lucid waters and lush mountains are invaluable assets, persist with a holistic and systematic approach to conserving and improving mountain, water, forest, farmland, grassland, and desert, and strengthen ecological environmental protection in an all-encompassing, territorial, and holistic manner.

We have implemented strong initiatives to take action. With unprecedented determination and strength, we are focusing on keeping our skies blue, waters clear, and lands clean, resolutely investigating and dealing with major typical cases of damage to the ecological environment, and resolving a number of prominent ecological and environmental problems that have been strongly reflected by the people. From 2013 to 2022, China's average concentration of fine particulate matter (PM2.5) has dropped by 57 percent and the quality of the ecological environment has significantly improved, despite the doubling of GDP. It has strengthened the protection of biodiversity, innovated and implemented the ecological protection red line system, built a nature reserves system centering on national parks, completed a cumulative total of more than 80 million mu of the holistic approach to protecting and restoring mountains, rivers, forests, farmlands, lakes, and grasslands, and has reforested a cumulative total of 1.02 billion mu in large-scale national land greening operations, with the area of planted forests ranking steadily at the top in the world. Carbon peak and carbon neutrality have been incorporated into the overall layout of ecological civilization construction, forming a "1+N" policy framework for carbon peak and carbon neutrality, accelerating the green and low-carbon transformation of energy sources, and vigorously developing renewable energy sources. The total installed capacity of renewable energy power generation has exceeded 1.3 billion kilowatts, and the total installed capacity has exceeded that of coal power. The number of new energy vehicles has reached 16.2 million, with more than half of the world's new energy vehicles traveling in China.

We actively participate in global environmental governance. Adhering to the concept of a community with a shared future for mankind, we have firmly practiced multilateralism and endeavored to promote the construction of a fair, reasonable and win-win global environmental governance system. We have promoted the conclusion, signing, entry into force, and implementation of *the Paris Agreement*, announced the "3060" goal of carbon peaking and carbon neutrality, announced that no new coal power projects would be built outside China, and launched and stabilized the operation of the world's largest carbon market. As the presidency, we promoted the successful outcome of the 15th Conference of the Parties (COP15) to the United Nations *Convention on Biological Diversity (CBD)* and adopted the historic Kunming-Montreal Global Biodiversity Framework, which opened a new chapter in global biodiversity governance. We have pushed forward the construction of the green development of Belt and Road and established BRI International Green Development Coalition, which has more than 170 partners from 43 countries. It has carried out South-South cooperation in addressing climate change, signed 46 cooperation documents with 39 developing countries, and helped them improve their capacity to address climate change through cooperation in building low-carbon demonstration zones and implementing material assistance projects.

While seeing the achievements, we are also keenly aware that, as a developing country with a huge population, China's ecological environmental protection work is uniquely arduous and complex. At present, the structural, root, and trend pressures on ecological environmental protection in China have not yet been alleviated at all, and the construction of an ecological civilization is still in a critical period of overlapping pressures and heavy loads.

We will be guided by Xi Jinping's thought on ecological civilization, stand in the height of the harmonious coexistence of man and nature to plan for development, integrate industrial restructuring, pollution control, ecological protection, and respond to climate change, and synergistically promote the reduction of carbon emissions, pollution reduction, expansion of green, growth, high-quality ecological environment to support high-quality development, and promote the development of the quality of the low-carbon transformation to achieve an effective enhancement of the quality of the quantity and reasonable growth. First, adhere to the realization of pollution reduction, carbon reduction, and synergistic efficiency as the general grasp; accelerate the promotion of a green and low-carbon economy and society. The second is to adhere to precise, scientific, and lawful pollution control and continue the nationwide battle to prevent and control pollution. Thirdly, persisting with a holistic and systematic approach to conserving and improving mountain, water, forest, farmland, grassland, and desert, and endeavors to enhance the diversity, stability, and continuity of ecosystems. Fourth, adhere to worst-case scenario thinking, strengthen the environmental risk warning, prevention and control and emergency response to emergencies, and build a solid ecological security barrier. Fifth, adhere to reform and innovation and strive to build a modern environmental governance system. Sixth, we will adhere to win-win cooperation, deeply participate in global environment and climate governance, and continue to contribute Chinese wisdom, Chinese approach, and Chinese power to global sustainable development.

The past year was the launching year of the seventh session of CCICED. Over the past year, overcoming the adverse impacts of the epidemic and international geopolitics, the CCICED has conducted joint Chinese and foreign researches on topics such as carbon peaking and carbon neutrality, collaborative promotion for carbon reduction, pollution reduction, green expansion and growth, and green development of the Belt and Road and have produced more than 30 policy research reports. CCICED has submitted annual policy recommendations to the State Council of China under the title of "Maintaining the Strategic Dual-Carbon Determination and Exploring the

Innovative Path of Multi-objective Synergy: Accelerating Green and Low-Carbon High Quality Development," organized more than 30 online and offline conferences and activities, and held 36 bilateral talks with partners, which have served the cause of China's environment and development, and contributed to China's solutions for global sustainable development.

Looking ahead, CCICED should be based on the needs of China's new era of ecological civilization and the United Nations 2030 Agenda for Sustainable Development and take high-level joint research between China and foreign countries as a means to serve China's modernization of accelerating the harmonious coexistence of human beings and nature, as well as the global green, inclusive, and high-quality development. Here, I would like to put forward three suggestions on the work of the CCICED.

First, CCICED should make new efforts to promote the Beautiful China initiative. Focusing on major topics such as actively and steadily promoting the "dual-carbon" work, accelerating the topics on transformation of the development mode, and deeply promoting ecological and environmental governance, should carry out fundamental, forward-looking, and innovative research to provide decision-making reference and intellectual support for China to synergistically push forward high-quality development and high-level protection, accelerate the comprehensive green transformation of the economy and society, and push forward the sustained improvement of the ecological and environmental.

Second, CCICED should achieve new results in promoting global environmental governance. Utilizing the unique advantage of CCICED in connecting China and foreign countries should conduct in-depth dialogues and seminars on global environmental crises such as climate change, loss of biodiversity, acidification of the oceans and plastic pollution, promote policy exchanges and practical cooperation between China and foreign countries, and work together to build a cleaner and more beautiful world.

Thirdly, new progress should be made in promoting the 2030 Agenda for Sustainable Development. From the perspective of the development of human civilization, CCICED should focus on the serious challenges to global sustainable development, such as imbalanced development, geopolitical conflicts, fair transition, and reverse globalization, study and propose comprehensive solutions and achievable paths, and promote dialogue, enhance trust and reduce misgivings and bridge differences among all parties, so that they can work together to achieve the goals of sustainable development.

As China's longest-established, highest-level, most fruitful, and most influential mechanism for high-level dialogue and cooperation between China and foreign countries on environment and development, CCICED has played a unique and important role in China's ecological civilization building and sustainable development process. I hope that you will continue to speak freely and exchange views with each other at this annual meeting and make greater contributions to the cause of promoting China's environment and development and building a clean and beautiful world!

Beijing, China Runqiu Huang

Acknowledgements

Chapter 1 Xie Zhenhua, Liu Shijin, Scott Vaughan, Arthur Hanson, Knut Alfsen, Zhou Guomei, Li Gao, Li Yonghong, Kate Hampton, Zou Ji, Wang Yi, Lei Hongpeng, Liu Qiang, Zhao Xiao, Qi yue, Xin Jianan, Gu Baihe, Han Wei, Meng Qi, Wang Chen, Li Jiaxin, Cai Jingjing, Zhang Xiaohan, Zhao Wenbo, Peng Linan, Zhang Huiyong, Liu Kan and Tang Huaqing.

Chapter 2 Dai Minhan, Jan-Gunnar Winther, Su Jilan, Liu Hui, Birgit Njåstad, Wang Juying, Cao Ling, Kate Bonzon, Yang Songying, Li Daoji, Tang Zhi, Cheng Danyang, Guan Dabo, Huang ShuoLin, Kristin Kleisner, Li Jianghui, Shen Wei, Zhang Weiwei, Sun Fang, Li Wei, Li Yanting, Kristian Teleki, Xie Xi, Liu Chunyu, Jiang Ziyu, Jin Shuaichen, Sun Chuanwang.

Chapter 3 Scott Vaughan, Gabrielle Dreyfus, Richard "Tad" Ferris, Christian Mielke, Danting Fan, Dimitri de Boer, Dominic Waughray, Huw Slater, Jiuhong Qi, Jun Ma, Junling Zhuo, Knut H. Alfsen, Kan Liu, Kang Sun, Liping Li, Maosheng Duan, Michael Zimonyi, Tara Sharma, Wei Li, Xiaolu Zhao, Ze Ran, Huaqing Tang, Meizhen Wang.

Chapter 4 He Kebin, Zhang Yongsheng, Zhang Hongjun, Lei Yu, Bian Shaoqing, Lu Xi, Ma Xiaoqin, Tong Dan, Zhang Qiang, Zhang Shaojun, Bo Yu, Feng Yueyi, Geng Guannan, He Dongquan, Sun Shida, Wang Yunshi, Yan Xizhe, Yu Xiang, Zhao Pei.

Chapter 5 Li Xiaojiang, Hu Jingjing, Zhang Yongbo, Xiao Yingguang, Liu Kunyi, Lu Xiaobei, Li Hao, Qin Yi, Du Xiaojuan, Peng Li, Lei Xia, Wu Kai, Yu Miao, Pan Jie, Liang Ce, Zhao Zhiqi, Wu Ke, Yin Jun, Zhao Xiang, Wu Chunfei, Zhou Pengfei, Pan Xiaodong, Qi Yuyao, Zhang Zunhao, Bai Jing, Li Linqing, Zhu Naixuan, Jiang Guoxiang. Fernando Miralles-Wilhelm, Hans Mommaas, Bob Tansey, Tjitte Nauta, Yang Bo, Kees Bons, Anna Koster, Wilfried ten Brinke, Xu Xin.

Chapter 6 Zhang Yongsheng, Zhu Chunquan, Ma Xiaoqin, Hu Yue, Xu Zhengxue, Jia Kejing, Ouyang Zhiyun, Kong Lingqiao, Xu Yinlong, Zhao Mingyue, Meng Han, Yao Lin, Kang Nannan, Yu Xiang, Zhang Zhuoqun, Dong Yaning, Tanja Ploetz, Eva Sternfeld, Ursula Becker, Linda Wallbott, Niels Theves, Dai Min, Jan-Hendrik Eisenbarth, Zhou Zhou, Zhang Li, Ling Songting, Qiao Hui, Zhang Bowen, Jan Bakkes, Karel van Bommel, Tamas Hajba, Liu Guoliang, Gim Huay Neo, Akanksha

Khatri, Scott Vaughan, Liu Shijin, Zhou Guomei Zhang Yujun, Li Yonghong, Zhang Huiyong, Liu Kan.

Chapter 7 Ke Gong, Dirk Messner, Gim Huay Neo, Antonia Gawel, Shijin Liu, Scott Vaughan, Lei Pei, Qiuping Li, Anna Rosenbaum, Gang Liu, Chaofeng Shao, Jie Liu, Xinwei Zhang, Mingxi Liao, Stephan Ramesohl, Felix Creutzig, Zhiguang Shan, Anna Zagorski, Guoyong Liang, Eric White, Daniel Hausmann, Xing Zhao, Chun Luo, Mattias Höjer, Marcel Dorsch, Hao Xu, Lisa Wee, Kitty Xia, Dabo Guan, Daoping Wang, Guanyi Wang, Ran Jing, Lizhi Zhao, Jan-Hendrik Eisenbarth, Dai Min, Niels Thevs, Markus Wypior, Maya Ben Dror, Na Na, Helen Burdett, Ting Peng, Tony Wu, Zheqiong Lei, Mario Canales, Marcel Dickow, Hans Baumgarten, Yufei Fang, Liuchuan Tong, Kristine St-Pierre, Rui Zhong, Huiyong Zhang, Kan Liu, Xiaoran Hao.

Chapter 8 Craig Hanson, Miaojie Yu, Anne Rosenbarger, Caroline Winchester, Xiaotian Fu, Liqing Peng, Tina Schneider, Rod Taylor, Wei Tian, Xinyu Chen, Zhuoyu Chen, Qishan Hu, Shijin Liu, Scott Vaughan, Huiyong Zhang, Kan Liu, Quan Mu, Brice Li, Samantha Zhang, Isaak Bowers, Yuyan Zhang, Hongjun Yu, Yishan Pan, Haobo Zhou, Shenggen Fan, Haiwei Jiang, Zhiyuan Li, Xiangjun Ma, Morgan Gillespy, Moazzam Malik, Tim Searchinger, Wei Zhao, Xin Yu, Jin Xu, Anne-Marie Belley, Bo Li, Sarah Stettner, Xiaoyu Fan, Yanping Zhang, Jian Wan, Haizi Hu, Jiaxi Meng, Ran Wei, Jinghan Xu, Yu Ji, Keyu Zhao, Shuang Wu, Jingyu Zhou.

Chapter 9 Liu Shijin, Scott Vaughan, Peng Wensheng, Manish Bapna, Jan Erik Saugestad, Zhang Jieqing, Li Yating, Wu Qi, Cheng Yuxin, Emine Isciel, Chen Chao, Zhang Junjie, Wang Yao, Cao Li, Guo Peiyuan, Liu Chunfa, Liu Junwei, Qi Xing, Li Yonghong, Zhang Huiyong, Liu Kan, Wang Ran, Samantha Zhang, Sarah Doughty, Alfonso Pating.

Chapter 10 Guo Jing, Kevin Gallagher, Erik Solheim, Zhang Yujun, Christie Ulman, Lei Hongpeng, Zou Ji, Zhang Jianyu, Zhang Jieqing, Dimitri De Boer, Guo Shenyu, Li Zhong, Xie Fei, Cecilia Han Springer, Niccolò Manych, Ishana Ratan, Tsitsi Musasike, Timothy Afful-Koomson, Kuda Ndhlukula, René Gomez-Garcia, Maria E. Netto, Paulo Esteves, Amar Bhattacharya, Jari Vayrynen, Farid Ahmed Khan, Diah Asri, Christoph Nedophil, Régis Marodon, Stephany Griffith Jones, Chris Humphre, Joe Thwaites, Bernice von Bronkhorst, Jeffrey Sachs, Jennifer Turner, Yunnan Chen, Liu Shuang, Rogèrio Studart, Samantha Attridge, Wei Shen, Daniel Kammen, Joanna I. Lewis, Lan Yan, Zhu Lin, Zhang Zhiqiang, Li Yan, Zhang Mengyan, Tao Ye, Li Dongya, Pang Xiao, Zhang Min, Yu Xiaolong, Ge Shaotong, Amilia Lee, Haishan Zhao.

Contents

Part II National Green Governance System

Part I
Innovation in Global Environmental Governance

Chapter 1
Pathways for Achieving Carbon Neutrality and China's Role in Global Climate Governance

1 Foreword

Since China proposed the "dual-carbon" targets, it has steadily pushed it forward, making climate actions a national strategy and incorporating it into the overall layout of the construction of an ecological civilization and the overall economic and social development set-up. In the meanwhile, China has actively participated in international climate change affairs, and is becoming an important participant, contributor and leader in the construction of a global ecological civilization.

However, carbon neutrality is a systematic project. At present, China's green and low-carbon transition still faces a series of challenges. On the one hand, the global journey to address climate change will remain long and challenging. On the other hand, the domestic short and medium-term structural transformation still faces a series of challenges in terms of just transition, technological innovation and policy mechanisms. Therefore, how to further improve China's "dual-carbon" policy system, how to scientifically respond to changes in the domestic and international situation, how to deepen policy actions in order to play China's key role in the climate change field, and how to promote international cooperation on climate change are the key issues that need to be studied at present.

Green Empowerment and High Quality Development,
https://doi.org/10.1007/978-981-96-4218-2_1

2 Judgment on the Domestic and International Situation of Green and Low-Carbon Transition

2.1 The Judgment on International Situation

2.1.1 The Current Global Situation Is Highly Uncertain, and the Deficit in Global Climate Governance Will Become More Prominent

In recent years, the international security situation has undergone profound and complex changes where great power competition and geopolitical competition are heating up. Under this circumstance, many have predicted that the global economy will grow negatively, and the deficit in climate governance will become prominent. The World Bank has significantly lowered the forecast of global economic growth rate for 2023 from 3.0 to 1.7% (see Fig. 1). Furthermore, in the report *World Economic Situation and Prospects in 2023* released by the United Nations, the expected world's economic growth rate in 2023 was reduced to 1.9%. The macro policies of the world's major developed economies are facing a dilemma of maintaining economic growth and restraining inflation [1].

Meanwhile, destabilizing factors in global climate governance still exist. The international community meet absence of strong leadership in Global climate change. The climate financing became insufficient, while the COVID-19 pandemic continues to have severe impacts globally, especially on developing countries.

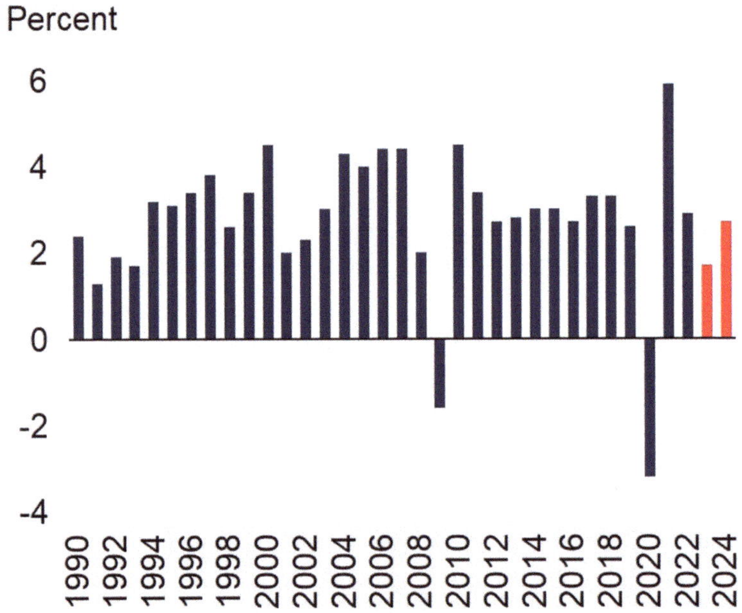

Fig. 1 The trend of global economy growth. *Source* World Bank

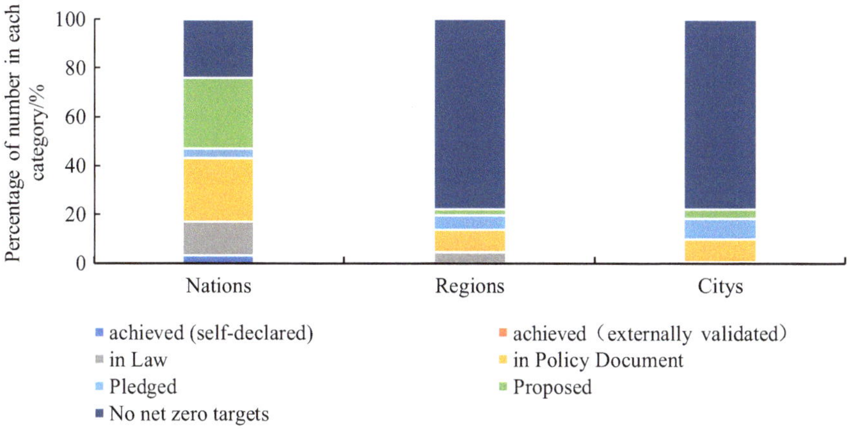

Fig. 2 Net zero target dates of Countries, Regions, and cities that are preparing to or have proposed carbon neutrality targets. *Source* NET-ZERO TRACKER

2.1.2 The Long-Term Trend of Global Energy Transition Is Clear, While the Short-Term Energy Crisis Shall Not Be Overlooked

The long-term trend of current global energy low-carbon transition is clear, and countries have proposed goals for carbon neutrality. So far, more than 130 countries and regions are preparing to or have proposed goals for carbon neutrality (see Fig. 2). In general, countries' strong objective to achieve carbon neutrality is clear, despite geopolitical conflict and reactivated coal power.

In the short term, energy crisis is still the focus of all countries. Europe has raised the energy security to a strategic position. Renewable energy production capacity is still insufficient whereas energy demand rises, and so there is still a certain degree of dependence on fossil energy [2], creating a short-term imbalance in energy supply and demand during energy transition process.

2.1.3 The Global Green Industrial Revolution Driven by Climate Change May Reshape the Competition and Cooperation System in the Fields of Global Trade, Technology, and Finance

Currently, a new round of global technological and energy revolution is accelerating, and great powers are accelerating the deployment of renewable energy as a new driving force for green growth. Global installed power capacity by renewable energy has reached 3372 GW as of the end of 2022 with a record of 9.6% (295 GW) increase in renewable energy stock [3]. The installed capacity of wind and solar power is increased by 15 times in terms of the accelerated and net zero, and increased by 9 times in terms of new momentum [4] (see Fig. 3). The unit cost of some low-emission technologies, such as solar power, wind power, and lithium-ion batteries, has continued to decline since 2010.The latest report from IPCC has shown that the costs of solar photovoltaic and onshore wind power generation are already far below fossil fuel power generation (see Fig. 4). In the future, the cost of new energy will be further reduced, and it will gradually replace fossil energy as the main force of the energy system and will play an important role in economic growth.

However, in the background of difficult recovery and deep adjustment of the global economy, many economies have introduced various trade restrictions and investment protection measures. There has been an obvious trend of localization and regionalization of industrial chains and supply chains, and the contradictions and problems in the supply of energy and bulk commodities are becoming more and more prominent, which seriously threatened the recovery and sustainable growth of the global economy [5]. On August 16, 2022, President Biden signed the *Inflation*

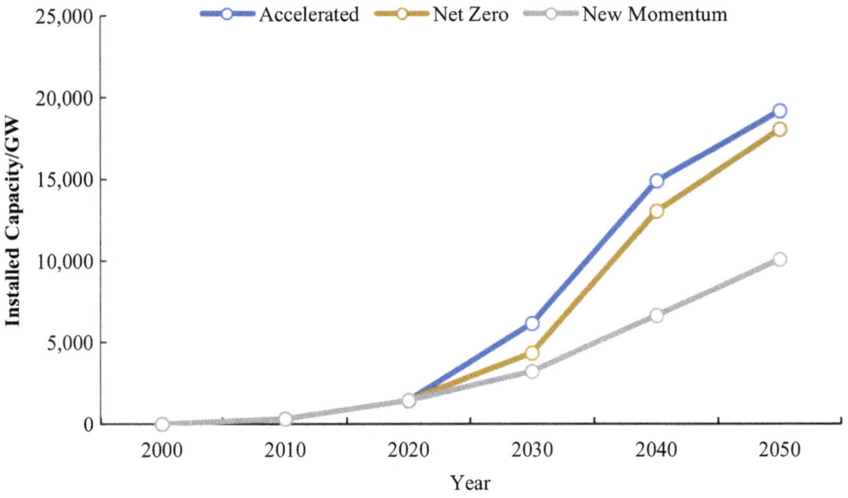

Fig. 3 Global installed capacity of wind and solar power. *Source* BP

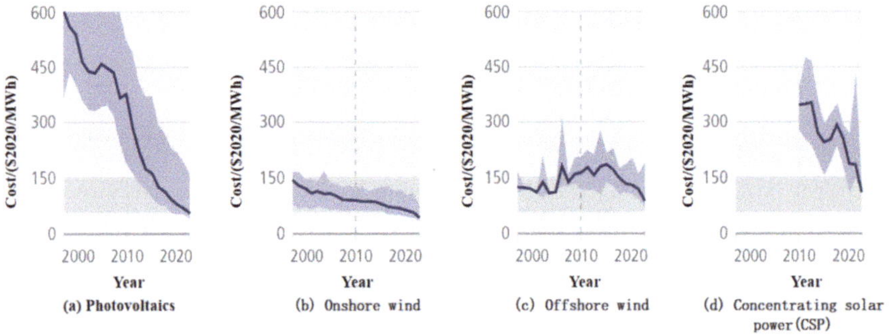

Fig. 4 Global trend of reduced costs of renewable energy. *Source* IPCC AR6

Reduction Act of 2022 (IRA) passed by both houses of Congress, which imposed restrictions on some supply chains, for example, automobiles must be assembled in North America, otherwise subsidies cannot be obtained; a certain proportion of key minerals must be mined or processed in the countries of North American Free Trade Agreement or have to be recycled in North America; a certain proportion of battery components must be manufactured in North America. The act caused many manufacturing companies to reinvest in the United States. In addition, the European Union established the Carbon Border Adjustment Mechanism (CBAM) and proposed to completely cancel free carbon emission quotas for related industries in the EU by 2035; meanwhile the European Commission proposed the *European Green Deal Industrial Plan*. The essence of these policies and actions is to protect domestic industries and accelerate the construction of their own green industry system through a great number of subsidies and incentives, which is a competition around the clean energy industry and green technology. Healthy competition will promote the accelerated deployment of new energy industries and technologies, but excessive trade barriers will increase the prices of new energy products, which is not conducive to the advancement of clean energy transition.

2.1.4 Southern Countries and Emerging Economies Are Actively Accelerating Energy Transition, but They Still Face a Series of Risks and Challenges

With the continuous innovation and popularization of renewable energy technologies, as well as the constant decline in costs, southern countries and some emerging economies have gradually begun to pay attention to the deployment and planning of renewable energy development (see Attachment Table 1). All countries are actively taking actions to accelerate the process of global energy green and low-carbon transition by formulating relevant coal reduction and renewable energy development goals. In 2020, China proposed the vision to reach carbon dioxide peak before 2030 and carbon neutrality before 2060, and basically formed the "1 + N" policy system for carbon dioxide peaking and carbon neutrality, mobilizing multiple subjects to participate extensively. India set targets to reduce greenhouse gas emission intensity by 50% from the level in 2005 and increase the share of non-fossil energy sources in the power sector to 45% by 2020. Brazil has proposed a 37% reduction of GHG emissions by 2025 and a 43% reduction by 2030 relative to the level in 2005. In the *ASEAN Plan Of Action For Energy Cooperation (APAEC) 2016–2025 Phase I: 2016–2020*, ASEAN set an overall regional target to increase the component of renewable energy to 23% by 2025 in the ASEAN energy mix, and the member countries set national targets accordingly.

However, these countries still have certain risks and challenges in accelerating energy transition. Firstly, due to the sluggish green recovery of the global economy after the pandemic, the interrupted industrial chains and supply chains has not yet fully restored, and the willingness of developed countries to invest in climate has weakened, so there is a large gap in climate funds. For example, Vietnam has planned and approved a number of renewable energy projects, but due to insufficient funds, the

Table 1 Global Top 10 countries with average annual net increase of forest area from 2010 to 2020

Ranking	Country	Net annual change	
		1000 ha/year	Change rate (%)
1	China	1937	0.93
2	Australia	446	0.34
3	India	266	0.38
4	Chile	149	0.85
5	Vietnam	126	0.90
6	Turkey	114	0.53
7	The U.S	108	0.03
8	France	83	0.50
9	Italy	54	0.58
10	Romania	41	0.62

Note The rate of change (%) is calculated as a compound annual rate of change
Source Global Forest Resources Assessment 2020[1]

conversion rate is very low. Secondly, some countries are highly dependent on coal power, and the accelerated transition of coal power will further increase the mismatch between energy supply and demand in these countries, which would trigger an energy security crisis. Moreover, the stranded capital cost from withdrawal of fossil energy is relatively high, which can easily lead to collapse of asset prices or huge debt defaults. The withdrawal of fossil energy may also cause a significant impact on traditional enterprises and employees which would lead to social turmoil. Thirdly, at the technical level, a large number of southern countries are insufficient in the innovation of renewable energy technologies, which restricts its development. Currently, the power grid infrastructure in southern countries and emerging economies is weak, and the power interconnection among countries is limited, which brings great difficulties to the integration of renewable energy into the grid. Failure to promote green and low-carbon transition in an orderly and smooth manner may lead to energy security risks.

2.1.5 Under the New Situation of Multiple Crises, the Urgency of International Climate Cooperation Has Further Increased

The current world political and economic landscape is full of uncertainties, and countries are facing multiple challenges including geopolitical conflicts, industrial and supply chain security, inflationary pressures, and energy crises. Under this circumstance, climate governance has become one of the important areas for international cooperation. According to the research by the United Nations Environment Programme, world faces a 2.8-degree warmer future by 2100 based on the current

[1] https://www.fao.org/forest-resources-assessment/past-assessments/fra-2020/zh/.

global climate policies. The urgency of human beings to deal with climate change is increasing day by day. To achieve the goals of the Paris Agreement, developed countries must take the lead in increasing emission reduction and take concrete actions to implement them. Developing countries have made a lot of efforts to deal with climate change, but limited by the technology and financial capabilities, they have not been able to make great progress; developing countries need financial and technical support from developed countries. Tackling climate crisis is a collective action that requires countries to actively implement corresponding measures. Strengthening solidarity and cooperation is the only way to tackle the challenge of climate change. But we must continue to uphold the principle of common but differentiated responsibilities, which is related to international fairness and justice. Deviation from this principle will seriously damage the solidarity and cooperation of the international community in tackling climate change.

Strengthening global climate governance requires the concerted efforts of the international community. It is necessary to further highlight the core position of climate issues in the international agenda, continue to promote bilateral and multilateral climate dialogues and cooperation, rebuild mutual trust among all parties, take concrete actions to deal with climate change, and promote comprehensive green transition and global cooperation for carbon neutrality. Meanwhile, the actions of great powers are crucial that they should proceed from the common interests of all mankind, seek more consensus on cooperation, prioritize climate issues in international cooperation, and actively implement the consensus and commitments reached under bilateral and multilateral frameworks, in order to lay a solid foundation for international climate cooperation [6].

2.2 *Analysis of the Domestic Situation*

2.2.1 Under the Downward Trend of the Economy, a Great Pressure Is Put on Green and Low-Carbon Transition, and the Green Transition of the Development Mode Should be Promoted Through Steady Progress

The downward pressure on the economy is obvious, as China is still in the recovery from COVID-19 pandemic. In 2022, the national per capita consumption expenditure of residents was 24,538 yuan, a decrease of 0.2% after deducting price factors [7]. According to the estimation of IMF, in 2022, the global economic growth rate was 3.4%, 2.6% points lower than the previous year, and will further lower to 2.8% in 2023 [8].

According to the *Government Work Report* for 2023, China's GDP growth target for 2023 is still around 5%, and economic development should prioritize "stability". In recent years, China has taken industrial structure as the main direction of economic structure optimization and upgrading. In 2021, the added value of China's strategic emerging industries continued to rise as a percentage of GDP, reaching 13.4%, an increase of 1.7% points over 2020 and a cumulative increase of 5.8% points over 2014 [9]. In addition, China's investment in the field of renewable energy has also ranked first in the world for many years in a row, and it reached $98 billion in the first half of 2022, accounting for 43% of the global total. China is the global leader in the field of renewable energy investment. Its investment in large-scale solar power projects was $41 billion, an increase of 173% over 2021. The investment in new wind power projects was $57.8 billion, a year-on-year increase of 107% [10].

2.2.2 China's Development Goals Have Shifted to Synergistically Promote Climate Response, Economic Growth, Energy Security, and Ecological Governance

The synergy effect of climate change and economic growth, environmental governance, and ecological protection has been scientifically proven. For example, air pollutants and greenhouse gases are homologous, and the realization of the synergistic governance of greenhouse gases and air pollutants will not only help reduce the total cost of air pollution control and greenhouse gas emission reduction, but also help avoid high carbon lock-in effects [11]. In addition, long-term economic gains will be achieved by tackling climate change. According to the research of OECD, compared with continuing existing policies, implementing a package of climate-compatible policies can increase the medium- and long-term GDP levels of G20 countries by an average of 2.8% in 2050. When the positive effects of preventing climate change are taken into account, the net effect on GDP of G20 developed and developing economies will rise to nearly 5% in 2050 [12].

The coordinated advancement of climate change, energy, economy, environmental ecology and other aspects is an inevitable choice for China's green and high-quality development. In the current new round of technology and industrial revolution, traditional boundaries between technologies and between industries are being gradually broken, adapting to the ever-changing world, and new-generation information technologies represented by artificial intelligence, Internet of Things, quantum computing, etc. are blending with low-carbon technologies represented by new-type industries and towns, renewable energy, green buildings and transportation to form new industries and new business forms [13]. Therefore, China will pay more and more attention to the innovation and development of integrated technologies, contribute effective solutions to the increasingly complex and changeable systemic challenges involving climate change, public health, pollution control, etc., and ultimately realize the multi-objective balanced and all-round and systematic development in economy, energy, environment and other fields.

2.2.3 China Has the Motivation for Green and Low-Carbon Transition and Development, and Has the Objective Conditions to Transform Climate Actions into High-Quality Economic and Social Development

Industrial transition is an objective requirement for China's green and low-carbon development. It is estimated that by the end of the century, the GDP at risk of the Shanghai and Guangzhou metropolitan areas due to climate change-induced sea level rise (in 2019 purchasing power parity terms) could exceed \$1.6 trillion and \$291 billion per year, respectively [14]. Moreover, with the popularization of green and low-carbon education, the current people, especially young people, are gradually more advocating green and environmentally friendly lifestyle.

China's manufacturing capacity and green financial system provide a solid guarantee for green and low-carbon transition. In 2021, in the four major photovoltaic supply chain links of polysilicon materials, silicon wafers, cells and photovoltaic modules, China's production capacity accounted for more than 70% of the world (see Fig. 5). The cost of wind power and photovoltaic power has dropped significantly. By the end of 2022, China's installed power capacity of renewable energy exceeded that of coal power, accounting for 47.3% of the national total installed power capacity (see Fig. 6). On the other hand, China's financial instruments and products such as green loans and green bonds are also developing vigorously. As of the end of June 2022, the balance of green loans in domestic and foreign currencies in China reached 19.55 trillion yuan, an increase of 40.4% year-on-year; the stock of green bonds was 1.2 trillion yuan, ranking second in the world [15].

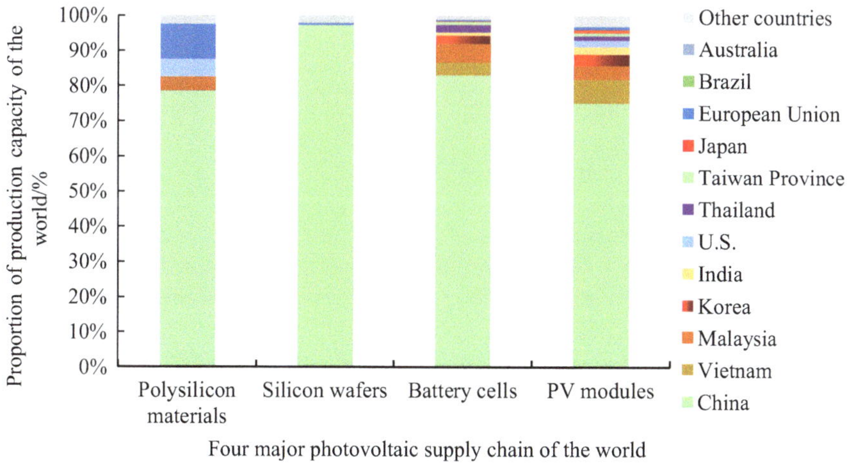

Fig. 5 Proportion of production capacity of major countries and regions in the four major photovoltaic supply chain links of polysilicon materials, silicon wafers, cells, and photovoltaic modules in 2021. *Source* CPIA, NREL, EIA, Solar Europe, IEA, BNEF

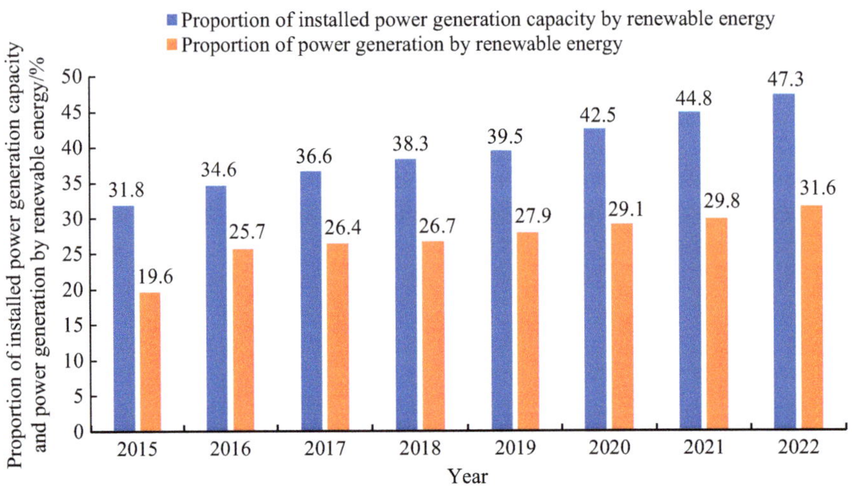

Fig. 6 Proportion of installed power generation capacity and power generation by China's renewable energy. *Source* National Energy Administration

2.2.4 The Long-Term Vision and Goal for Carbon Neutrality Are Clear, and the Short-to-Medium-Term Structural Transition Will Face Challenges in Terms of Just Transition, Technological Innovation and Policy Mechanisms

China has about 30 years from carbon dioxide peaking to carbon neutrality, compared to 60–70 years of developed countries. As a big country, the requirements for economic structural transition, technological innovation, and capital investment in the emission reduction process are also higher.

In the process of promoting structural transition, China will face problems such as cost increase and just transition. On one hand, coal is the primary energy of China, the transition of energy will inevitably lead to the handling of related stranded asset risks and structural unemployment [16]. On the other hand, some northern and western provinces are less developed in economy and more dependent on coal and heavy industry, and will face greater pressure, and the transition may expand the spatial imbalance of income and welfare [17, 18]. Technological innovation, market mechanism and other aspects will also pose related challenges in China's promotion of structural transition.

2.2.5 China Continues to Promote High-Level Opening Up, and Actively Carries Out International Cooperation to Advance Global Green Recovery and Low-Carbon Transition

Since 2016, China has launched 10 low-carbon demonstration zones, 100 climate change mitigation and adaptation projects, and 1000 climate change training quota cooperation projects in developing countries, and implemented more than 200 foreign aid projects to address climate change. Moreover, China has jointly launched the Belt and Road green development partnership initiative with 31 co-construction countries, and established the Belt and Road energy partnership with 32 co-construction countries [19].

China have jointly signed relevant climate agreement documents and joint statements with other key partners, such as the *U.S.-China Joint Glasgow Declaration on Enhancing Climate Action in the 2020s, China-U.S. Joint Statement Addressing the Climate Crisis, Joint Declaration between the People's Republic of China and the French Republic, Beijing Call for Biodiversity Conservation and Climate Change, China-EU Leaders' Statement on Climate Change and Clean Energy*, and *Joint Statement On the Implementation of the China-EU Cooperation on Energy*, laying a solid foundation for cooperation between China, the United States and Europe in the fields of energy, transportation, and construction.

China's overseas investment in the energy field is gradually becoming cleaner. According to statistics, China's investment in coal power in the Belt and Road countries is almost zero in 2021.

In 2021, China's green trade volume reached $1,161.09 billion, surpassing the European Union to become the world's largest green trading country, accounting for 14.6% of the world, an increase of 1.5% points compared to 2020 [20].

3 The Analysis on Progress of China's Carbon Dioxide Peaking and Carbon Neutrality Actions

3.1 Analysis on the Progress of China's Low-Carbon Transition

In recent years, China has placed climate change at the forefront of national governance and development. Under the guidance of the "dual-carbon" goals, China has made remarkable achievements in clean and efficient use of energy, etc.

3.1.1 The Energy System Is Developing Toward Cleanliness and High Efficiency

The clean and low-carbon transition of the energy consumption structure is accelerating, and non-fossil energy is developing rapidly. In the process of clean and low-carbon transition of the energy system, China insists on establishing before breaking down, and gradually realizes the orderly replacement of fossil energy on the basis of vigorously developing non-fossil energy. From 2013 to 2022, the proportion of China's non-fossil energy consumption increases from 10.2 to 17.5%, while the proportion of coal consumption decreases from 67.4 to 56.2% (see Fig. 7).

The development of China's renewable energy field has been acknowledged throughout the world. China has planned to build 450 million kilowatts of large-scale wind power photovoltaic bases in deserts, Gobi, and desert areas, and the construction of 100 million kilowatts projects has started [21]. By the end of 2022, China's installed power capacity by renewable energy reached 1.213 billion kilowatts, accounting for 47.3% of the country's total installed generation power capacity (see Fig. 8), surpassing the installed coal power capacity. Among them, the installed power capacity of wind power, solar power, and conventional hydropower reached 365, 393, and 368 million kilowatts respectively. In 2022, China's renewable energy power generation capacity was 2.7 trillion kWh, accounting for 31.6% of the electricity consumption of the whole society (see Fig. 9). In terms of proportions in the world, the installed power capacity and power generation of China's renewable energy accounted for 32.8 and 30.9% of the world's total in 2022, surpassing the sum of the United States and the European Union (see Fig. 10).

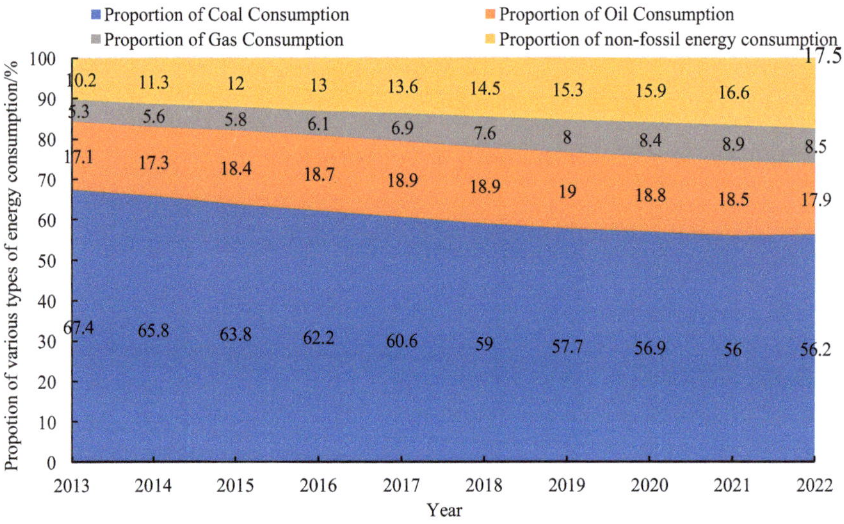

Fig. 7 Changes in China's overall energy consumption structure from 2013 to 2022. *Source* National Bureau of Statistics, National Development and Reform Commission

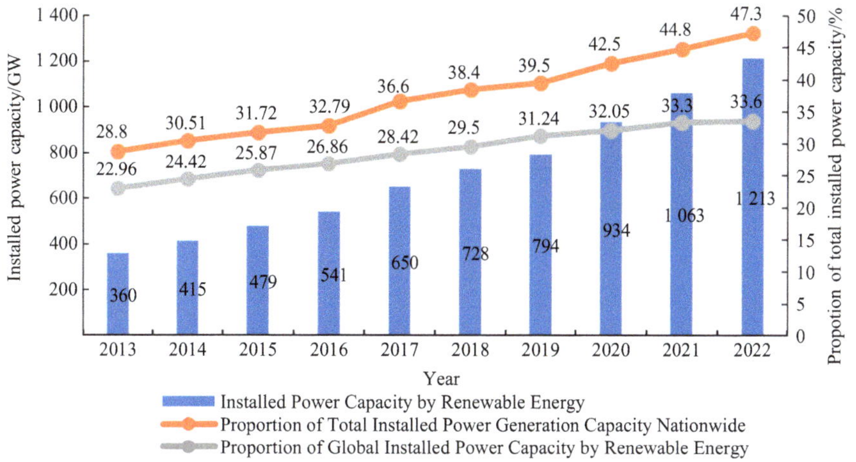

Fig. 8 Changes in installed power capacity by China's renewable energy from 2013 to 2022. *Source* National Energy Administration, China Electricity Council, IRENA, IEA

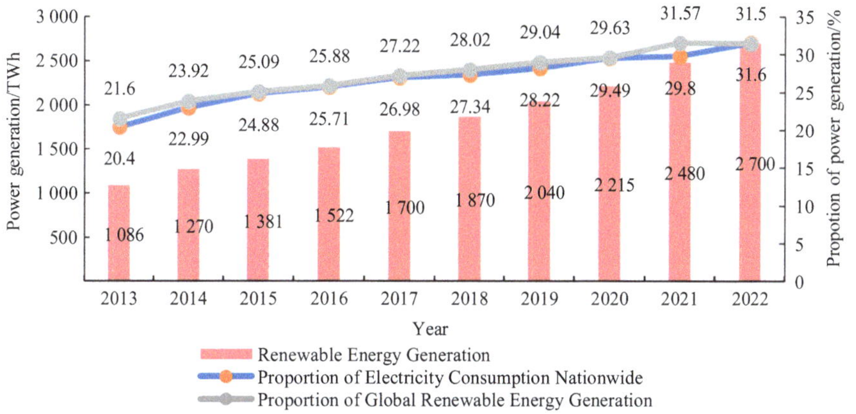

Fig. 9 Changes in power generation by China's renewable energy from 2013 to 2022. *Source* National Energy Administration, China Electricity Council, IRENA, IEA

By the end of 2021, China's large coal mines with an annual output of more than 1.2 million tons accounted for about 85% of the output [22]; bulk coal consumption was cut by about 440 million tons, a 58.7% drop from 750 million tons in 2015 [23]. By the end of 2021, China had phased out and shut down more than 100 million kilowatts of outdated coal power production capacity [24]; China had accumulatively implemented the energy saving and carbon reduction transition of nearly 900 million kilowatts of coal-fired generating units [22].

The energy consumption intensity is significantly reduced. China's energy consumption per unit of GDP in 2021 was 26.4% lower than that in 2012 [25].

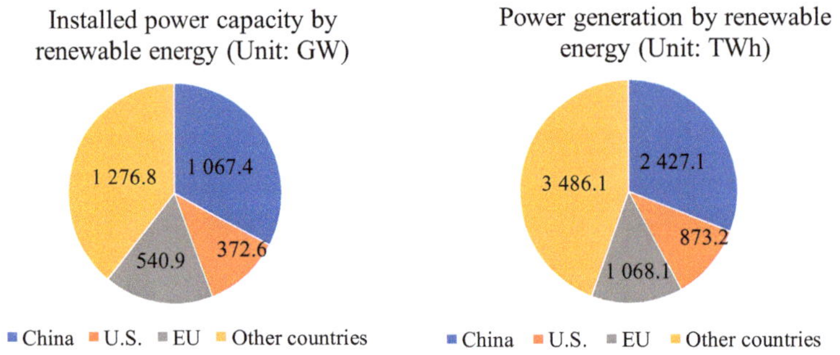

Fig. 10 Installed power capacity and power generation by renewable energy in China, U.S. and EU in 2021. *Source* IEA

Meanwhile, China is also actively developing new clean process transitions such as hydrogen energy steelmaking and green hydrogen to replace coal at the source, so as to promote the gradual green and low-carbon transition of terminal energy consumption in various industrial fields.

3.1.2 Green and Intelligent Transition of Industrial Structure

In recent years, China's industrial structure has been continuously upgraded and optimized, and the proportion of the added value of the tertiary industry in GDP has gradually increased. In 2022, the added value of China's tertiary industry was 63,869.8 billion yuan, a year-on-year increase of 2.3%, accounting for 52.8% of the country's total GDP, an increase of 5.9% points over 2013.

Green and low-carbon industries are booming. In 2022, China's new energy vehicles showed explosive growth, with production and sales of 7.058 million units and 6.887 million units, respectively, a year-on-year increase of 96.9% and 93.4%, respectively. In 2021, the output value of China's energy-saving environmental protection industry reached 8 trillion yuan, with an average annual growth rate of more than 10% [26].

Digital information technology enables green and low-carbon upgrading of traditional industries. Catalyzed by a new round of global technology and industrial revolution, emerging technologies such as big data, artificial intelligence, and 5G are booming in China. Studies have shown that by 2030, with the continuous improvement of the digitalization level of various industries, digital technology will enable China's entire society to reduce carbon by about 12–22%, and enable various industries to grow by 10–40% [27].

The green and low-carbon transition of traditional industries is accelerating. By 2020, China's carbon dioxide emissions per unit of industrial added value dropped by about 22% compared to 2015 [28]. By the end of 2022, China has accumulatively built 2783 green factories, 223 green industrial parks, 296 green supply chain enterprises, and released more than 20,000 green design products [29]. In the field of transportation, in 2021, China's freight volumes of railway and waterway ware 4.774 billion tons and 8.240 billion tons, respectively, a year-on-year increase of 4.9% and 8.2%, and their proportions in the total social freight volume increased from 7.8% and 14.1% in 2017 to 9.2% and 15.8%, respectively [30, 31]. By 2022, China has accumulatively built 5.21 million charging piles and 1973 battery swapping stations, of which 2.593 million new charging piles were built in 2022, almost the sum of previously built charging piles [32].

3.1.3 Synergy of Carbon Reduction, Economic Development, and Pollution Control

China actively promotes the coordinated governance of carbon reduction and pollution reduction, and give full play to the synergistic benefits between the two. In 2022, China introduced the *Implementation Plan for Synergizing the Reduction of Pollution and Carbon Emissions* to explore the establishment of a "source-process-end" whole-process synergy system for pollution reduction and carbon reduction. An separate CCICED study is development to address this topic.

3.1.4 The Carbon Sink Capacity of Ecosystems Such as Forests and Grasslands Has Been Significantly Enhanced

In recent years, China has implement projects for the protection and restoration of ecosystems with important ecological impacts such as shelterbelts, natural forest protection and restoration, and returning farmland to forests and grasslands to promote the continuous increase in the area of important ecosystems such as wetlands and soils. From 2000 to 2017, the global green area increased by 5%, to which China contributed about 25% [33], and the carbon sink capacity of the ecosystem has been significantly enhanced.

In terms of forest carbon sinks, in 2022, China's forest area was 231 million hectares, the forest coverage rate reached 24.02%, and the forest stock volume reached 19.493 billion cubic meters. Not only has the forest coverage rate and forest stock volume maintained "double growth" for more than 30 consecutive years, but China also has the fastest growing forest resources in the world (see Fig. 11 and Table 1). In terms of grassland carbon sinks, at present, the annual carbon sequestration capacity of China's grasslands can reach 100 million tons [34].

In terms of wetland carbon sinks, in the past ten years, China has implemented more than 3400 wetland protection projects, adding and restoring more than 800,000 ha of wetlands [35]. At present, the total aboveground carbon sequestration of herbaceous swamp vegetation in China is about 22.2 million tons, and the total soil organic carbon reserve of swamp wetlands is 9.9 billion tons [36].

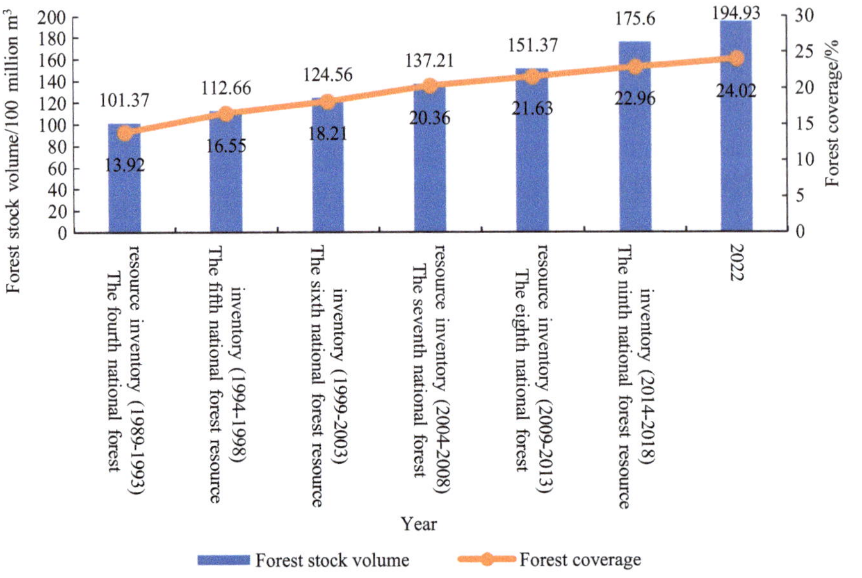

Fig. 11 Changes in forest stock volume and forest coverage in China in the past 30 years. *Source* National Forestry and Grassland Administration

3.1.5 Green and Low-Carbon Life Has Gradually Become a New Fashion

Under the guidance of the dual carbon goals, China has continued to carry out long-term themed publicity activities such as National Energy Conservation Publicity Week, National Low-Carbon Day, and World Earth Day to citizens to actively popularize climate change knowledge and strengthen green and low-carbon awareness. Meanwhile, China has also issued the *Implementation Plan for the Construction of the National Education System for Green and Low-Carbon Development*, which clearly proposes to guide young people to firmly establish the concept of green and low-carbon development, and build a national education system for green and low-carbon development with distinctive features, coherence, and rich content. In addition, China has also issued the *Ten Codes of Conduct for Citizens' Ecological Environment*.

3.2 *China's "1 + N" Policy System for Carbon Dioxide Peaking and Carbon Neutrality*

3.2.1 China Has Established a "1 + N" Policy System for Carbon Dioxide Peaking and Carbon Neutrality with Clear Goals, Reasonable Division of Labor, Effective Measures, and Orderly Connections

In October 2021, China issued the *Working Guidance For Carbon Dioxide Peaking And Carbon Neutrality In Full And Faithful Implementation Of The New Development Philosophy* (hereinafter referred to as the *Working Guidance*) and the *Action Plan for Carbon Dioxide Peaking Before 2030* (hereinafter referred to as the *Action Plan*), as the overall programmatic documents which is the "1" in the "1 + N" system, while "N" is the implementation plans for key areas and key industries and related supporting plans for key areas such as energy, industry, transportation, reduce carbon emissions and pollution, urban and rural construction, agriculture and rural areas, circular economy, pollution and carbon reduction, implementation plans for key industries such as coal, oil, gas, building materials, electrical power, steel, non-ferrous metals, petrochemicals, and building materials, as well as supporting plans for scientific and technological support, financial support, energy security, statistical accounting, inspection supervision (as shown in Fig. 12).

Meanwhile, we must also pay attention to the long-term and arduous nature of the carbon dioxide peaking and carbon neutrality work, and its path and policy need

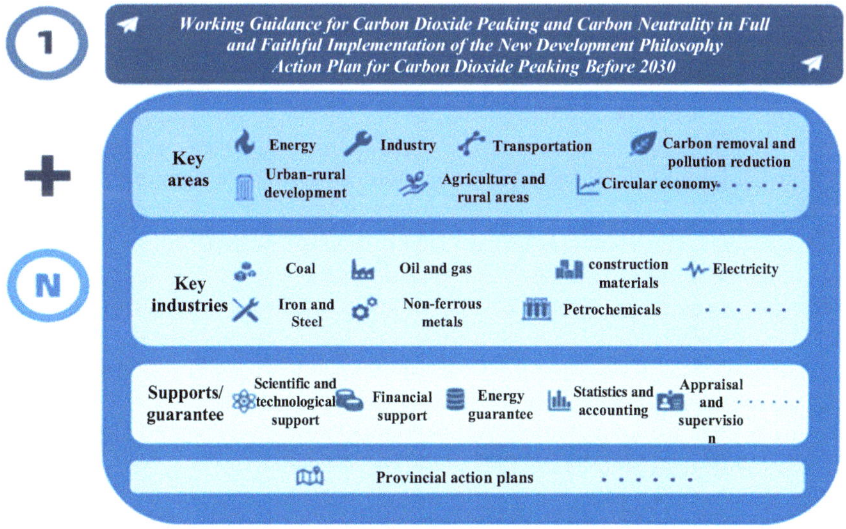

Fig. 12 Diagram of "1 + N" policy system for carbon dioxide peaking and carbon neutrality

to be constantly and dynamically adjusted according to the changes in the domestic and foreign situation.

3.2.2 The "1 + N" System for Carbon Dioxide Peaking and Carbon Neutrality Implements the Mid-to-Long-Term "Dual-Carbon Goal" Through Practical Actions in Multiple Fields

The "1 + N" policy system for carbon dioxide peaking and carbon neutrality clarifies the goals of carbon dioxide peaking before 2030 and carbon neutrality before 2060, providing a clear vision and transition signal for the whole society, and implementing the goals through actions in multiple fields.

In the field of energy, China will accelerate the planning and construction of a new energy system. The *Action Plan* pointed out that during the 14th Five-Year Plan period, the growth of coal consumption will be strictly controlled, and coal consumption will be gradually reduced during the 15th Five-Year Plan period (2026–2030).

In the industrial field, the *Development Plan for Green Industry During the 14th Five-Year Plan Period* pointed out that during the 14th Five-Year Plan period, the carbon dioxide emissions per unit of industrial added value will be reduced by 18%, and the energy consumption per unit of added value of industrial enterprises above designated size will be reduced by 13.5%.

In the field of transportation, China will accelerate the formation of green and low-carbon transportation methods. It is planned to increase the proportion of new energy and clean energy-powered vehicles to about 40% by 2030, and vigorously promote the application of electric trains and hydrogen cell vehicles.

In the field of construction, China will improve the quality of green and low-carbon development of buildings, and control the excessive growth of new construction land; improve the level of building efficiency to increase the energy efficiency of new residential buildings and public buildings in cities and towns by 30% and 20% respectively [37] by 2025, and promote the full electrification of more than 20% of buildings by 2030 [38].

In the fields of agriculture, forestry and ecological carbon sinks, studies have shown that the global greening area increased by 5% between 2000 and 2017, of which 25% came from China [33]. Forests and agricultural land contributed 42% and 32%, respectively.

In the field of circular economy, by 2025, the comprehensive utilization rate of bulk solid waste will reach 60% [39], the comprehensive utilization rate of construction waste will reach 60%, and the output value of the resource recycling industry will reach 5 trillion yuan [40]. By 2030, the resource utilization rate of urban solid waste will increase to 65% [41].

3.2.3 The "1 + N" Policy System for Carbon Dioxide Peaking and Carbon Neutrality Provides a Solid Guarantee for Low-Carbon Transition Through Systematic Institutional Construction and Capacity Building

China has introduced the *Implementation Plan for Accelerating the Establishment of a Unified and Standardized Statistical Accounting System for Carbon Emissions*, indicating that it will continuously strengthen capacity building in terms of statistical basis, accounting methods, technical means, and data quality of carbon emissions, and promote the formation of an orderly and standardized carbon emission statistics and accounting system covering the national, local, industrial, enterprise and key product levels, so as to provide comprehensive, reliable and scientific data support for China's carbon dioxide peaking and carbon neutrality work.

The *Implementation Plan for Science and Technology Support for Carbon Dioxide Peaking and Carbon Neutrality (2022–2030)* proposes the direction of China's green and low-carbon scientific and technological innovation and provide scientific support for realizing carbon dioxide peaking and carbon neutrality goals.

In terms of carbon market construction, the *Carbon Emissions Trading Management Measures (Trial)* introduced that China will clarify the entry threshold of the national unified carbon market, gradually enrich the trading varieties and trading methods in the existing carbon market pilots, and promote local pilot carbon markets to gradually transition to the national carbon market, and finally form a national unified carbon market system. In terms of electricity market construction, the *Guidelines on Accelerating to Build a Unified National Power Market System* pointed out that China should improve the multi-level and cross-regional electricity market system.

The *Work Plan for Strengthening the Construction of the Higher Education Talent Training System for Carbon Dioxide Peaking and Carbon Neutrality* pointed out that China should strengthen the forecast of talent demand in key industries related to "dual carbon" targets, and promote the upgrading and transition of carbon related majors in universities such as energy, transportation, and management.

3.2.4 The "1 + N" Policy System for Carbon Dioxide Peaking and Carbon Neutrality Plays the Role of Multiple Subjects in Multiple Fields to form a Good Pattern of Joint Efforts of the Whole Society to Reduce Emissions

At the central level, the CPC Central Committee and the State Council play a key role in overseeing and coordination. Local governments formulate action plans and implementation plans according to local conditions to effectively promote the decomposition and implementation of the "dual-carbon" goals and tasks at the regional level. At present, most provinces in China have issued action plans and implementation plans for carbon dioxide peaking and carbon neutrality.

At the enterprise level, the State-owned enterprises at national level formulate action plans for carbon dioxide peaking in the principle of "one policy for each enterprise" and play a leading role. In December 2021, the State-owned Assets Supervision and Administration Commission of the State Council issued the *Guidelines on Promoting the High-quality Development of Central Enterprises and Performing the Carbon Dioxide Peaking and Carbon Neutrality Work*, and made arrangements for national-level stated-owned enterprises to formulate and implement the action plans for carbon dioxide peaking.

At the public level, the public consciously participates in green and low-carbon activities, strengthens green and low-carbon education, and effectively promotes green and low-carbon life to become a new social trend. On October 26, 2022, the Ministry of Education issued the *Implementation Plan for the Construction of the National Education System for Green and Low-Carbon Development*, which clearly proposed to build a national education system. China is also exploring the development of an innovative voluntary emission reduction mechanism—carbon inclusiveness.

3.2.5 The "1 + N" Policy System for Carbon Peaking and Carbon Neutrality Shows That China Will Continue to Uphold the Image of a Responsible Major Country with an Open and Inclusive Attitude, and Continue to Promote the Construction of a New Pattern of Multi-party Win–Win Climate Cooperation

China will provide assistance and support within its capacity to developing countries in coping with climate change. China will continue to promote the establishment of green climate partnerships with developing countries through the Belt and Road South-South Cooperation Initiative on Climate Change, the Belt and Road Energy Partnership, and the BRI International Green Development Coalition as important intermedia. The *Guideline on Promoting Green Development under the Belt and Road Initiative* was issued which clearly pointed out 9 key cooperation areas such as green infrastructure, green energy, etc.

China will also continue to actively seek opportunities to cooperate with developed countries in Europe and America on a larger scale, in a wider field, and at a deeper level.

3.3 Brief Summary

3.3.1 China Has Made Progress in the Field of Carbon Peaking and Carbon Neutrality

To achieve carbon peaking by 2030 and carbon neutrality by 2060 are major strategic decisions made by the CPC Central Committee after careful consideration. The central government has incorporated them into the overall layout of economic construction and ecological civilization construction and the local governments have responded positively and put forward their respective schedules and roadmaps [42]. In 2022, China's energy consumption per unit GDP accumulatively decreased by 44.1% when compared with that in 2005. The proportion of the consumption of non-fossil energy in the total energy consumption was 17.5%.

3.3.2 China Needs to Further Deepen and Perfect Its Policies and Actions for Carbon Peaking and Carbon Neutrality

It is a tightly scheduled and arduous task for China to achieve the goals of carbon peaking and carbon neutrality. To complete its transition from the largest emitter in the world to a country featured by carbon neutrality within 30 years, China needs to reduce emissions faster and more vigorously than developed countries such as Europe and the United States did so as to achieve comprehensive green and low-carbon transitions of economic and social development patterns. This requires China to further deepen and improve its policies and actions for carbon peaking and carbon neutrality.

4 China's Policy Direction of Deepening Green and Low-Carbon Transition

4.1 To Take Carbon Reduction as a Guide and Promote Overall Green and Low-Carbon Transition of Whole Economy and Society

The goals of carbon peaking and carbon neutrality are inherent requirements for high-quality development. Development remains the key to solving all the problems in China facing the future and the necessity and urgency of transforming the development mode have become even more prominent. A circular economy oriented to carbon emission reduction should be developed. We should give full play to the synergistic effects of carbon reduction, pollution reduction, and green growth through green design as well as green and low-carbon full product lifecycle and supply chain and further improve the extended producer responsibility system. We should also deepen the pilot work of the program of "waste-free cities" and expand it to regional and urban agglomerations.

4.2 To Promote Green Investment, Low-Carbon Consumption, and Trading of Low-Carbon Products for the Purpose of Injecting New Impetus into Economic Growth

By 2050, the direct investments oriented to carbon neutrality can reach at least CNY 140 trillion. In consideration of related investments, the actual investment potential will be much greater than this scale. Based on the analysis of Energy Foundation China, during the 14th Five-Year Plan, the total investment potential in the fields of digital economy and digital upgrading and green transition of traditional industries, green and low-carbon urbanization and modern urban construction, green and low-carbon consumption, and construction of renewable energy, environmental-friendly energy, and power systems can reach CNY 44.6 trillion, with an annual average investment of about CNY 8.9 trillion, accounting for about 1/6 [43] of the total social investments in 2021 (see Table 2). From the perspective of consumption needs, although only less than 30% of China's population is currently in the middle-income category, this proportion will continue to increase and the consumption of this part of the population is the main growth point of the Chinese consumer market.

In terms of low-carbon trading, China also has great potential, with comparative advantages mainly including renewable-energy equipment and related parts, high-speed trains, ultra-high-voltage direct current transmission equipment, electric vehicles, low-carbon electrical appliances, and high-energy efficiency refrigeration and other equipment. China also has a comparative advantage in the manufacturing of green and low-carbon equipment and can supply products with competitive prices to the global market to reduce energy transition costs globally.

4.3 To Optimize the Spatial Structure of the Territory and Construct a New Spatial Pattern Meeting the Requirements for Carbon Peaking and Carbon Neutrality

It is recommended to take into account regional differences, comprehensive resource endowments, new-energy developments, carbon neutral industrial layout of industrial transformation strategies and the integration of the three major spaces of the regional low carbon transformation partnership, use controls and dual carbon targets, formulate a mixed land use policy in favor of large-scale and high-proportion developments of renewable energy sources, promote the protection, development and utilization of biomass energy, reduce carbon emissions, increase carbon sink, and be adapted to and protect biodiversity synergistically. We should establish a system of natural

Table 2 Key areas of the 14th five-year plan and green stimulus measures (accumulative investments of CNY 44.6 trillion)

Category	Priority area	Investment scale (2021–2025)	Source/ channel of funding
Information infrastructure	5G base station	CNY 2.5 trillion	Based on public + market debts and loans
	Artificial intelligence and big data center	CNY 2 trillion	Based on market debts, loans, and stocks
	Industrial Internet	CNY 800 billion	
Renewable energy/ environmental-friendly energy/power system	Centralized/distributed renewable energy, power system flexibility, and smart grid, etc.	CNY 4.7 trillion	Based on public + market debts and loans
Green and low-carbon urbanization and modern cities	Public service facilities, low-carbon buildings, clean heating and cooling, charging piles for electric vehicles, and high-speed railway and intercity transportation of urban agglomerations	CNY 7.8 trillion	Based on market debts, loans, and stocks
Digital upgrading and green re-construction of traditional industries	Digital applications of specific scenarios; electrification of specific departments and processes; re-structuring of integrated supply chain of medium and small enterprises in specific regions and urban agglomerations; and improvement of environmental quality and ecological remediation (in consideration of carbon emission)	CNY 16.5 trillion	Based on market debts, loans, and stocks
Expansion and re-shaping of green consumption	Consumption of green and low-carbon products: high-efficiency electric appliances and electric vehicles Low-carbon lifestyle in smart cities: medical treatment, supporting for old people, sports, education/training, and entertainment	CNY 5.5 trillion	Based on public + market debts, subsidies, and loans
Innovative infrastructure	Significant infrastructure for science and technology, science and education, and innovation of industrial technologies	CNY 300 billion	Based on public + market debts and loans

Data source From green stimulus and the 14th five year plan to China's modernization: writing a new growth story around natural capital, by Energy Foundation China reserves with national parks as the main body and stabilize the carbon sequestration function of existing forest, grassland, wetland, ocean, soil, permafrost, karst, etc. [41].

4.4 To Complete the Mechanism and Institution for Management Over Carbon Peaking and Carbon Neutrality and Focus on Building the Local Capacity to Achieve the Dual Carbon Goals

We should improve the management and coordination mechanism for various departments and fields and promote the participation of various stakeholders. We should give full play to the role of the National Leading Group on Climate Change, Energy Conservation, and Emission Reduction and enhance the overall planning. Then, we should further complete the duties and the working procedures of the related departments, reach a wider consensus and a normalized coordination mechanism. We should also strengthen the building of the capacities of various regions in addressing climate change and low-carbon transition issues from the perspectives of institution, human resources, and law enforcement and should promote communication and coordination between departments. We should break through the departmental barriers in fundamental institutional systems, and strengthen the local awareness of carbon peaking and carbon neutrality.

4.5 To Push Forward the Transition and Transformation of Key Mechanisms and Accelerate the Transition from the Control Over Both of the Total Amount and the Intensity of Energy Consumption to the Control Over Those of Carbon Emissions

We should further complete the system for the control over both of the total amount and the intensity of energy consumption. The elasticity of the total amount of energy consumption should be enhanced and the relevant statistics and assessment systems should be improved so as to ensure the energy use of raw materials or new renewable energy will not be included in the total amount of energy consumption for control.

We should establish and improve relevant laws and regulations, promote the updating, improvement, and development of the supporting system for renewable energy. A mechanism for connecting renewable energy projects with carbon market should be set up as well.

4.6 To Speed Up the Carbon Pricing and the Carbon Market Mechanism Building of China

China should consolidate the political, legal, policy, and management mechanism foundation for the construction of the carbon market, and establish a total carbon emission constraint mechanism for enterprises on the carbon market and set an annual proportion of increase or decrease in the total carbon emission of an industry involved in the carbon market based on the characteristics of the industry.

China should promote the connection and coordination between the carbon market and other markets and policies, by enhancing the compatibility and cooperation of the carbon market with market-oriented reforms such as electricity market reform and related policy mechanisms and fully explore and leverage their synergistic effects [44].

4.7 To Focus on Achieving just Transition at the Industrial and the Regional Levels

China should formulate a series of policies and measures relating to financial assistance for coal and related enterprises. Make a package of policies and measures relating to compensation, job replacement, and retraining for unemployed laborers in the coal and related industries. Lay down relevant policies and measures to promote the diversified economic development of coal-intensive and underdeveloped areas and enhance their "hematopoietic function".

4.8 To Push Forward the Perfection of the Investment and Financing Policies Relating to Climate so as to Promote the Transition to Carbon Neutrality

Establish a natural capital accounting system that can fully reflect the scarcity of natural resources as well as the corresponding financial incentives and regulatory models. Improve the taxonomy for investment and financing projects relating to climate and gradually promote the synergy of the taxonomy with those of developed economies in Europe and America. At present, China's construction of the taxonomy for investment and financing projects relating to climate is still in the initial stage.

Strengthen the assessments over environmental and climate risks of the financial industry and especially climate risks in industries presenting high energy consumption such as coal, heavy industry and thermal power, and effectively filter projects with high risks of stranded assets and/or high financial risks.

4.9 To Lead a New Global Green Development Pattern and Continue to Deepen the International Cooperation in Climate Changes

China has been always performing the responsibilities of a major power with the idea of building a community with a shared future for mankind, and will continue to uphold the principle of openness and inclusiveness, deepen and expand climate cooperation with developing and developed countries in policy, technology, and other aspects, and promote the construction of a new global climate cooperation pattern for a win–win situation.

China should continue to engage in dialogues and cooperation via the main channel of the governance mechanisms under the United Nations Framework Convention on Climate Change (UNFCCC) and the *Paris Agreement*, explore multilateral climate cooperation outside the UNFCCC, such as making use of the G20 to discuss green finance and green financial policies for promotion of green and low-carbon recovery and to explore cooperative investments in low-carbon infrastructure. China should also establish a working group mechanism within the framework of the WTO, and actively share best practices and experience during transition with developing countries. Also, promote pragmatic cooperation between China, the United States, and Europe in green industries and technologies.

5 Analysis of Gender Mainstreaming

Women and girls represent half of the population of the world. Gender equality is a fundamental human right as well as an important prerequisite for fully unleashing human potential, promoting sustainable development, and ultimately achieving a peaceful society. In China, gender equality is a basic state policy for promoting national and social development. The *Outline for Development of Women in China* (2021–2030) has identified Women and Environment as one of its eight major themes. However, in the fields of environment and climate, China still has a gap with the international community. Adopting the gender mainstreaming analysis in the work of carbon peaking and carbon neutrality, fully considering the impacts on different gender groups during policy formulation and implementation, and ensuring equal benefits for men and women can not only end or reduce inequality but also give full play to women's abilities and potential and further promote the realization of sustainable development goals, to finally provide a multiplier effect for guaranteeing sustainable development.

5.1 To Give Full Play to Women's Leadership in Climate Change-Related Affairs and Improve Women's Participation and Representativeness in Climate Decision Making

Women and girls are disproportionately affected by climate change. At present, women's participation level in climate-related work is even less than a half, which is not directly proportional to the impact of climate change upon them. In 2020, only 15% of the leaders of the environmental departments throughout the world were women [45] and the average proportion of female employees was only one-third [46]. Therefore, more women should be promoted to participate in the making of decisions about mitigation of and response to climate change.

Under the framework of the United Nations Framework Convention on Climate Change (UNFCCC), a practical and effective measure of promoting women's leadership in the climate-related work is to elect China's National Gender and Climate Change Focal Point (NGCCFP).[2] If the Chinese government designates some persons of the Ministry of Ecological Environment to act as its gender and climate change liaisons responsible for coordinating gender-related work, participating in international discussions, and promoting capacity building, etc., it will be a simple, practicable, and effective step to enhance gender considerations in the work relating to carbon peaking and carbon neutrality.

5.2 To Actively Drive just Transition and Promote Women's Equal Employment Through the Opportunities of Climate Transition

Based on the research by the International Labor Organization, without the policies for fair transition, even in emerging green industries, the existing gender employment stereotypes will continue [47]. According to the data provided by the International Labor Organization, if actions are taken in the energy sector to control the global warming within 2 °C by the end of twenty-first century, approximately 24 million job opportunities can be created [48]. The majority of new employment opportunities in emerging industries, which are undergoing low-carbon transition, are concentrated in industries currently dominated by men, such as renewable energy, manufacturing, and construction. Relevant departments need to take the following pertinent measures to narrow the gender gap in these industries and ensure that women can also participate and contribute.

China should include gender considerations in macroeconomic and industrial development policies. Ide the goal of fair transition or gender considerations in

[2] National Gender and Climate Change Focal Points. https://unfccc.int/topics/gender/resources/list-of-gender-focal-points-under-the-unfccc.

the *Nationally Determined Contributions*. Make full use of the role of climate-related investment and financing, guide funds towards projects of fair transition, ensure the direct participation and profitability of women in project design, decision-making, and implementation, and take into account the social disciplines and the other challenges to be faced by women during transition.

5.3 To Improve' Women's Ability of Being Adapted to Climate Change and Achieve a Multi-win Situation Featured by Adaptation to Climate Change, Increased Benefits for Low-Income Groups, and Uplifted Gender Equality

The special needs of women should be included in the decision-making mechanisms for formulating policies on adaptation to climate change. The first is to introduce a gender perspective when formulating a policy relating to adaptation to climate change, fully consider the needs and roles of women, use women's experience in adaptation to climate change as a reference for decision-making, and promote women's participation in the making of decisions about adaptation to climate change. The second is to ensure that women can sustainably utilize natural resources [49], adapt themselves to changes in natural resources to be brought about by climate change, and enhance their economic empowerment during pre-disaster preparedness, post-disaster relief, and long-term recovery and reconstruction processes. The third is to provide financing mechanisms for women (especially impoverished women) to be adapted to climate change at the local level, or to support females through special funds for gender equality, and prevent them from being trapped in further poverty and avoid exacerbating gender inequality due to climate change.

5.4 To Consider the Social Impacts of Overseas Green Investments and Aids in a Strengthened Way, Promote Gender Equality, and Give Play to the Leading Role of China in Global Climate Governance

Currently, in China's overseas investment and aid projects (and especially green overseas investment and aid projects), the environmental impacts of these projects have been fully considered and recognized but the recognition and management of their social impacts have been still in the initial stage. China should complete the standards and guidelines for overseas investments and aids by taking gender equality into account, consider incorporating gender issues into the information disclosure dimension of an overseas investment project through the self-discipline pact on the overseas investment project, and to provide training and technical support for women in combination of the characteristics of local industries and culture, and strengthen international cooperation, promote knowledge sharing and gender mainstreaming as well as encourage more stakeholders to participate in gender equality issues.

6 Policy Recommendations

6.1 Accelerate Coordinated Efforts in Green Economic Development, Energy Security, and Combating Climate Change

Green and low-carbon investment is the connecting point of cross-cycle and countercyclical measures, which can organically combine economic growth and carbon neutral transition. Approximately 140 trillion RMB of green investment will be needed to achieve carbon neutrality by 2060, with power, transportation, and building sectors as the major green investment targets. From 2020 to 2022, China's new installed renewable capacity exceeded 120 million kW annually. With a powerful governance of the system, party, and the government, and the strong manufacture sector and market, China's renewable energy is expected to develop with momentum and huge potential. Electric vehicle as well has entered the fast lane, with 13.1 million units by the end of 2022 and expected growth of more than 10 million units in 2023 alone. While other products face challenges in export in the current international landscape, EV export still demonstrates a strong momentum. Also, 90% of the production capacity of heat pump is in China. Exports of solar modules, high-efficiency cooling equipment, and ultra-high voltage DC transmission equipment have become a new highlight in China's international trade business, and the export mix and global market share of China are becoming more matched with China's economic performance. With the recovery of China's economic growth, the overlap among energy security, climate change, and environment targets will increase, therefore enhancing the coordination between China's modernization and achieving the dual carbon targets.

6.2 Take Comprehensive Consideration of Various Factors (Such as Energy Security, Economic Costs, Etc.) and Gradually Promote Coal Power Transition and Accelerate the Development of a New Energy System Dominated by Renewable Energy Based on China's Situation

President Xi Jinping has announced that China will strictly control coal power, and will peak coal consumption in the 14th five-year-plan (FYP) period and phase down coal consumption in the 15th FYP period. Globally, coal phase-out is also an irreversible trend. Firstly, emergency power safety and long-term supply guarantee cannot be equated. It is important to understand that the renewable-dominated new power system is the long-term basis of China's energy and power security. Secondly, comprehensive comparison among different coal power transition paths in terms of security, cost-effectivenss, and sustainability should be made, thus to formulate better and more economic systematic transition plan. Thirdly, the overall time frame of coal power transition needs to be clearly planned, and the opportunity window of coal power transition and exit needs to be well managed in consideration of the "three cycles"[3] the lifespan of coal power units, technology iteration for new energy system, and fixed asset investment to solve employment and other just transition issues in the process. Fourthly, we need to accelerate the development of renewable energy, promote investment on energy storage, grid interconnection, and long-distance power transmission, and accelerate R&D on V2G policy and technologies to coordinate EV development with construction of the new power system.

6.3 Improve Green and Climate Finance System that Supports Low-Carbon Transition and Innovation

Firstly, to enhance carbon pricing through a comprehensive set of instruments, including taxes, prices, subsidies, procurement, etc. to motivate diversified green climate investment and financing mechanisms. Secondly, to improve taxonomies for climate investment and financing, and gradually enlarge the common ground with those of Europe, the United States, and other regions. Thirdly, to promote financial sector to enhance environmental and climate risk assessment, and effectively filter out high risk projects to avoid stranded asset and to avoid "green washing". Fourthly, to promote mandatory disclosure of climate-related information for financial institutions and establish green standards for oversea financing and investment projects and compliance and accountability mechanisms for financial institutions.

[3] The first cycle is the lifespan of coal power units. The average service time of coal power units is about 20 years nationwide, and about 1200 GW coal power plants will be retired around 2040. By then, the increase of coal power and other fossil energy should be strictly controlled and replaced by non-fossil energy. The second cycle is cycle of technology iteration for new energy systems. Based on the technology of renewable energy, energy storage, power transmission, smart grid, and demand-side management, a modern energy system with high penetration of renewable energy will be established in the next two decades. In the last decade, 90% of the cost of electricity generation from renewable energy has reduced. With technology iteration and market development under the technology cycle, a more significant economic scale will be reached. The third cycle is the investment boom cycle under the economic environment. Under the general trend of resisting recession, China's relatively loose fiscal and monetary policies, as well as flexible bonds and commercial loans, will together create a favorable investment environment. Over the next two decades, development of renewalbe energy production and consumption will be accelerated, and the increase of fossil energy will be strictly controlled. During the "15th Five-Year Plan" period, the incremental electricity demand is expected to be fully supported by renewable energy.

Fifthly, raise the standards for overseas green investment and establish mechanisms for compliance and accountability.

6.4 Accelerate the Transformation of Key Policies and Mechanisms, Especially from Energy Dual Control to Carbon Dual Control

Firstly, to establish a sound system of laws and regulations on carbon peaking and carbon neutrality and formulate a framework law on addressing climate change and/or promoting carbon neutrality as soon as possible. Secondly, adhere to the principle of "establish before break down" and define the short-, medium- and long-term roadmap for institutional transformation. It is recommended during the mid and late 14th Five-Year period, select some provinces/cities and key emission departments/industries as soon as possible to pilot carbon dual control in parallel of energy dual control and study the implementation process; in early 15th Five-Year period, implement nation-wide carbon dual control, with the intensity as the binding target and the total amount as an indicative target. After 2030, refine the climate policy system with the carbon cap as the core measure and carbon/energy intensity benchmarks embedded in industry and product standards. Thirdly, further improve China's ETS through establishing the property right of carbon assets, including carbon intensive industries such as iron and steel into the ETS, and coordinating the study on CBAM with domestic ETS development.

6.5 Continue to Deepen International Cooperation on Climate Change, and Promote the Global Climate Governance for Win–Win Cooperation, Including Sustainable Supply Chains

Firstly, to promote global green transition and carbon neutrality process under the *United Nations Framework Convention on Climate Change* and its *Paris Agreement*, and strengthen multilateral climate cooperation on other key platforms such as G20 and WTO; Secondly, to reform global finance architecture to support developing countries facing poly-crises, and to increase support for climate actions, debt restructuring, etc. Thirdly, to actively establish cooperation partnerships with developing countries on energy transition and continue expanding and enlarging the fields, forms, and scale of cooperation. For the African countries who boast significant potential in terms of renewable energy generation and low-emission manufacturing, China can lead to unlock Africa's sustainable development by facilitating them to move up the global value chains, building on the China-Africa Strategic Partnership. Fourthly, to actively seek for opportunities of cooperation with the United States, Europe and other developed countries in technology transfer, talent, policy, and manufacturing, thus to carry out fair, reasonable, effective, and win–win cooperation by complementing to each other. Fifthly, to enhance efforts to optimise global supply chains, accelerate the development of strategy and policy to shape new multinational corporations, and explore partnerships and collaborative solutions to the increasing competition in clean energy manufacturing and supply, across critical minerals, materials, and modules. It is also important to increase new production capacity of bulk soft commodities in regions with rich renewable resources. These collaborative bilateral and regional partnerships should cover enhanced and cooperative green standards, certification, and traceability; enhanced security; and work to reduce carbon intensity of supply chains for more secured supply chains.

References

1. China Development Network, 2023. (2023, February 15). *In the Spring of Hope—Remarks of the Expert Symposium on "How to See the World Economy in 2023"*, April 10, 2023. https://baijiahao.baidu.com/s?id=1757869289742659148&wfr=spider&for=pc
2. People's Forum. (2022, December 16). *Possible impacts and implications of the European energy crisis*, April 14, 2023. http://www.rmlt.com.cn/2022/1216/662687.shtml. (in Chinese).
3. International Renewable Energy Agency. (2023). *Renewable capacity statistics 2023*. Abu Dhabi.
4. BP. (2023). *BP Energy Outlook* (2023 ed.).
5. China Youth Net. (2022, September 20). *Maintaining the stability of the global industrial chain supply chain*, April 10, 2023. https://baijiahao.baidu.com/s?id=1744465088297889045&wfr=spider&for=pc. (in Chinese).
6. People's Daily Online. (2023, June 05). *The international community needs to work together to address the climate crisis*, June 10, 2023. http://v.people.cn/n1/2023/0605/c177969-40006809.html. (in Chinese).
7. National Bureau of Statistics of China, 2023. (2023, February 28). *Statistical communiqué of the People's Republic of China on the 2022 National Economic and Social Development*, April 15, 2023. http://www.stats.gov.cn/sj/zxfb/202302/t20230228_1919011.html
8. International Monetary Fund. (2023). *World economic outlook*. Washington, DC.
9. China Government Network, 2023. (2022, August 31). *National Bureau of Statistics explains China's new momentum index for economic development in 2021*, June 10, 2023. https://www.gov.cn/xinwen/2022-08/31/content_5707546.htm
10. Bloomberg New Energy Finance. (2022). *Renewable Energy Investment Tracker 2H 2022*. New York.
11. Wang, C. et al. (2020). The outlook for research on synergistic management of greenhouse gases and air pollutants. *Chinese Journal of Environmental Management, 12*(04), 5–12. https://doi.org/10.16868/j.cnki.1674-6252.2020.04.005. (in Chinese).
12. OECD. (2017). *Investing in climate, investing in growth*. Paris.
13. Energy Foundation. (2020, July 07). *From green stimulus and the 14th Five-Year Plan to China's modernization: Building a new growth story around natural capital*, Febuary 27, 2023. https://www.efchina.org/News-zh/EF-China-News-zh/news-efchina-20200703-zh. (in Chinese).
14. Bernard, S., & Shepherd, C. (2021, June 12). *China's sea-level rise raises threat to economic hubs to extreme*, March 12, 2023. https://www.ft.com/content/4dd9860b-664e-4ca0-86a4-5a935d2a22f1
15. World Bank. (2022). *China country climate and development report*. Washington.

16. Zhang, Y. (2018). Challenges and countermeasures for China's coal transformation. *Environmental Protection, 46*(02), 24–29. https://doi.org/10.14026/j.cnki.0253-9705.2018.02.005. (inChinese)
17. IEA. (2021). *An energy sector roadmap to carbon neutrality in China.* Paris.
18. Zhu, T. (2023). Reflections on international rules and domestic mechanisms for "carbon reduction." *Wind Energy, 03*, 8–11. (in Chinese).
19. State Council Information Office. (2023). *China's green development in the New Era.* Beijing.
20. Chinese Academy of International Trade and Economic Cooperation. (2022). *China green trade development report.* Beijing. (in Chinese).
21. Ministry of Ecology and Environment. (2022). *Annual report on responding to climate change: China's policies and actions.* Beijing.
22. National Bureau of Statistics. (2022). *Energy Transition Continues to Advance and Remarkable Results have been Achieved in Saving Energy and Reducing Consumption—The 14th Report in a Series of Reports on Achievements in Economic and Social Development Since the 18th CPC National Congress.* Beijing. http://www.stats.gov.cn/xxgk/jd/sjjd2020/202210/t20221008_1888971.html
23. Natural Resources Defense Council. (2022). *China comprehensive coal management research report.* Beijing. (in Chinese).
24. National Energy Administration. (2022). *Summary of Reply to Proposal No. 02486 of the Fifth Session of the 13th CPPCC National Committee.* Beijing. http://zfxxgk.nea.gov.cn/2022-08/09/c_1310668930.htm
25. www.gov.cn. (2022, October 08). *National Bureau of statistics: Over the past 10 years, China's energy consumption per unit of GDP has dropped by an average of 3.3% per year*, March 08, 2023. http://www.gov.cn/shuju/2022-10/08/content_5716737.htm
26. www.gov.cn. (2022, August 24). *The scale of China's ecological environmental protection industry continues to expand*, April 26, 2023. http://www.gov.cn/xinwen/2022-08/24/content_5706583.htm
27. China Financial Information Network. (2023, March 12). *Digital economy continues to drive high-quality economic development and its share of GDP expected to rise to 41% by 2022*, April 26, 2023. https://www.cnfin.com/hg-lb/detail/20230312/3821292_1.html
28. Drc.net. (2023, February 16). *The green manufacturing system is moving towards a new journey of high-quality development*, April 26, 2023. https://h5.drcnet.com.cn/docview.aspx?version=emerging&docid=6755168&leafid=28071&chnid=5553
29. Sina Finance. (2022, September 15). *Ministry of housing and urban-rural development: Intensify the promotion of building energy efficiency, green buildings and green construction*, April 26, 2023. https://finance.sina.com.cn/esg/2022-09-16/doc-imqmmtha7455132.shtml?cre=tianyi&mod=pcpager_news&loc=16&r=0&rfunc=2&tj=cxvertical_pc_pager_news&tr=174&wm=#!/index/1#250579937
30. Ministry of Transport. (2022). *Statistical bulletin on the development of the transportation industry in 2021.* Beijing.
31. www.gov.cn. (2023, December 30). *National Development and Reform Commission Press Conference to Introduce the "14th Five-Year Plan of Modern Logistics Development"*, April 26, 2023. https://www.gov.cn/xinwen/2022-12/30/content_5734915.htm
32. www.gov.cn. (2023, January 24). *China's new energy vehicle production and sales have ranked first in the world for 8 consecutive years*, April 26, 2023. https://www.gov.cn/xinwen/2023-01/24/content_5738622.htm
33. Chen, C., Park, T., Wang, X., et al. (2019). China and India lead in greening of the world through land-use management. *Nature Sustainability, 2*(2), 122–129.
34. National Forestry and Grassland Administration. (2022, June 14). *Grasslands have important functions as carbon pools—relevant persons in charge of the Management Department of the State Forestry and Grassland Administration explain the functions of the "four pools" of grasslands*, April 26, 2023. http://www.forestry.gov.cn/main/586/20220614/083858613965364.html

35. Ministry of Natural Resources. (2022, November 07) *Building Consensus on Cherishing Wetlands, Promoting Cooperation and Facing the Future—Written on the Occasion of the 14th Conference of the Parties of the Ramsar Convention*, April 26, 2023. https://m.mnr.gov.cn/dt/pl/202211/t20221107_2763886.htm
36. Northeast Institute of Geography and Agroecology, Chinese Academy of Sciences. (2022). *China wetland research report*. Changchun. (in Chinese).
37. Ministry of Housing and Urban-Rural Development. (2022). *Development plan for building energy-saving and green building during the 14th five-year plan period*. Beijing.
38. Ministry of Housing and Urban-Rural Development, National Development and Reform Commission. (2022). *Plan for carbon peaking implementation in urban and rural construction*. Beijing.
39. National Development and Reform Commission. (2021). *Guiding opinions on comprehensive utilization of bulk solid waste during the 14th five-year plan period*. Beijing.
40. National Development and Reform Commission. (2021). *Development plan for circular economy during the 14th five-year plan period*. Beijing.
41. State Council of the People's Republic of China. (2021, October 24). *Action plan for carbon dioxide peaking before 2030*, April 26, 2023. https://www.gov.cn/zhengce/content/2021-10/26/content_5644984.htm. (in Chinese).
42. National Development and Reform Commission. (2021, October 29). *Integration of carbon emission peaking and carbon neutrality into the overall layout of economic and social development and ecological civilization construction*, April 20, 2023. https://www.ndrc.gov.cn/xxgk/jd/jd/202110/t20211029_1302188_ext.html. (in Chinese).
43. Fan, Y., Mo, J. L. (2015). Major issues and suggestions for the top-level design of China's carbon market. *Bulletin of Chinese Academy of Sciences, 30*(04), 492–502. https://doi.org/10.16418/j.issn.1000-3045.2015.04.008. (in Chinese).
44. UN. (2022, March 08). *Gender equality today for a sustainable tomorrow: why women are key in the race against climate change*, April 26, 2023. https://china.un.org/en/174134-gender-equality-today-sustainable-tomorrow-why-women-are-key-race-against-climate-change
45. UNDP. (2021). *Gender equality in public administration*. New York.
46. China Women's News Daily. (2021, June 22). *Gender mainstreaming and women's participation in practices of carbon peaking and carbon neutrality*, April 26, 2023. https://www.women.org.cn/art/2021/6/22/art_25_166547.html
47. International Labour Organization. (2021, October 22). *Frequently asked questions on just transition*, April 26, 2023. https://www.ilo.org/global/topics/green-jobs/WCMS_824102/lang--en/index.htm
48. International Labour Office. (2018). *World employment and social outlook 2018: Greening with jobs*. Geneva.
49. Liu, B. H., & Wang, X. B. (2011). Social gender and climate change. *Journal of Shandong Women's University, 06*, 1–9.

Chapter 2
Pathways and Policies of Blue Economy in Supporting Carbon-Neutrality Target

1 Framing the Issue

As the ocean economy takes off in new tangents beyond traditional areas, it is critical that the short-term growth in the ocean economy should not come at the expense of the long-term prosperity of the ocean, including the key roles the ocean plays in regulating our climate and providing critical habitats for a diverse array of marine animals and plants. In December 2022, the parties to the United Nation (UN) Convention on Biological Diversity reached an agreement to set a global target to effectively conserve and manage at least 30% of the world's lands, inland waters, coastal wetlands, and oceans, with emphasis on areas of particular importance for biodiversity and ecosystem functioning and services. In March 2023, the UN agreed on text to ensure the conservation and sustainable use of marine biological diversity in areas beyond national jurisdiction (BBNJ). June 2023 also saw the second round of the formal negotiations for a UN treaty to end plastic pollution—an important steppingstone in the process—which aims to have a treaty ready for adoption in 2024.

Recognizing that a healthy ocean environment is a prerequisite for the growing ocean economy, an integrated ocean management (IOM) approach is proposed to strike a balance between environmental, economic, and societal goals, and between short-term economic gains and long-term prosperity based on marine ecosystem services. In the past, IOM has been an overarching concept for the CCICED Special Policy Studies on Ocean governance. This report is a first building block of the Ocean SPS's contribution to CCICED work in phase 7 and this, as well as following work on ocean governance, will continue to take a comprehensive and sustainable approach and will strive to address climate change and work to balance trade-offs between growth of the ocean economy and environmental protection.

Existing and potential new economic activities related to oceans, seas, and coasts—the so-called ocean economy—cover a wide range of interlinked established and emerging sectors. The value of the global ocean economy today is an

Green Empowerment and High Quality Development,
https://doi.org/10.1007/978-981-96-4218-2_2

estimated US $2.5 trillion annually [1], equivalent to the size of the world's seventh-largest economy. China's ocean industry has been estimated to be around RMB 3.8 trillion (US $0.5 trillion) in 2021 and RMB 3.9 trillion in 2022 (China Marine Economic Statistics Bulletin), accounting for approximately 3% of China's overall gross domestic product (GDP). According to projections from the Organisation for Economic Co-operation and Development (OECD), the blue economy could by 2030 outperform the growth of the global economy as a whole, both in terms of value added and employment. The long-term potential for innovation, employment, and economic growth offered by the ocean economy is promising.

The ocean offers a wide array of potential ocean-based climate mitigation options that can contribute to carbon neutrality goals, including, but not limited to, the grooming of carbon-efficient ecosystems ("blue forests" or "blue carbon"), the use of the ocean's inherent energy potential, minimizing the carbon footprint of ocean-based activities such as shipping, protecting and potentially enhancing the ability of ocean sediments to store carbon (carbon capture and storage, or CCS), as well as reorienting food policy and fisheries management to value aquatic foods from fisheries and aquaculture as key sources of low-carbon ocean-based protein and micronutrients.

It should be noted that the suite of policy areas and topics covered in this report on the synergies between ocean economies and carbon neutrality connect to and are relevant for topics and discussion taking place within other CCICED Special Policy Studies. Consequently, a holistic view and approach are required across the CCICED agenda to build policy recommendations that utilize the benefits of these synergies.

2 Carbon Neutrality as an Opportunity for Transformation into Sustainable Blue Economy

2.1 Introduction

The ocean "economy" or "blue economy" covers a wide range of interlinked established and emerging sectors, such as marine energy, seafood production, coastal tourism, and marine biotechnology that have a direct or indirect link to the oceans, the seas, and the coasts. They are typically categorized into two pillars, the sum of the economic activities of ocean-based industries, and the assets, goods, and services provided by marine ecosystems.

The proliferation of the blue economy in political discourse has gained traction in recent years; however, there remains no standardized definition [2]. To emphasize the sustainable component of the blue economy and to differentiate from components of the ocean economy that are at least partially prone to being unsustainable (e.g., offshore oil and gas industry), we adopt the term "Sustainable Blue Economy (SBE)" throughout the report.

The ocean already significantly moderates our planet's climate [3]. It has absorbed the majority of the heat generated from increased emissions over the past century and about one quarter of the CO_2 emissions [4]. Together this has greatly impacted the ocean, leading to increased temperature and acidity, changes in ocean circulation, reduced oxygen levels, and the loss of biodiversity [5].

2.2 *Status*

2.2.1 Carbon Neutrality and Ocean-Based Solutions

The ocean is vital in meeting this carbon neutrality goal at both the international and national levels because it is the primary and sustained carbon sink, accounting for an overall uptake of around 37% of the fossil fuel CO_2 emissions, or around 25% of the combined fossil fuel burning and emissions due to changes in land use between 1850 and 2019 [6]. Hoegh-Guldberg et al. [7] further projected that ocean-based mitigation options could reduce the "emissions gap" by up to around 21% for a 1.5 °C pathway and by around 25% for a 2.0 °C pathway.

2.2.2 Ocean Economy

China is a vital player in the global production, consumption, and trade of maritime products. In 2022, China's gross ocean industry product reached 3.8 billion yuan and RMB 3.9 trillion in 2022 (China Marine Economic Statistics Bulletin), accounting for approx. 3% of China's GDP, the same proportion as the previous year.[1] China is the leading aquaculture and ship producer in the world, accounting for approx. 58% (2020) and around 45% (2021) of the global total seafood and ship production (gross tonnage), respectively [1, 8].

In this chapter, we briefly provide overviews of five substantial sub-sectors of the ocean economy, specifically marine mineral resources and offshore oil and gas, maritime transport, ocean renewable energies (ORE), food production and other supply chain issues, and offshore carbon capture, utilization, and storage (CCUS), which may provide significant opportunities to transform into an SBE.

[1] Note that China's gross marine product reached 9462 billion yuan in 2022, an increase of 1.9% over the previous year, accounting for 7.8% of China's GDP. However, in the gross marine product statistics are included a significant amount of the upstream and downstream industry related to ocean industry, as well research, education, and governmental management contribution to the ocean economy. We provide the statistics for the more limited ocean industry for comparability and relevance in context of the policy study.

Marine Mineral Resources and Offshore Oil and Gas

Mining the seabed is an important potential component of China's future maritime economy. Current activity within China is predominantly focused on marine aggregate extraction. The oil and gas subsector is a highly capitalized industry. A large portion of current China's oil and gas production takes place offshore, mainly in the Bohai Sea and the northern South China Sea. The generation of hydrogen offshore along with offshore oil and gas extraction has a number of advantages. Within the oil and gas sector, companies are increasingly investing in digital and environmentally friendly solutions. Suppliers from the oil and gas industries play an important role in this transition. There are also great opportunities in both the mineral resources sector and the oil and gas sector to play a crucial role in the transition to an SBE.

Maritime Transport

Maritime transport has been a significant driving engine of the global economy. It is reported that the greenhouse gas emissions associated with the shipping sector increased from 977 Mt in 2012 to 1076 Mt in 2018, which will continue to increase by about 130% by 2050 [9]. Around 15% of global anthropogenic NOx and 5–8% of global SO_x emissions are from oceangoing ships [10]. The shipbuilding industry is known as a high-energy consumption, high-material consumption, and high-pollution industry [11].

Ocean Renewable Energies

Ocean renewable energy (ORE) is one of the emerging sectors of the ocean economy of international interest. Offshore wind (both bottom-fixed and floating), tidal, wave, solar, and hydrogen represent the most viable opportunities to significantly expand renewable energy capacity for many coastal and island countries. In 2021, China's offshore wind power added 16.9 million kW of grid-connected capacity, 5.5 times the previous year, and the cumulative installed capacity jumped to the world's first [12]. Facilitating and sustainably deploying ORE can thus significantly contribute to the decarbonization of the energy system, which is essential for achieving China's carbon neutrality goals. However, there remain potential ecological and social risks facing the large-scale development of ORE.

Food Production and Other Supply Chain Issues

Parker et al. [13] estimated that globally, marine capture fisheries generate around 179 million tonnes of CO_2-equivalent GHGs annually (~ 4% of global food production), of which fuel combustion accounts for over 70%. Advances in fishing technology have spurred the development of more powerful engines that have increased the demand for fossil fuels, and it has been estimated that fuel costs can account for up to 60% of total fishing costs [14]. Additionally, the use of particular gear types, such as trawl nets, contributes substantially to GHG releases, directly through emissions from fishing vessels and indirectly via the disturbance of bottom sediments that hold carbon.

Aquaculture is becoming an increasingly important source of seafood in many countries and regions [15]. Results of quantitative research analyzing both the entire globe and China (as the leading producer) consistently illustrated that feed production was the main source of GHG emissions from aquaculture [16, 17]. Key improvements in the aquaculture sector include reducing the use of energy-intensive feeds, improving feed management, and, where appropriate, considering the development of regenerative aquaculture practices that may help mitigate climate impacts and provide substantial co-benefits (e.g., biodiversity and habitat enhancement, livelihood improvements).

Offshore CCUS

CCUS may contribute significantly to cutting emissions, especially in hard-to-abate sectors with limited or no other alternatives. China has installed the first offshore CCUS project "Enping" in the northern South China Sea. The Asian Development Bank and Chinese Academy of Sciences have estimated a theoretical storage potential of 500–800 billion tonnes of CO_2 in geological structures on China's offshore sedimentary basins.

2.3 Challenges and Opportunities

2.3.1 Science and Technology

Currently, there are critical gaps in our understanding of whether most ocean-based CDR techniques would offer significant drawdown potential of CO_2 and their effect on overall GHG fluxes.

MRV (Monitoring, Reporting, Verification) Mechanism

A significant challenge for all ocean-based CDR pathways is monitoring, reporting, and verification (MRV) of the quantity and durability of carbon stored. There are also substantial uncertainties surrounding emissions of other GHGs, including methane, from mangroves and salt marshes; in some cases, these emissions could severely limit the climate mitigation potential of these ecosystems [18]. Overall, ocean-based CDR is a nascent field and is garnering a lot of attention, but cannot substitute for rapid and deep cuts in greenhouse gas emissions.

2.3.2 Policy

Blue Finance

The World Bank's International Bank for Reconstruction and Development (IBRD) has issued two sets of "sustainability bonds" (World Bank and Credit Suisse Sustainability Bond and World Bank and JP Morgan Sustainability Bond) with strong considerations of the need to safeguard marine and coastal ecosystems. In addition, some coastal countries and regions have issued blue bonds, such as the "Seychelles Blue Bond," "Nordic-Baltic Blue Bond," "The Nature Conservancy Blue Bonds for Conservation," and "Fiji Blue Bond." While these blue bonds could be considered a part of blue finance, there is a need for frameworks or resources to detail how to better utilize these tools to facilitate the transformation of ocean economy and drive conservation of critical marine ecosystems.

Evaluating Marine Ecosystem Services

The Paris Agreement reaffirms "the importance of protecting and enhancing, as appropriate, the sinks and reservoirs of greenhouse gases referred to in the Convention" (Preamble). Australia and the United States have also begun to include blue carbon in their numerical reduction targets. Japan has implemented several blue carbon offset credit projects for seagrass meadows and is piloting carbon crediting projects for natural and cultivated macroalgae beds.

International policies and programs that aim to assess and protect multiple marine ecosystem services other than carbon sinks are centred around the UN SDG of restoring marine ecosystems for a healthy and productive ocean (SDG 14).

Assessing the Impact of Benefits When Implementing Specific Interventions/Solutions

The policy for assessing the impact of benefits when implementing specific interventions/solutions is derived from the technical standards of the UN Clean Development Mechanism (CDM). Some non-governmental organizations (NGOs) have also developed assessment policies based on these standards, i.e., the Verified Carbon Standard (VCS), which is a more detailed version.

Policies on Including Blue Carbon in NDC Commitments

Blue carbon ecosystems (coastal ecosystems that sequester carbon) are valued for their climate mitigation and adaptation benefits. Countries have begun to properly account for blue carbon ecosystems in their nationally determined contributions (NDCs), national greenhouse gas (GHG) inventories, national adaptation plans (NAPs), and other high-level climate-related policies.

2.3.3 Legal Framework

Legal Framework for Maritime Transport

The International Convention for the Prevention of Pollution from Ships (MARPOL) is the dominant legal instrument for the prevention and control of environmental damage caused by pollutant discharges from ships. The International Maritime Organization (IMO) has also introduced mandatory constraint regulations and policies for shipping carbon emission reduction, mainly through the revision of relevant contents of MARPOL.

At the national level, the Maritime Bureau of China issued the Measures on the Management of Energy Consumption Data and Carbon Intensity of Ships in November 2022, which stipulates the requirements for the management of China's ship energy consumption data and carbon intensity, and applies to ships of Chinese nationality with 400 gross tons or more and foreign ships entering and leaving China's ports.

Legal Framework for Offshore Renewable Energies

At the international level, none of the treaties that China has acceded to are directly related to this field. At the national level, the Renewable Energy Law of China should be applied as a fundamental legal instrument. However, the Renewable Energy Law was adopted in 2005 and revised in 2009. Therefore, the Renewable Energy Law lacked beneficial guidance for the exploration and utilization of ocean renewable energy in the context of carbon neutrality strategy.

Legal Framework for Fisheries Management

At the international level, the United Nations Convention on the Law of the Sea (UNCLOS), the CBD, the WTO Agreement on Fisheries Subsidies (the WTO Fishery Subsidies Agreement), and the Agreement under the United Nations Convention on the Law of the Sea on the Conservation and Sustainable Use of Marine Biological Diversity of Areas beyond National Jurisdiction (the BBNJ Agreement) should be applied to China.

Legal Framework for Offshore CCUS

At the international level, the UNCLOS, the CBD, and the Convention on the Prevention of Marine Pollution by Dumping of Wastes and Other Substances (the 1972 London Convention) and its 1996 protocol should be applied to regulate the implementation of the offshore CCUS projects. At the national level, the Marine Environment Protection Law is mandated to consider the environmental impact of offshore CCUS projects. In addition, there is no systematic legal framework for the full life cycle of offshore CCUS in domestic law.

2.4 Chapter-Specific Recommendations

High-level recommendations:

- Set SBE as a strategic development goal of the nation and as part of Carbon Peak and Neutrality goals and include it in international cooperation frameworks.
- Evaluate the use of global ocean technologies, in particular, digital technologies, to support the growth of the sustainability measures and promote carbon neutrality in blue economy. Adopt policies and measures on tax incentives, industrial matching, entrepreneurial support, talent attraction and training, etc. More

importantly, promote the industrialization and large-scale application of such emerging technologies vigorously.

- Refine management systems to account and balance for both ecological and socio-economic goals. Establish a multi-level integrated ocean management system covering the central to local authorities from the socio-economic-nature complex ecosystem perspective.
- Develop frameworks and metrics for holistically accounting for sustainability and socio-economic outcomes, and strengthen financial support for the blue economy. Assess the existing national "Green Industry Guidance Directory" and green financial policies and examine the need to establish a new framework of blue finance. Encourage financial institutions to develop diversified financial products to support the low-carbon transformation of the ocean economy. Improve the role of government-guided funds in promoting the blue transformation of the ocean economy.
- Seek opportunities for international cooperation in scientific and economic research. Promote comprehensive international cooperation in marine and carbon neutrality science, technology, education, investment, trade, etc., through bilateral and multilateral platforms and mechanisms, such as the BRI, as well as Boao Forum for Asia (BFA), World Economic Forum (WEF), etc.

Specific recommendations:

- Develop a research code of conduct for ocean-based CDR that addresses fundamental principles of scientific integrity (e.g., transparency and dissemination of results), fairness and equity (e.g., public consultation), and responsible research (e.g., minimization of potential harms and assignment of responsibility) across all ocean-based CDR methods. Recipients of federal grants should be required to follow this code of conduct and there should be plans to incentivize uptake by scientists performing CDR research supported by private funding. Furthermore, translate the fundamental principles into domestic laws at the right time to strengthen the regulation of ocean-based CDR.
- Accelerate the transformation to low-carbon marine industrial practices.

3 Blue Economy and Marine Plastic Reduction

3.1 Introduction

Global plastic production and consumption have grown exponentially since the 1950s and are set to triple by 2060 if business continues as usual [19]. Since most of the discarded plastic products end up in the ocean, plastic pollution is recognized as a severe anthropogenic issue in coastal and marine ecosystems across the world. This chapter reviews the pollution status and sources of marine plastic in China and provides an overview of the impact of plastic litter on marine ecosystems.

3.2 *Status*

3.2.1 Scale of the Marine Plastics Debris in China

According to the "National Urban and Rural Construction Statistical Yearbook 2017" issued by the Ministry of Housing and Urban–Rural Development of China, the proportion of mismanaged waste of large and small cities in China are 1.00% and 3.89%, respectively, which is close to the level of developed countries. A recent study showed that harmless treatment rate of urban domestic waste had reached 99.7% in 2020, where treatment mainly involves sanitary landfill and incineration [20]. It is estimated the amount of mismanaged plastic waste in coastal areas could be about 55,000 metric tons, and about one-third of some plastic waste may leak into the ocean [21]. A study indicated that from 2011 to 2019, the amount of plastic waste entering the sea in China showed a rapid downward trend as a whole [21].

3.2.2 Source Analysis of the Marine Plastics Debris

Riverine Input

Rivers have long been considered to be the major source of ocean plastic waste. The latest monitoring research shows that the annual amount of plastic waste entering the ocean from the Yangtze River, the largest river in China, is now at about 10–20% of previous estimation by Lebreton et al. [22–25]. Moreover, the total amount of plastic waste entering the sea from all other rivers in China is only 50–60% of that of the Yangtze River.

Coastal Household Garbage Spill

The leakage of coastal plastic waste into the ocean in China is mainly from people's daily activities, tourism, and leisure activities. Relevant monitoring in 2021 shows that the average density of floating garbage in the Chinese coastal waters is about 3.6 kg/km^2.

Leakage from Activities Such as Maritime Commercial Activities

Varied substantially by region, sea-based sources could contribute 32–60% of the total marine litter [26]. However, the amount of plastic waste generated by those activities has not been rigorously quantified. It is important it become a focus of future marine plastic waste management in China.

Transfer Across Oceans

There is still a knowledge gap to know about the amount of plastic waste transported along the seabed. The key pathways of microplastics from the coasts of Bohai, Yellow, and East China Seas were detected by applying the Lagrangian particle tracking method in a hydrodynamic model [26]. It was found that less than 18%

of terrestrial microplastics was eventually transported from the coast to the Pacific Ocean [26].

3.3 Challenges and Opportunities

3.3.1 Impacts of Marine Plastics Debris on the Marine Ecosystem

Plastic pollution has reached almost every part of the ocean, from the sea surface to the deep ocean floor. Physical impacts of marine plastic include entanglement, plastic ingestion, colonization of plastic items by marine life, and contact or coverage (e.g., smothering) of organisms with plastics, with effects including restrained movement, injury, suffocation, mortality, dispersal of organisms by rafting, and spread of pathogens.

Impacts of Plastic Pollution on Species

Impacts of plastic pollution on main marine species groups including endangered species and commercial species:

- A spatial risk analysis for seabirds [27] concluded that 59% of seabird species and 29% of seabird individuals had ingested plastic between 1962 and 2012.
- A global analysis of sea turtles estimates that 52% (340,000 individuals) of all turtles have already ingested plastics [28]. Among the thousands of sea turtles that strand every year, 6% were found entangled in marine debris, of which 91% were dead [29].
- Whales, dolphins, and porpoises have been subjected to both entanglement and ingestion of macroplastics. The necropsies of whales stranded between 1990 and 2015 along Irish coasts revealed that 9% of whales had eaten plastic debris [30].
- However, ingestion of microplastic can make zooplankton fecal matter more buoyant [31] and reduce zooplankton's ingestion rate [32], growth and reproduction [18], thus affecting the functioning of ocean as a carbon sink.
- In China, a study on commercial marine fish collected from Yangtze Estuary, East China Sea, and South China Sea concluded that all 21 species sampled ingested micro- or meso-plastics, with plastic fibre being the most common morphotype found in their stomach and intestines [33].

Effects of Plastic Pollution on Marine Habitats

Plastic pollution also affects major marine habitat types and impairs the ecosystem functions, including their ability to sequester carbon dioxide through primary production. One third of the investigated 159 coral reefs in the Asia–Pacific region were polluted with macroplastics [34]. Macroplastic debris is prone to be trapped in coral reefs and can smother large parts of coral colonies and promote coral disease [35, 36]. In the South China Sea, microplastic abundances are of up to 45,200 items/m^3 in coral reef surface waters, 5738.3 items/kg in mangrove sediments, and 927.3 items/kg in seagrass bed sediments were reported [37].

3.3.2 Impacts of Marine Plastics Debris on the Blue Economy

As a typical problem of marine ecological pollution, marine plastic waste poses substantial challenges to the sustainable development of the blue economy.

Impact on Coastal Tourism

Marine plastic waste is concentrated on coastlines and beaches where recreational activities are the main function, destroying the original natural scenery of these coastlines and beaches, and causing negative perceptions and experiences for tourists [38]. For example, marine litter accumulation events, including marine plastic litter, caused approximately $29–$37 million USD in lost tourism revenue to Geoje Island, South Korea, in 2011 [39]. Surveys of tourists at coastal tourist attractions in China have shown that the amount of beach litter significantly affects tourists' willingness to pay [40, 41].

Impact on Marine Fisheries

Marine plastics have had an unpromising impact on the development of capture fisheries and aquaculture [42, 43]. In a survey of 21 major marine species off the coast of China, all of them were found to contain plastic of varying degrees [33]. Whereas a global consolidation of regional studies showed that a total of 323 out of 494 fish species tested were recorded to contain plastic, and more than 262 out of 391 commercial fish species were also detected to contain plastic [44].

Microplastics in the marine environment can enter the bodies of aquatic organisms through ingestion (Fig. 1) and have a series of negative effects on various physiological processes. In addition, microplastics can also act as a stable adherent and a carrier of toxic and harmful substances, such as pathogens, in the water column [45, 46].

Food is also one of the factors contributing to the increase in plastic levels in fish. The food here includes both artificial fish feed in the farming process and the natural prey of wild fish. Studies have shown that the plastic content of fishmeal ranges from 0 to 526.7 n/kg, with Chinese fishmeal (337.5 $\pm$ 34.5 n/kg) having a relatively high plastic content [47]. While the harmful effects of marine plastics on fisheries development are obvious, fisheries themselves are also an important source of marine plastics. In a study of floating raft cultivation systems in the Maowei Sea of Guangxi, China, it was estimated that approximately 3840 tonnes of plastic waste would be discharged into the sea within the next 4 years if left unchecked [48]. Many aquaculture facilities contain plastic components, the loss of farming facilities and fishing gear is one of the main sources of marine microplastics. In Australia, the main loss areas of plastic fishing gear are ropes (47%), tank components (30.7%), and floats (22.3%) [49].

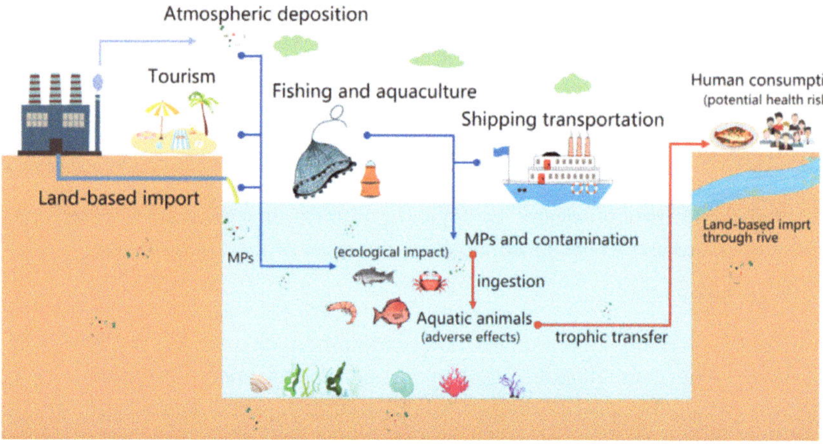

Fig. 1 Marine plastic pathways in aquaculture systems

Impact on the Marine Shipping Industry

Depending on the distribution of marine plastic litter, the main impact on shipping is caused by floating litter on the sea surface. Vessel losses and mass casualties due to floating debris entangled in propellers have been documented in established studies [50], but economic losses due to repair and maintenance costs are more common [49]. Domestically, a study suggests that China should improve the marine plastic waste management system, reduce plastic waste from the source into the ocean, strengthen the principle of tripartite governance among the government, producers, and consumers, strengthen the harmless disposal and recycling of used fishing nets and fishing gear, and encourage public participation in marine waste management, to contribute Chinese wisdom to the global governance process [50, 51].

3.3.3 Knowledge and Policy Gaps for the Life Cycle of Plastic Governance

A more comprehensive approach is urgently required for its governance beyond the narrow focus of the plastic litter or pollution in the marine ecosystem. Governing marine plastic pollution and litter hence needs "outside-the-box" thinking There are two general trends of research on life-cycle plastic governance. One is the attempt to construct a "circular" plastic economy, and the other is to establish a global or regional convention or treaty on plastics [52]. More studies should be advanced in addressing the following knowledge and policy gaps and challenges.

3.4 Chapter-Specific Recommendations

- Actively engage in the ongoing multilateral negotiations around the global plastics treaty. Support the international negotiation contributing to a treaty that is scientifically based, includes effective measures with specific, implementable, and efficient global rules at the most appropriate stage in the life cycle of plastic with close consideration of gender equality in order to address the transboundary marine plastic problem.
- Deploy appropriate policy instruments upstream of the plastic production industry (such as effective extended producer responsibility) that internalize the full cost of plastics and incentivize waste reduction, implementation of reuse models, the creation and use of recycled plastic over new plastic, and the development of viable alternatives to plastic that have smaller environmental footprints.
- Strengthen the control of plastic pollution in fishing activities. Establish a production licence system for plastic fishing gear in accordance with industry standards, strengthen the promotion of environmentally friendly plastic fishing gear, implement policy and a subsidy scheme for enabling and speeding up replacement of eco-friendly fishing gear; establish a collection and recycling mechanism for discarded fishing gear by providing financial benefits, and encourage fishermen to salvage "ghost fishing gear" from the sea, so as to better promote the industrialization of the value chain of plastic fishing gear recycling.
- Facilitate cooperation between industries, civil society groups, and government. Establish a systems-based approach that addresses plastic production, consumption, waste management, and recycling as a singular and coherent system.
- Develop an effective knowledge dissemination plan on waste sorting and collection among citizens.
- Work actively to establish a dynamic marine plastic pollution monitoring and accounting system at the global scale.
- Support multidisciplinary and collaborative research that involves new technologies (such as digital, AI, and satellite-based models) and action-based research in order to identify the most polluting plastic objects and sectors.

4 Enhancing Blue Carbon and Reducing Carbon Footprint Through Fishery Governance

4.1 Introduction

Marine capture fisheries and mariculture bring abundant food and nutrient supply to human society and provide basic livelihoods for coastal populations, thus becoming an indispensable pillar of the blue economy. For capture fisheries, fuel use are the largest contributor to the carbon footprint of this sector. According to recent estimates, global marine fishing is responsible for 179 million tonnes of CO_2-equivalent emissions per year, of which fuel consumption contributes more than 70%, with the highest emissions from fishing crustaceans and the lowest from fishing small pelagic fishes [13]. The carbon footprint of global mariculture has not been fully measured, but recent studies indicate that global marine and freshwater aquaculture together contribute 263 million tonnes of CO_2 equivalent emissions per year, with feed use being the largest source of carbon footprint and crustacean aquaculture being the most intensive due to high energy consumption [16].

4.2 Status

4.2.1 Current Status and Challenges Facing the Development of Fisheries that Are Climate-Resilient and Promote Carbon Neutrality

Carbon Footprint of Capture Fisheries and Mariculture

For marine capture fisheries, the fuel consumption of fishing vessels constitutes the largest part of the carbon footprint. Currently, global marine fishing consumes approximately 40 billion litres of fuel per year, directly generating 132 million tonnes of CO_2-equivalent emissions; when fishing vessel construction and maintenance, gear manufacturing, and cold chain logistics are included, the total annual emissions are 179 million tonnes of CO_2-equivalent emissions, accounting for approximately 0.5% of global anthropogenic carbon emissions [13].

The carbon footprint of mariculture is more complex and includes three components: on-farm, upstream (represented by feed production), and downstream (represented by processing and transport), where the upstream and downstream emissions are often greater than the farming process itself [53]. Some specific categories of marine aquatic foods (both captured and cultured products) have the highest level of climate efficiency among major animal protein sources [54], and there are increasing calls to shift human diet structure from land to sea.

In the field of mariculture, the biggest opportunity lies in strengthening research on the potential for macroalgae and shellfish farms to act as carbon sinks and accelerating the implementation of incentive policies, such as fisheries carbon trading.

Carbon Sequestration by Capture Fisheries and Aquaculture

Aquaculture

China's mariculture is dominated by non-fed shellfish and macroalgae culture, which led to concept of fisheries carbon sink [55]. Fishery carbon sinks can also be called "removable carbon sinks" and "industrializable blue carbon" [56]. Recent studies recognize the carbon sink function of macroalgae and suggest that macroalgae can have multiple roles in climate change mitigation, producing large amounts of detritus, particles and dissolved organic carbon during their growth, a small amount of which can accumulate in the rocky substrates where the algae themselves grow, and the majority of which transported to the deep sea and its sediments by currents, thus being sequestered for a long time [57–59].

Compared with macroalgae, the source and sink effects of filter-feeding shellfish in the ecosystem are more complex. Tang et al. [60] systematically discussed the characteristics of four carbon pools, namely, carbon used, carbon removed, carbon stored, and carbon released by aquaculture and their quantitative relationships, and then confirmed that shellfish aquaculture enhanced the carbon sink capacity of the aquatic ecosystem and was a carbon sink.

Fisheries carbon sinks have received wide attention, but so far there are no international and national standards for their monitoring and measurement, and it is impossible to comprehensively and systematically assess their carbon sink capacity and tradable volume.

Capture Fisheries

Fish and large marine mammals contribute to the global carbon cycle through five main pathways: (1) they can act as a short-lived reservoir of carbon by storing it in their biomass, (2) through redistribution of carbon and nutrients throughout the ocean (especially to the deep sea) through vertical or horizontal migrations, (3) by mixing of water or resuspension of sediments (i.e., bioturbation), (4) by exporting carbon directly from the surface ocean to the deep ocean when the dead organisms sink to the bottom, and (5) in some fish, via intestinal precipitation of calcium carbonates, followed by export of large amounts of particulate inorganic carbon in fish feces to the deep ocean. The elemental carbon in marine fish biomass is estimated to range from 120 million to 1.9 billion tonnes [61–65].

Fisheries management typically aims to sustain or rebuild depleted stocks back to target levels, such as maximum sustainable yield (MSY) or maximum economic yield (MEY). At present, it is premature to attempt to set fisheries goals to assist in more rapid carbon drawdown given the large uncertainties in the knowledge of the five processes described above.

Challenges and Development Principles for Climate-Resilient Fisheries

Climate change amplifies uncertainty about the effectiveness of fisheries management and poses significant direct challenges to fisheries managers and practitioners around the world. In recent years, China has been working to improve the sustainability and resilience of its fisheries, and the 13th Five-Year Plan, implemented in 2016, provides a robust policy platform for the conservation and restoration of China's marine ecosystems and fisheries [66]. A notable example of recent progress is the launch of several Total Allowable Catch (TAC) pilots in Zhejiang, Shandong, Fujian, Guangzhou, and other coastal provinces and cities. A report entitled "Progress of China's TAC System: Evaluation Report for Zhejiang and Fujian Pilots" was completed in 2021 and proposed 29 specific policy recommendations for further implementation of sound TACs in China.

Illegal, Unreported and Unregulated Fishing and Other Harmful Practices

Illegal, unreported, and unregulated (IUU) fishing is one of the greatest challenges to global fisheries governance. The annual catch of the IUU fishery ranged from 11 to 26 million tonnes, with an annual value of $1–$23.5 billion at the time [67]; meaning that, on average, one out of every five fish caught in the global ocean may be from IUU fishing, and in certain regions, this ratio may be as high as one-half [68].

Because IUU fishing is difficult to trace and often employs harmful practices, its presence inevitably leads to the failure of ecosystem-based fisheries management (EBFM).

High economic returns, lack of governance mechanisms, and weak enforcement are considered to be the main reasons for the persistence of IUU fishing. Currently, standardized fishing logbooks and inspection and enforcement data are shared electronically among EU member states, greatly improving the effectiveness and efficiency of traceability and reducing human interference with information quality, but catch legality certificates are still paper-based.

Equal Rights in Fisheries: The Role and Contribution of Small-Scale Fisheries and Women in Fisheries

Small-Scale Fisheries and Their Governance

In the context of climate change, small-scale fisheries (SSF) are more vulnerable SSF contribute around 40% of the world's seafood production. In 2016, more than 60 million people worldwide were employed in small-scale fisheries, representing 90% of all employment in capture fisheries.

In China, small-scale fisheries are an essential part of the blue economy. Xiong et al. [69] examine SSF in Shengsi County in Zhejiang Province and note Chinese government could improve the management of SSF by (1) more clearly defining SSF to set management goals, (2) by developing multidisciplinary data collection and monitoring systems, (3) working to develop cooperatives, and (4) strengthen the coordination among government departments at various levels.

Women in Small-Scale Fisheries

Climate change has disparate influences on different socio-economic groups and will have more negative impacts on women than men [70]. In climate disaster events, men carry more of the work in post-disaster recovery construction, while women and children suffer more. About 50% of SSF workers are women [71] and about 90% of workers in the seafood processing industry are women [72]. In SSF, women are responsible for gleaning, pre-harvest preparation (e.g., repairing fishing gear, preparing bait and food for trip), post-harvest work (sorting and processing the catch), and selling the catch. Women's contribution in SSF is often disregarded.

Although women fishers are numerous, they are typically excluded from decision-making processes regarding the allocation of these resources. This is partly due to the traditional perception of the fishing industry as a "male-dominated industry".

To promote women's meaningful participation, the following recommendations are suggested:

(a) Increase sex-disaggregated data collection to include pre-harvest, post-harvest, and household fishing activities undertaken by women within fisheries data to further understand the gender-specific contributions of fishers in small-scale fisheries for targeted policy development.
(b) Carry out gender-specific training to upgrade the productive skills and knowledge of female fishery workers and to increase the capacity of women to cope with natural disasters and other changes.
(c) Carry out gender-inclusive governance reforms, increase women's participation in fisheries management decisions and research, fully incorporate women's experience and wisdom, promote fair distribution of resource use and management rights, ensure equal pay for equal work, and protect women's rights and interests through laws, regulations, and policies.
(d) Raise awareness of women's lack of agency in resource management and promote effective measures to address it, including the creation of more non-male-centred communication spaces and opportunities for women, especially in small-scale fisheries.

4.2.2 Existing National and International Policy Frameworks on Marine Biodiversity Conservation

Conservation and Sustainable Use of Marine Biodiversity Beyond National Jurisdiction

Marine areas beyond national jurisdiction (ABNJ), including the high seas and the international seabed area, account for 64% of the world's oceans and seas, and the development, conservation, and sustainable and equitable use of ABNJ biodiversity resources and the protection of marine ecosystems and biodiversity has become a key issue in global ocean governance for the international community. The legally binding international instrument on the conservation and sustainable use of BBNJ under UNCLOS was a response to the growing importance of this issue (see text box for background).

The main topics of BBNJ negotiations included:

- Marine genetic resources (MGRs), including benefit-sharing issues.
- Area-based management tools (ABMTs), including MPAs.
- Environmental Impact Assessment (EIA). This includes EIA initiation thresholds and criteria, decision making and implementation, internationalization, monitoring and review, and relationship with other IFBs' EIA.
- Capacity building and transfer of marine technology.

It is recommended that China actively participate in the science and technology bodies and committees established by the agreement and increase the power input to gain more voice and influence.

WTO Fisheries Subsidies Agreement

After 21 years of negotiations, the WTO reached an Agreement on Fisheries Subsidies (referred to as the Agreement in the following) at its 12th Ministerial Conference (MC12) in June 2022. The Agreement consists primarily of three subsidy disciplines and seven notification requirements that apply to exclusive marine fishing and subsidies for fishing-related activities at sea and do not apply to non-exclusive subsidies, inland water fishing and aquaculture, and intergovernmental payments through access agreements. The subsidy disciplines under the WTO Fisheries Subsidies Agreement are broad-based to curb subsidies harmful to fisheries, e.g., prohibition of fishing for overfished fish stocks and includes a comprehensive notification mechanism to improve transparency. *It is recommended to seize the window period for the agreement to enter into force, accelerate the adjustment of offshore and distant-water fisheries management and subsidy policies, strengthen research and mechanisms to support compliance, and closely follow the negotiations, so that the agreement supports promoting the high-quality and sustainable development of China's fisheries.*

Regional Fisheries Management Organizations or Arrangements

Regional fisheries management organizations or arrangements (RFMO/As) has been developed rapidly since the 1990s. At present, 15 RFMOs, 2 single-species management organizations, and 3 RFMAs with high seas fisheries management functions have been established globally, covering almost all regions of the global ocean except the Southwest Atlantic.

RFMO/A fisheries management is basically guided by the following principles: science-based fisheries management, which requires that fisheries management be based on the best available scientific evidence or information; application of the precautionary approach; fisheries management that considers ecosystems; non-contradiction (compatibility) between conservation and management measures; and development and use of selective fishing gear.

RFMO/A has become the implementation body of international fisheries in a practical sense, especially high seas fisheries management, and its role in international fisheries governance will become increasingly important. At present, China has joined 8 RFMOs and is a member party of 2 RFMAs. *It is recommended that China should increase and deepen its participation in RFMO/A so that RFMO/A can become a fundamental platform for China to participate in international ocean governance and enhance China's voice and influence in equitable international ocean equity governance.*

4.3 Opportunities and Challenges

To ensure the maximum benefit and synergies between a sustainable blue fishery economy and achieving carbon neutrality, governance of marine fisheries should be effectively incorporated into the holistic and integrated ocean management (IOM) system.

4.3.1 Current Progress and Potential Challenges

The use of marine fisheries resources is the oldest and most extensive form of exploitation of the oceans, and the sustainability of marine fisheries development has long been a global concern.

IOM aims to achieve sustainable development of the ocean by coordinating various ocean development activities, balancing conservation and exploitation of marine resources using EBFM [73], and supporting livelihoods and employment while maintaining the health and resilience of marine ecosystems [74]. Internationally, IOM has already achieved advanced application that is worthy of reference for China. It is important to note, however, that IOM always requires locally tailored strategies.

In China, there is still a lot of room to advance IOM. Since the institutional reform of the State Council in 2018, changed the long-standing two-segment governance of land and sea in China, and shifting to a new pattern of integrated land-sea governance [42].

In both marine capture and mariculture, China is the world's largest producer, and includes various scales of production, marine fisheries account for only 5% of the country's gross marine product, although they are responsible for the livelihoods of a large population.

4.3.2 Future Trends and Work Paths

Throughout the world, the concept of IOM has been widely recognized and implemented in many countries. China has just undergone a round of institutional reform.

In order to promote IOM in China's fisheries, *it is recommended that*:

(a) At the central government level, drawing on the common "leadership group" working model in China, establish a coordinating working group composed of multiple marine-related departments (National Development and Reform Commission, Ministry of Natural Resources, Ministry of Ecology and Environment, Ministry of Agriculture and Rural Affairs, Ministry of Science and Technology).
(b) At the local level, drawing on the practical experience represented by Xiamen, build a locally adapted implementation framework for IOM.
(c) Strengthen the synergy of law enforcement agencies.
(d) For the WTO Fisheries Subsidies Agreement, FAO PSMA and other international cooperation matters that have far-reaching impacts on China's fisheries

governance system, further integrate the management forces of relevant administrations to form a professional, multidisciplinary and inter-departmental work team.

4.4 Chapter-Specific Recommendations

Throughout the review and analysis presented in this chapter, it should be fully recognized that reducing carbon footprints and increasing climate resilience are not only necessary for the sustainable development of marine fisheries and mariculture but are also integral components of efficient ocean use to achieve carbon neutrality goals. This requires high-level policy-makers in major marine aquatic production countries, represented by China, to bring together the concerns and demands of different stakeholder groups at the domestic level, to lead government departments at all levels to implement strong policy governance, and to actively advocate and lead multilateral cooperation at the international level with the concept of a community with a shared future for humanity. The key principles that policy-makers need to implement include: (1) reduce carbon emissions from fishing vessels and incentivize the development of carbon sink fisheries; (2) further regulate harmful and carbon-intensive production practices, such as IUU fishing; (3) focus on the equal rights of marginalized groups in production and decision making and fully incorporate their experiences and wisdom in fisheries governance; and (4) incorporate fisheries governance into the strategic framework of IOM. Based on these principles, this chapter establishes a number of priority actions to achieve synergy between the high-quality development of marine fisheries and carbon neutrality.

Specifically, it is recommended to:

(1) Avoid fisheries management practices that spawn competition in fishing capacity, phase out harmful fishing vessel fuel subsidies, reduce excess fishing capacity, and promote a shift from fuel-intensive marine fishing gear and practices to those with lower carbon footprints.
(2) Promote research on the process and mechanism of fisheries as a carbon sink, and promote "negative-emission mariculture" that has the ability to sequester carbon, such as the cultivation of macroalgae and filter-feeding shellfish.
(3) Strengthen fisheries supervision and enforcement, apply big data technology to build a legality tracing mechanism for marine catches, and combine relevant international collaborative frameworks to combat IUU fishing and other harmful fishing practices.
(4) Build on the best available scientific knowledge to enhance the climate resilience of marine fisheries and the marine ecosystems on which they depend and to ensure fishing opportunities for small-scale fisheries.
(5) Carry out gender-inclusive fisheries governance reforms, promote the equitable distribution of resource use and management rights to fishers of all genders, and fully safeguard the rights and interests of women in fisheries.

(6) Increase input in participation in international agreements and processes related to marine living resources conservation and fisheries governance to ensure balanced voices and build an efficient global collaborative governance system.
(7) Bridge the functional boundaries between the various ocean management administrations, fully incorporate the stakeholder groups in different fields, promote the linkage between science and technology and management decisions, and build an integrated ocean management framework tailored to local conditions.

Abbreviations

ABMT	area-based management tools
ABNJ	areas beyond national jurisdiction
BBNJ	maritime biological diversity in areas beyond national jurisdiction
BRI	Belt and Road Initiative
CBD	Convention on Biological Diversity
CCICED	China Council for International Cooperation on Environment and Development
CCS/CCUS	carbon capture and storage/carbon capture, utilization, and storage
CDR	carbon dioxide removal
CPUE	catch per unit of fishing effort
EBFM	Ecosystem-Based Fishery Management
EDF	Environmental Defense Fund
EIA	Environmental Impact Assessment
EMFAF	European Maritime, Fisheries and Aquaculture Fund
EMFF	European Maritime and Fisheries Fund
FAO	Food and Agriculture Organization of the United Nations
GDP	gross domestic product
GHG	greenhouse gas
IGC	Intergovernmental Conference
IMO	International Maritime Organisation
IOM	integrated ocean management
IUU	illegal, unreported, and unregulated
MCS	monitoring, control, and surveillance
MEY	maximum economic yield
MGR	marine genetic resources
MPA	marine protected area
MSY	maximum sustainable yield
NDC	nationally determined contribution
NGO	non-governmental organization
NRDC	Natural Resources Defense Council
OECD	Organisation for Economic Co-operation and Development

PSMA	Port State Measures Agreement
RAS	Recirculating Aquaculture Systems
RDOC	recalcitrant dissolved organic carbon
RFMO/A	regional fisheries management organizations or arrangements
SBE	Sustainable Blue Economy
SDG	Sustainable Development Goals
SSF	Small-scale fisheries
TAC	total allowable catch
UN	United Nation
UNFCCC	United Nations Framework Convention on Climate Change
WTO	World Trade Organization

References

1. UNCTAD. *Advancing the potential of sustainable ocean-based economies: Trade trends, market drivers and market access*. Available from: https://webaplicacion.apn.gob.pe/proyecto/wp-content/uploads/2021/10/Advancing-the-potencial-of-sustainable-ocean-based.pdf
2. Wuwung, L., Croft, F., Benzaken, D., et al. (2022). Global blue economy governance—A methodological approach to investigating blue economy implementation. *Frontiers in Marine Science, 9*.
3. Gattuso, J. P., Magnan, A., Bille, R., et al. (2015). Contrasting futures for ocean and society from different anthropogenic CO_2 emissions scenarios. *Science, 349*(6243), aac4722.
4. Doney, S., Bopp, L., & Long, M. (2014). Historical and future trends in ocean climate and biogeochemistry. *Oceanography, 27*(1), 108–119.
5. Doney, S. C., Ruckelshaus, M., Duffy, J. E., et al. (2012). Climate change impacts on marine ecosystems. *Annual Review of Marine Science, 4*(1), 11–37.
6. Friedlingstein, P., O'Sullivan, M., Jones, M. W., et al. (2020). Global carbon budget 2020. *Earth System Science Data, 12*(4), 3269–3340.
7. Hoegh-Guldberg, O., et al. (2019). *The ocean as a solution to climate change: Five opportunities for action*. D.W.R.I. Available from: http://www.oceanpanel.org/climate
8. FAO. *Fisheries and aquaculture: Global production by production source quantity (1950–2020)*. FAO. Available from: https://www.fao.org/fishery/statistics-query/en/home
9. IMO. *IMO fourth greenhouse gas study 2020*. Available from: https://wwwcdn.imo.org/localresources/en/OurWork/Environment/Documents/Fourth%20IMO%20GHG%20Study%202020%20-%20Full%20report%20and%20annexes.pdf
10. Corbett, J. J., Winebrake, J. J., Green, E. H., et al. (2007). Mortality from ship emissions: A global assessment. *Environmental Science and Technology, 41*, 8512–8518.
11. Rahman, A., & Karim, M. M. (2015). Green shipbuilding and recycling: Issues and challenges. *International Journal of Environmental Science and Development, 6*(11), 838–842.
12. Available from: https://www.gov.cn/xinwen/2022-06/07/5694511/files/2d4b62a1ea944c6490c0ae53ea6e54a6.pdf
13. Parker, R. W. R., Blanchard, J. L., Gardner, C., et al. (2018). Fuel use and greenhouse gas emissions of world fisheries. *Nature Climate Change, 8*(4), 333–337.
14. Greer, K., Zeller, D., Woroniak, J., et al. (2019). Global trends in carbon dioxide (CO_2) emissions from fuel combustion in marine fisheries from 1950 to 2016. *Marine Policy, 107*.
15. Poore, J., & Nemecek, T. (2018). Reducing food's environmental impacts through producers and consumers. *Science, 360*(6392), 987–992.

16. MacLeod, M. J., Hasan, M. R., Robb, D. H. F., et al. (2020). Quantifying greenhouse gas emissions from global aquaculture. *Science and Reports, 10*(1), 11679.
17. Xu, C., Su, G., Zhao, K., et al. (2022). Current status of greenhouse gas emissions from aquaculture in China. *Water Biology and Security, 1*(3).
18. Wieczorek, A., Croot, P., Lombard, F., et al. (2019). Microplastic ingestion by gelatinous zooplankton may lower efficiency of the biological pump. *Environmental Science & Technology, 53*(9), 5387–5395.
19. OECD. (2022). *Global plastics outlook: Economic drivers, environmental impacts and policy options*. OECD Publishing.
20. Huang, L., & Yang, L. (2022). Status and prospect of urban waste treatment in China. *Sustainable Development, 12*.
21. Bai, M., Zhu, L., An, L., et al. (2018). Estimation and prediction of plastic waste annual input into the sea from China. *Acta Oceanologica Sinica, 37*(11), 26–39.
22. Lebreton, L. C. M., van der Zwet, J., Damsteeg, J.-W., et al. (2017). River plastic emissions to the world's oceans. *Nature Communications, 8*(1), 15611.
23. Zhao, S., Wang, T., Zhu, L., et al. (2019). Analysis of suspended microplastics in the Changjiang Estuary: Implications for riverine plastic load to the ocean. *Water Research, 161*, 560–569.
24. Mai, L., Sun, X., Xia, L., et al. (2020). Global riverine plastic outflows. *Environmental Science & Technology, 54*(16), 10049–10056.
25. Meijer, L., van Emmerik, T., van der Ent, R., et al. (2021). More than 1000 rivers account for 80% of global riverine plastic emissions into the ocean. *Science Advances, 7*(18).
26. GESAMP. *Sea-based sources of marine litter*. In K. Gilardi (Ed.). IMO/FAO/UNESCO-IOC/UNIDO/WMO/IAEA/UN/UNEP/UNDP/ISA Joint Group of Experts on the Scientific Aspects of Marine Environmental.
27. Tekman, M. B., Walther, B. A., Peter, C., Gutow, L., & Bergmann, M. (2022). *Impacts of plastic pollution in the oceans on marine species, biodiversity and ecosystems*. Zenodo.
28. Koelmans, A., Bakir, A., Burton, G., et al. (2016). Microplastic as a vector for chemicals in the aquatic environment: Critical review and model-supported reinterpretation of empirical studies. *Environmental Science & Technology, 50*(7), 3315–3326.
29. Wilcox, C., Heathcote, G., Goldberg, J., et al. (2015). Understanding the sources and effects of abandoned, lost, and discarded fishing gear on marine turtles in northern Australia. *Conservation Biology, 29*(1), 198–206.
30. Schuyler, Q., Wilcox, C., Townsend, K., et al. (2016). Risk analysis reveals global hotspots for marine debris ingestion by sea turtles. *Global Change Biology, 22*(2), 567–576.
31. Duncan, E., Botterell, Z., Broderick, A., et al. (2017). A global review of marine turtle entanglement in anthropogenic debris: A baseline for further action. *Endangered Species Research, 34*, 431–448.
32. Lusher, A., Hernandez-Milian, G., Berrow, S., et al. (2018). Incidence of marine debris in cetaceans stranded and bycaught in Ireland: Recent findings and a review of historical knowledge. *Environmental Pollution, 232*, 467–476.
33. Cole, M., Lindeque, P., Fileman, E., et al. (2015). The impact of polystyrene microplastics on feeding, function and fecundity in the marine copepod *Calanus helgolandicus*. *Environmental Science & Technology, 49*(2), 1130–1137.
34. Cole, M., Lindeque, P., Fileman, E., et al. (2013). Microplastic ingestion by zooplankton. *Environmental Science & Technology, 47*(12), 6646–6655.
35. Jabeen, K., Su, L., Li, J., et al. (2017). Microplastics and mesoplastics in fish from coastal and fresh waters of China. *Environmental Pollution, 221*, 141–149.
36. Lamb, J., Willis, B., Fiorenza, E., et al. (2018). Plastic waste associated with disease on coral reefs. *Science, 359*(6374), 460–462.
37. Harris, P., Westerveld, L., Nyberg, B., et al. (2021). Exposure of coastal environments to river-sourced plastic pollution. *Science of the Total Environment, 769*.
38. Zheng, X., Sun, R., Dai, Z., et al. (2023). Distribution and risk assessment of microplastics in typical ecosystems in the South China Sea. *Science of the Total Environment, 883*.

39. Pervez, R., Wang, Y., Jattak, Z., et al. (2021). The distribution and composition of litter on the Aoshan Beach Qingdao, China. *Journal of Coastal Conservation, 25*(4).
40. Stenger, K., Wikmark, O., Bezuidenhout, C., et al. (2021). Microplastics pollution in the ocean: Potential carrier of resistant bacteria and resistance genes. *Environmental Pollution, 291*, 118130.
41. Yu, X., Du, H., Huang, Y., et al. (2022). Selective adsorption of antibiotics on aged microplastics originating from mariculture benefits the colonization of opportunist pathogenic bacteria. *Environmental Pollution, 313*, 120157.
42. Tudor, D. T., & Williams, A. T. (2006). A rationale for beach selection by the public on the coast of Wales, UK. *Area, 38*, 153–164.
43. Kutralam-Muniasamy, G., Pérez-Guevara, F., & Shruti, V. (2022). (Micro)plastics: A possible criterion for beach certification with a focus on the Blue Flag Award. *Science of the Total Environment, 803*.
44. Lima, A., Silva, A., Pereira, L., et al. (2022). Anthropogenic litter on the macrotidal sandy beaches of the Amazon region. *Marine Pollution Bulletin, 184*.
45. Chen, G., Li, Y., & Wang, J. (2021). Occurrence and ecological impact of microplastics in aquaculture ecosystems. *Chemosphere, 274*.
46. Sussarellu, R., Suquet, M., Thomas, Y., et al. (2016). Oyster reproduction is affected by exposure to polystyrene microplastics. *Proceedings of the National Academy of Sciences of the United States of America, 113*(9), 2430–2435.
47. Greven, A., Merk, T., Karagöz, F., et al. (2016). Polycarbonate and polystyrene nanoplastic particles act as stressors to the innate immune system of fathead minnow (*Pimephales promelas*). *Environmental Toxicology and Chemistry, 35*(12), 3093–3100.
48. Mahamud, A., Anu, M., Baroi, A., et al. (2022). Microplastics in fishmeal: A threatening issue for sustainable aquaculture and human health. *Aquaculture Reports, 25*.
49. Walkinshaw, C., Tolhurst, T., Lindeque, P., et al. (2022). Detection and characterisation of microplastics and microfibres in fishmeal and soybean meal. *Marine Pollution Bulletin, 185*.
50. Skirtun, M., Sandra, M., Strietman, W., et al. (2022). Plastic pollution pathways from marine aquaculture practices and potential solutions for the North-East Atlantic region. *Marine Pollution Bulletin, 174*.
51. van Bijsterveldt, C. E. J., van Wesenbeeck, B. K., Ramadhani, S., et al. (2021). Does plastic waste kill mangroves? A field experiment to assess the impact of macro plastics on mangrove growth, stress response and survival. *Science of the Total Environment, 756*, 143826.
52. Napper, I., & Thompson, R. (2020). Plastic debris in the marine environment: History and future challenges. *Global Challenges, 4*(6).
53. Froehlich, H., Gentry, R., & Halpern, B. (2018). Global change in marine aquaculture production potential under climate change. *Nature Ecology & Evolution, 2*(11), 1745–1750.
54. Cheung, W., Lam, V., Sarmiento, J., et al. (2009). Projecting global marine biodiversity impacts under climate change scenarios. *Fish and Fisheries, 10*(3), 235–251.
55. Gephart, J., Henriksson, P., Parker, R., et al. (2021). Environmental performance of blue foods. *Nature, 597*(7876), 360–365.
56. Zhang, J., Liu, J., Zhang, Y., & Li, G. (2021). Strategic approach for mariculture to practice "ocean negative carbon emission." *Bulletin of Chinese Academy of Sciences, 36*(3), 252–258.
57. Tang, Q., Zhang, J., & Fang, J. (2011). Shellfish and seaweed mariculture increase atmospheric CO_2 absorption by coastal ecosystems. *Marine Ecology Progress Series, 424*, 97–105.
58. Tang, Q. S., Jiang, Z. J., & Mao, Y. Z. (2022). Clarification on the definitions and its relevant issues of fisheries carbon sink and carbon sink fisheries. *Progress in Fishery Sciences, 43*(5), 1–7.
59. Hill, R., Bellgrove, A., Macreadie, P., et al. (2015). Can macroalgae contribute to blue carbon? An Australian perspective. *Limnology and Oceanography, 60*.
60. Sebille, E., Spathi, C., & Gilbert, A. (2016). *The ocean plastic pollution challenge: Towards solutions in the UK*. Imperial College London, Grantham Institute.
61. Heisterkamp, I., Schramm, A., de Beer, D., et al. (2010). Nitrous oxide production associated with coastal marine invertebrates. *Marine Ecology Progress Series, 415*, 1–9.

62. Tang, Junchao., Wu, Y., Zhang, Y., et al. (2022). A Brief Introduction on the Industrial Development Mode of Photovoltaic Agriculture. *Chinese Agricultural Science Bulletin* (038–011).
63. Anderson, T., Martin, A., Lampitt, R., et al. (2019). Quantifying carbon fluxes from primary production to mesopelagic fish using a simple food web model. *ICES Journal of Marine Science, 76*(3), 690–701.
64. Bar-On, Y., Phillips, R., & Milo, R. (2018). The biomass distribution on Earth. *Proceedings of the National Academy of Sciences of the United States of America, 115*(25), 6506–6511.
65. Bianchi, D., Carozza, D., Galbraith, E., et al. (2021). Estimating global biomass and biogeochemical cycling of marine fish with and without fishing. *Science Advances, 7*(41).
66. Wilson, R., Millero, F., Taylor, J., et al. (2009). Contribution of fish to the marine inorganic carbon cycle. *Science, 323*(5912), 359–362.
67. Proud, R., Handegard, N., Kloser, R., et al. (2019). From siphonophores to deep scattering layers: Uncertainty ranges for the estimation of global mesopelagic fish biomass. *ICES Journal of Marine Science, 76*(3), 718–733.
68. Cao, L., Chen, Y., Dong, S., et al. (2017). Opportunity for marine fisheries reform in China. *Proceedings of the National Academy of Sciences of the United States of America, 114*(3), 435–442.
69. Trebilco, R., Melbourne-Thomas, J., & Constable, A. (2020). The policy relevance of Southern Ocean food web structure: Implications of food web change for fisheries, conservation and carbon sequestration. *Marine Policy, 115*, 103832.
70. Xiong, M., Wu, Z., Tang, Y., et al. (2022). Characteristics of small-scale coastal fisheries in China and suggested improvements in management strategies: A case study from Shengsi County in Zhejiang Province. *Frontiers in Marine Science, 9.*
71. UNFCCC. *Dimensions and examples of the gender-differentiated impacts of climate change, the role of women as agents of change and opportunities for women.* Available from: https://unfccc.int/documents/494455
72. FAO. (2017). Towards gender-equitable small-scale fisheries governance and development—A handbook. In *Support of the implementation of the voluntary guidelines for securing sustainable small-scale fisheries in the context of food security and poverty eradication.* Nilanjana Biswas.
73. FAO. (2012). *The state of world fisheries and aquaculture 2012 (SOFIA).* Food and Agriculture Organization of the United Nations (FAO). Available from: https://www.fao.org/3/i2727e/i2727e00.htm
74. Galappaththi, M., Armitage, D., & Collins, A. (2022). Women's experiences in influencing and shaping small-scale fisheries governance. *Fish and Fisheries, 23.*

Chapter 3
Innovative Technologies for Greenhouse Gas Emissions and Carbon Sequestration Monitoring

The 20th National Party Congress pointed to the importance of improving "carbon emissions statistics and accounting systems." China's green transition and climate plan similarly highlighted the need to improve "carbon emissions verification, accounting, and reporting standards for regions, industries, businesses, and products and establish a unified, well-regulated carbon accounting system."[1]

In recent years, there has been a sharp increase in the demand for different types of high-quality climate data as the basis of credible GHG monitoring and reporting. Climate data has also become an important feature of recent climate risk disclosure and environmental, social, and corporate governance (ESG) standards for green finance.

There are well-established international standards for GHG monitoring and reporting, notably those under the UN Framework Convention on Climate Change (UNFCCC). It is expected that the UNFCCC's forthcoming Enhanced Transparency Framework and Global Stocktake will increase attention to quality GHG monitoring, reporting, and verification (MRV) systems.

This study examines four topics: (i) GHG data monitoring and reporting for mandatory carbon markets based on China's sector-based reporting standards; (ii) methods and practices related to carbon sequestration measurement; (iii) metrics and measurement standards for current and emerging financial sector climate risk disclosure, and (iv) innovative new monitoring. It begins by discussing the characteristics of data quality.

Given the pace of change in climate data MRV systems within China and internationally as well as across mandatory and voluntary systems, this scoping study recommends that CCICED should examine evolving best practices and standards for GHG monitoring and reporting on a regular basis.

---

[1] A recent high-level report on China's carbon neutrality policy framework, co-authored by Zhu and Stern among others, notes: "An essential prerequisite to the effective delivery of [carbon neutrality] is a statistical and accounting system capable of supporting efficient carbon-neutral policy design and tracking its implementation."

Green Empowerment and High Quality Development,
https://doi.org/10.1007/978-981-96-4218-2_3

This study notes significant changes and innovations underway in how climate data is generated. Such changes are advancing rapidly and should be welcomed. At the same time, there are risks associated with moving too quickly from familiar and established practices to new systems.

Accordingly, this study recommends a transition plan in which current bottom-up self-reporting systems continue and complement continuous monitoring systems and other top-down systems. As a bridging approach, hybrid monitoring systems are recommended. A pilot system that combines bottom-up, top-down and hybrid methods should be launched and tested for 1–2 years, building on the recent five-sector CEMS pilot. The goal of an integrated bottom-up and top-down multi-sector pilot is to enhance the efficiency and accuracy of data needed to support a well-functioning national carbon market.

This study notes current best practice in bridging bottom-up and top-down monitoring of methane emissions and recommends that lessons from methane gas monitoring should inform approaches to other GHG emission monitoring, including other powerful non-CO_2 greenhouse gases.

This study explores some evolving monitoring and reporting systems for carbon sequestration. This study recommends that domestic and international case studies of credible carbon offset project-level monitoring and reporting be maintained and updated, with examples illustrating carbon sequestration characteristics of different ecological natural CDR systems, like forests, wetlands, grasslands, etc., as well as emerging CCUS engineered practices. This study also recommends that China's carbon offset metrics and measurement standards align and are interoperable with international best practices.

In the area of green finance, this study recommends that evolving domestic standards for climate risk disclosure and reporting align with comparable climate data metrics and measurement standards recommended in international jurisdictions as well as through the new International Sustainability Standards Board (ISSB) standards.

1 Data Quality

Accurate, timely, and authoritative data are of central importance to the implementation of climate change policies. As countries advance toward 2030 and mid-century or 2060 greenhouse gas emissions (GHG) reduction targets, climate data is essential both in tracking progress and attributing emission reductions to specific measures to determine if the right tools are in place.

The standard benchmark for official national-level climate data remains National Inventory Reports (NIR), designed to track national GHG emissions and report via the UNFCCC. The standard NIR methodology remains that of the IPCC. Virtually all net-zero frameworks are built around sector-based monitoring.

The common feature of national GHG monitoring and reporting system is the goal of providing quality data. The most common definition of quality data comes from national statistical agencies, which produce weekly, monthly, quarterly, and annual economic statistics and, as noted below, are increasingly issuing climate data.

At the heart of these six attributes is accuracy. Accurate data is described as information that is a truthful measure, indicator or estimate of actual conditions or in the case of climate policy, GHG emissions.[2] Since statistics are rarely 100% accurate all of the time, data providers disclose the level of confidence concerning the accuracy of the data they produce.

While accuracy is the most important feature of data systems, the six data principles should be seen together: for example, data that is accurate but inaccessible, irrelevant or untimely does not meet data quality thresholds. For many national targets under the Paris Climate Agreement closely focused on interim and 2030 targets, data that is timely becomes more critical.

1.1 How Much Does Good Data Cost? What Is It Worth?

Quality data is expensive to produce, both for governments and increasingly for companies that are required to disclose it. Experts in this study noted that increasing accuracy levels from 95 to 99% can escalate data costs sharply.

At the same time, the returns on quality climate data investments are substantial. Real-time and accurate data can signal which sectors or measures are underperforming and need to be strengthened, adjusted, or replaced. Accurate data is essential to enable markets to realize the resource allocation efficiency of market-based measures, which in turn can help unleash innovation through competition.

For decades, the main sources of public climate data have been national environmental agencies, such as Sweden's Environmental Protection Agency to Environment and Climate Change Canada. Each has its own QA standards to ensure quality data. The European Environment Agency (EEA) reports that its QA philosophy entails aligning the "degree of agreement between the true value and the data used to represent the value" of emissions.[3]

A recent and welcome step has seen national statistical agencies begin to issue their own climate statistics, usually on a more frequent basis than the NIRs. In addition, a whole new set of net-zero implementation indicators are being reported within the decarbonization plans of the United Kingdom, Scotland, France, the European Union, Sweden and others.

China has taken important steps in creating a coherent and accurate set of national carbon statistics. In April 2022, the National Bureau of Statistics, together with MEE and NDRC, issued the "Implementation Plan for Accelerating the Establishment of a Uniform and Standardized Carbon Emission Statistics and Accounting System."

[2] Quality data terms related to accuracy include definitions of "trustworthiness, authenticity or consistency," as well as the disclosure of acceptable margins of error in estimates and the disclosure of uncertainties.

[3] European Monitoring and Evaluation Programme (EMEP); EEA Guidebook, 2016.

The Plan aims to support China in achieving its 30/60 climate goals by identifying four key tasks and five safeguard measures.

Key tasks:

- Establish national- and local-level carbon emission statistics and accounting systems;
- Improve the current carbon emission accounting mechanism, which works at a sectoral and enterprise level;
- Establish and improve carbon emission accounting methodologies for key products;
- Improve the National Greenhouse Gas Inventory Compilation Mechanism.

Safeguard measures:

- Lay a solid statistical foundation;
- Build emission factor database;
- Apply advanced technology;
- Carry out methodology research;
- Improve supporting policies.

Drawing from the implementation of this Plan and emerging international examples, NDRC should consider issuing quarterly climate data reports.

Sector-based emissions data: China's national climate data system to monitor GHG emissions at the economic sector level was launched in 2013. Its coverage has progressively expanded to comprise power generation, power grids, iron and steel, chemical production, petrochemicals, aluminum, magnesium, glass, cement, paper and pulp, and civil aviation. More recent reporting standards include mining, public buildings, the food and beverages sector, and road transport.

Like other national monitoring systems, China's climate data system is based on self-reporting by businesses at the company and facility levels. MEE plays a key role in issuing QA guidelines, checking data misstatements or fraud, conducting site inspections, providing training modules and other steps. Company and facility-level reporting is via the competent provincial-level MEE authorities. Data is reported to the National Pollutant Discharge Platform, which includes, in addition to air and other pollutants, GHG emissions data.

As in other bottom-up, self-reporting systems based on Pollutant Release and Transfer Register (PRTR) approaches, discrepancies or anomalies in data accuracy should be reported directly by companies or through third-party consulting companies engaged by enterprises to calculate emissions.

An important new and potentially game-changing phase in sector-based GHG monitoring and reporting was launched last year. In 2021, MEE initiated a pilot to test the efficacy of a continuous GHG monitoring system. The new pilot program involves five sectors: thermal power, steel, oil and gas, coal mining, and waste treatment. The pilot entails monitoring CO_2 emissions through a Continuous Emissions Monitoring System (CEMS) to monitor criteria air pollutants including particulate matter (PM), NO_X, and SO_2.

As its name suggests, the key characteristic of the CEMS system is the provision of real-time—usually hourly—air pollution emission information. These measurements are at the facility and smokestack level. Empirical data is overlaid with emissions factors and other information, whereby CEMS appears to enhance the spatial–temporal resolution of emission factor estimates.[4]

The CEMS pilot is designed to monitor CO_2 emissions. It will be important to augment that initial focus with other non-CO_2 GHG emissions. Methane is an especially potent GHG, as well as the focus of a number of innovative monitoring and reporting initiatives.

Fossil fuel methane emissions: Since methane is a relatively short-lived GHG, reducing emissions is crucial in slowing warming in the near term. Globally, the oil and gas (O&G) sector comprises the largest proportion of methane emissions, while in China, coal mining accounts for the largest share.

Measuring methane accurately is key to meeting China's dual-control goals.

Bottom-up inventories are a primary tool for monitoring methane emissions from the oil and gas sector. Bottom-up inventories provide essential information on potential sources of emissions and are critical for planning mitigation approaches. However, a major challenge is the dependence of this bottom-up approach on emission factors that are developed for equipment and components at normal operation. When compared with atmospheric measurements, these inventory-based approaches are found to systematically underestimate emissions.

Atmospheric monitoring approaches combine measurements of atmospheric concentrations (CH_4 mole fraction in the atmosphere) with transport and dispersion models that use meteorological inputs to convert detected concentrations to emissions. These models rely on initial input assumptions about sources (priors) and provide the most accurate results when detailed facility and activity data are available.

Dedicated testing facilities like the Methane Emissions Technology Evaluation Center (METEC) and TotalEnergies Anomaly Detection Initiatives (TADI) are enabling the development of international standards for leak detection and quantification of methane emissions.[5]

Bottom-up monitoring systems are being complemented by a new generation of top-down systems led by remote sensing, satellite-monitoring platforms.

[4] CEMS has been used for years in China and other jurisdictions. For example, the U.S. EPA system to monitor acid rain and wider criteria air pollutants focuses on monitoring large emission sources like coal power plants. The EPA CEMS system has been augmented to include CO_2 monitoring. This builds on earlier CO_2 monitoring used to verify whether corresponding SO_2 and NO_x emissions for a given facility based on emission factors or historical trends were accurate. (This use of CO_2 emissions to help determine the quality of NOx and SO_x emissions predates GHG inventory self-reporting reporting requirements under the EPA for large emission sources; smaller combustion sources continue to rely on average emission factors such as energy and fuel use.)

[5] US Department of Energy (4 April 2023) *Joint Statement by the U.S. and EU following the 10th U.S.-EU Energy Council.*

China's 2022–2023 pilot GHG monitoring test includes methods to improve methane monitoring in the oil and gas sector, and, in particular, to combine ground-level and remote sensing monitoring to assess fugitive emissions (leakage, hot flaring and abnormal operations). The results and recommendations of this pilot will be important given the unique challenges methane emission monitoring pose.

Recommendations: Tracking CO_2 on a continuous basis has the potential to transform the current MRV system in an innovative way. At the same time, it will take several years to determine the level of accuracy of new systems, as well as their overall advantages, as well as any shortcomings and limitations. Therefore, a transition plan should continue in which roles, responsibilities and clear protocols are developed. Transitions are also a good time for intentional system redundancies, which can also be used to benefit quality assurance. It is recommended that there be a pilot test of the interplay between the current self-reporting system with the use of a new CEMS system and the use of top-down systems.

The results of MEE's pilot methane monitoring system should provide valuable lessons in the wider application of bottom-up and top-down monitoring. China should also consider deepening its partnership with two UN initiatives, UNEP's IMEO and the Climate and Clean Air Coalition (CCAC) to address other potent short-lived climate pollutants.

2 Carbon Sequestration

Carbon sequestration is usually divided into two categories: ecosystem-based or nature-based sequestration derived from forests, wetlands, grasslands, and oceans, and "technological" or engineered solutions, such as carbon removal technologies.

The vast majority of public and private carbon sequestration projects involve nature-based climate solutions.

As noted, markets and governments are increasing investments in carbon sequestration. Part of this increased demand comes from exaggerated claims of carbon sequestration of large-scale initiatives such as large-scale afforestation.

Subsequent work at the scientific and implementation levels points to the significant challenges associated with deriving fungible carbon credits from carbon sequestration projects. A central challenge relates to monitoring and reporting. The IPCC has issued numerous and more detailed methodologies to measure carbon sequestration. Thousands of companies in turn have been launched, providing services to governments and companies.

The methodological complexities of monitoring and reporting LULUCF have long been noted. Among the measurement challenges associated with estimating carbon sequestration include comparing over time the sequestration rates of diverse ecosystems against average baselines. Carbon stocks and sequestration rates differ across ecosystems and geographies. Estimating above-ground carbon stocks requires data on the extent and health of standing forests and other ecosystems and the extent of other factors, like dead and decaying organic matter. Estimating below-ground carbon stocks is more complex, requiring on-site testing that is costly and intrusive. Other challenges include estimating the duration or permanence of offsets in lieu of disruptions due to land-use change, wildfires, floods, and other impacts.[6]

The importance of ensuring co-benefits of carbon sequestration activities were examined in a CCICED Special Policy Study on Nature-Based Solutions.

Given these and other complexities, it is not surprising that there has been long-standing criticism of carbon credits claimed from carbon sequestration projects.

There have been steps to improve the efficacy of carbon offset projects and markets.

At the multilateral level, there have been welcome steps to define and benchmark high-quality carbon sequestration projects, especially as they related to nature-based solutions. The final text of COP 27 referenced for the first time Nature Based Solutions, underscoring the climate-nature nexus, while the Kunming-Montreal Global Biodiversity Framework notes linking climate and nature through nature-based solutions and/or ecosystem-based approaches (Target 8).

At the operational level, the finalization of Article 6 rules at COP 26 sets out the framework governing international trading in carbon credits. While much work needs to be done to operationalize these new rules, it is expected that the rules will clarify how carbon sequestration can be measured, traded, and accounted for. Various pilots have been initiated by governments around Article 6, while the World Bank has launched the Climate Action Data Trust to support carbon sequestration registries digitally.

At the market level, there have been several efforts to set out standards covering voluntary carbon markets. In response to a number of extraordinarily deficient carbon sequestration transactions, the Voluntary Carbon Market Integrity Initiative (VCMI) is formulating safeguards and standards to ensure the integrity of carbon credits.

An additional development has been emerging standards and rules related to carbon offset credits set out in recent mandatory and voluntary climate risk disclosure standards that include recommendations on how carbon offsets can be claimed.

Recommendation: Considerable uncertainties remain in estimating carbon sequestration outcomes. It is important that bodies in China overseeing offset standards keep up to date with the latest science and methodologies in the international voluntary carbon market, across the full range of sequestration activities.

---

[6] A recent audit from Canada's Commissioner of the Environment and Sustainable Development concludes that the government's two-billion tree program intended to enhance carbon stocks will remain a net source of GHG emissions to 2031, due to start-up site preparation and tree-planting activities.

3 Climate Risk Disclosure Data

As green finance continues to expand, so too have rules and standards covering metrics and measurement protocols specifying the kind of data to be used for climate risk disclosure.

Recent rules and standards are based on evolving recommendations of the Task Force on Climate Risk Disclosure (TCFD), such as its October 2021 Guidance on Metrics, Targets and Transitions.

The GHG Protocol is an initiative of two CCICED partners—WRI and WBCSD—first launched in 1998. Akin to the role of the International Organization for Standardization vis-à-vis the World Trade Organization, the Protocol has become the data quality standard for the international financial sector.

Carbon footprint: TCFD further recommends companies calculate their carbon footprint. These estimates are typically overlaid with data from emission factor estimates developed by the IPCC for the given sector.

Converging standards: While TCFD has been the foundation for climate data, there has been a high degree of fragmentation in how those standards are interpreted.

This is why recent steps by many national jurisdictions to set out new rules based on TCFD are so important.

Certainly, there are some differences among recent regulations, particularly regarding Scope 3 emissions and the EU's inclusion of double-materiality. At the same time, there has been a notable convergence around climate data that can be used, in particular, their common reference to GHG Protocol and Carbon Footprint as the basis of climate data metrics.[7] Such convergence is a key recommendation of the 2022 G20 Sustainable Finance Working Group's call for greater comparability and interoperability across green finance markets.

Green finance has been at the forefront of China's agenda for nearly a decade. It is expected that by 2025 the current system will switch to mandatory climate risk reporting.

Recommendation: China's green finance standards and company practices should align with emerging international norms and standards, including new ISSB ESG and climate risk disclosure standards.

---

[7] Even with recent mandatory regulations, work continues to drill down into greater details of the type of climate data most appropriate to specific market segments. For example, the Global GHG Accounting and Reporting Standard was launched by the Partnership for Carbon Accounting Financials (PCAF) to define information related to mortgages, commercial vehicle loans, and real estate. In October, 2022, the UK Financial Conduct Authority (FCA) launched new Sustainability Disclosure Requirements (SDR), based on TCFD and ISSB, arguing that even with new reporting standards, the risk of greenwashing remains significant. Among its proposals are a series of sustainable investment labels designed to inform consumers of different financial products.

4 Continuous GHG Monitoring Systems

There has been a sharp increase in the deployment of new, innovative climate data systems that promise to provide near-continuous, highly granular, and spatially referenced carbon source and sinks data.

Climate Trace presents country-level maps that identify all major facility-level sources of major GHG emissions sources, notably oil and gas—the largest sector-based source of GHG emissions globally—electrical power stations, cement factories or airports, based on satellite observations, ground-level observations, and AI. The platform received considerable attention at COP 27. However, it has also been criticized for how it aggregates different types of climate data that may not be comparable.

There has also been a notable increase in climate data intended to support Article 6 rules. In addition to the World Bank's new portal noted above, the Climate Warehouse was recently launched with a similar goal of improving the quality and transparency of climate data to support Article 6 in several developing countries.[8]

Given the challenges in estimating net above- and below-ground carbon stocks discussed in Sect. 2, there have also been impressive new measurement techniques that include quickly evolving, near-continuous spatial measurements. Finally, as in other areas, there has been an increased use of AI to augment top-down, continuous GHG data systems, with ongoing research and company applications. These new hybrid systems promise to transform how climate data is generated and accessed via smartphones and other platforms.

Recommendation: Given the central role that digitization and technological innovation play in China's high-quality development and 14th FYP, the MEE can be a world leader in new hybrid climate data systems. A pilot system, noted above, that tests a holistic bottom-up, top-down and hybrid system should be tested with the objective of improving China's availability of high-quality climate data for enhancing climate governance and decision making at all levels.

[8] The pilot involves Bangladesh, Bhutan, Chile, Ghana, Kazakhstan, Japan, Peru, Rwanda, Senegal, Singapore, Sweden, Switzerland, Ukraine, and Namibia, well as UNDP and other entities.

Part II
National Green Governance System

Chapter 4
Collaborative Mechanism for Carbon Reduction, Pollution Reduction, Green Expansion and Growth

1 Introduction

The report of the 20th National Congress of the Communist Party of China has made a significant strategic deployment to promote green development and harmonious coexistence between human beings and nature. The report emphasizes the construction of a beautiful China, reducing carbon dioxide (CO_2) emissions, curbing air pollution, and fostering low-carbon economic growth, all while giving priority to ecological conservation alongside green development.

Promoting the synergy of pollution and carbon reduction is a powerful tool to implement this new development concept and drive the comprehensive green transformation of economic and social development. It is also an inevitable choice to achieve the construction of a beautiful China and the goal of "dual carbon." The concept is based on the fact that air pollutants and carbon emissions are highly interconnected and largely consistent in terms of control approaches, management strategies, tasks, and measures.

Conventional air pollutants (NO_x, $PM_{2.5}$, SO_x and other particulate matter) and greenhouse gases (CO_2, CH_4, N_2O, etc.) have a high degree of homology, and synergistic control can effectively integrate resources, improve efficiency, reduce costs, improve public health, and at the same time promote the application of new technologies, direct investment to better choices, and provide new impetus for economic growth.

Green Empowerment and High Quality Development,
https://doi.org/10.1007/978-981-96-4218-2_4

2 Promoting Economic Growth Through Emissions Reductions

The relationship between the "dual-carbon" goals and economic growth is, in essence, the relationship between the environment and development. However, emissions reductions have long and wrongly been considered a burden on economic growth. Today, the debate about the relationship between the "dual-carbon" goals and economic growth places more emphasis on how to strike a balance between the two. But increasingly, evidence shows that decarbonizing the economy is a major domestic opportunity and a driving force for promoting the high-quality development of China's economy.

The key to recognizing and seizing the opportunity lies in whether we can completely, accurately and comprehensively understand and put into practice a new development philosophy. A simple criterion to distinguish between old and new development philosophies is whether the environment and development are deemed to be mutually reinforcing or contradictory. In advancing both the "dual-carbon" goals and the need to cut conventional pollutants, it is essential to abandon the traditional outdated and disproven idea that decarbonization is a net cost to the economy.

2.1 Why Carbon Neutrality Is a Major Opportunity for China

2.1.1 Carbon Neutrality Is an Opportunity for China to Create a Sustainable Development Paradigm

The "dual-carbon" goals involve not only energy but also a shift in development paradigm; carbon neutrality is not only China's action, but also reflects worldwide consensus and action. So far, more than 150 countries have committed to carbon neutrality in various forms. These countries account for around 90% of the world's carbon emissions and economy.[1] Why have so many countries committed to carbon neutrality? First, of course, because the climate crisis is getting worse and, if unabated, the humanitarian burdens on China and the planet will be staggering. Second, experts and policymakers increasingly see the huge opportunities behind emissions reduction.

This is creating an international trend. About 70% of the 150 countries that are committed to carbon neutrality are developing countries. The global consensus and action on carbon neutrality signifies that the traditional development paradigm is becoming a thing of the past, and a new green development paradigm is on the horizon.

[1] https://zerotracker.net.

China possesses unique advantages in green development. The first advantage is its new development philosophy. The second advantage is its strong government commitments. The third advantage lies in its market The fourth advantage lies in technology. In sum, the industrial revolution-built prosperity at the expense of the environment. The green development paradigm turns the trade-off between environment and development into synergy and will determine the future.

2.1.2 Carbon Neutrality Is a Major Opportunity for China's Industries to "Change Lanes"

Carbon neutrality is the most comprehensive and profound shift in development paradigms since the Industrial Revolution, and this means that many of the industries established in the traditional industrial era will be modified or even rebuilt. This transition process is a major challenge, but it will also bring a wealth of new opportunities. China has already developed a competitive edge in many realms that can help it shift its development paradigm.

2.2 How the "Dual-Carbon" Goals Are Promoting Economic Growth

2.2.1 Are the "Dual-Carbon" Goals Dragging or Driving Economic Growth?

The "dual-carbon" goals can be either a drag on or a driver of economic growth: the difference depends on the *pattern* of economic growth. If the traditional pattern of high emissions and high resource consumption, namely "digging coal, opening mines and running factories," established after the Industrial Revolution was adopted, then the "dual-carbon" goals would impede economic growth; if a green transition pattern was adopted, which features a modern concept and content of development, then the "dual-carbon" goals and prosperity would reinforce each other.

In fact, China has an increasingly sophisticated view of the relationship between environmental protection and growth. In the early days of its industrialization, China considered emissions reduction to be a burden on economic growth because production requires energy inputs, and for China especially, that has meant huge amounts of fossil energy. This pathway offered two possible outcomes from carbon reduction: less production, which impacts economic growth, or higher production costs. In both cases, the "dual-carbon" goals would become a burden on economic growth. Logically, then, under this obsolete model, each country perceived carbon emissions as a so-called right to development, with the hope that other countries would reduce more while it could reduce less.

After the 18th CPC National Congress, China made a fundamental shift in its understanding of the relationship between the environment and development and related action, from "being told to reduce emissions" to "reducing emissions."

The "dual-carbon" goals are "what we want to do ourselves" rather than "what others want us to do." The core reason behind this is that the traditional development model is no longer sustainable, and China's economy must undergo a green transition. The traditional development model has not only brought on an unsustainable environmental crisis but also often left behind this core purpose of development. The new development philosophy has the core goals of a better life and people-centred benefits. It is the essence of the fundamental purpose of development.

Once such a green transition takes place, the environment and development reinforce each other. Under the new structure of wind, solar, and other renewables, cutting emissions is accelerating clean energy technology, and bringing about numerous new economic opportunities. In this way, cutting emissions can become a process of "creative destruction," which will enable the economy to leap from an old structure to a more competitive new one.

With traditional development theories and thinking, one can scarcely perceive the profound shift in the development paradigm behind the "dual-carbon" goals and the huge opportunities for growth brought about by such a shift.

2.2.2 Shifting Away from the Old Sources of Growth in the Old Development Paradigm

The first old source of growth in the old development pathway is, based on consumerism and excessive consumption, to encourage people to wastefully consume more physical goods.

The second old source of growth is to create new market demand by stimulating people's desires through marketing and "innovation." Most of these desires do improve people's well-being, but many are useless and may even have negative side effects.

2.2.3 New Development Paradigm: A Return to the Original Purpose of Economic Development

Unfortunately, the so-called modern economic growth is, to a great extent, moving toward the first or second pathway above. A significant portion of the so-called modern economic activities are essentially "ditching and filling" in the Keynesian sense. Green transition is the third pathway. Indeed, the "smiling curve" could be a pathway for an individual firm or country to foster economic growth, but what we need now is a green transition for the whole world economy. Hence, a green transition is more fundamental.

The new development pathway is to satisfy people's demand for well-rounded development—that is, translating the demand for cultural and other services beyond materials into a driver of economic growth after core material needs are met. This means a change in what is produced and consumed. Economic growth is increasingly built upon intangible resources, such as knowledge, technology, environment, culture, and experience, rather than the input of material resources, as in the past.

Today, China's development strategy is changing dramatically, from GDP-oriented development to well-being-oriented development or people-centered development. This could be seen from the progress in the Human Development Index and life satisfaction in China.

2.2.4 Analysis of the Impact of Emission Reduction on Economic Growth[2]

Promote Economic Growth and Emission Reduction at the Same Time by Improving the Economic Efficiency of Traditional Sectors, Including by Reducing Carbon Intensity Through Technological Innovation, Organizational Innovation, and Management Innovation

Compared to the U.S., the European Union and the world average levels, there is still a big gap for China to improve, including to decrease its share of electricity generation from fossil fuels. China, at this stage, is much lower than the international level in the absolute level but is catching up quickly.

Internalizing External Costs as Much as Possible to Make Green Development More Cost-Effective

If the external and social costs are taken into account, the seemingly efficient traditional development model is actually more costly. Instead, green development might be more cost-effective.

We use air pollution as an example to show the potential for China to decrease the external cost of its traditional growth model. Though the share of deaths attributed to air pollution from 1990 to 2019 keeps dropping, it is still much higher than the world average level.

How Emission Reduction Contributes to the Economy's Leap to a New, More Sustainable Structure, Which Is Called Creative Destruction—For Example, a Leap from the "Fossil Fuel-Powered Vehicles" Structure to the "NEVs" One

As we can see from the dramatic decline in new energy prices and increase in new energy generation, the economy is experiencing a leap to a new, more competitive structure.

---

[2] Further Quantitative Research Yet to Be Conducted.

The Economy Can Be More Dematerialized, and Growth Can Be More Dependent on the Input of Non-material Resources—For Example, Technology, Knowledge, Culture, Ecological Environment, Creativity ... To Achieve Higher Productivity, Greater Sustainability, and Better Well-Being

We use "the number of people using the Internet" as a proxy to show China's potential to achieve a green economy in the digital era. The number in China is almost equal to the number in Europe and North America combined. As Smith [1] pointed out, the division of labor is the source of economic growth, and the division of labor is limited by the extent of the market. The large number of people using the Internet is a unique opportunity for China in developing its green economy in the digital era.

2.3 *The "Dual-Carbon" Goals Are Becoming a New Driver of China's Economic Growth*

2.3.1 The "Dual-Carbon" Goals Create New Drivers of Growth

In fact, the "dual-carbon" goals have not affected China's economic growth. After the 18th CPC National Congress, China made a fundamental shift in its understanding of the relationship between the environment and development and has taken unprecedented actions to protect the environment without sacrificing economic growth.

From the perspective of new energy and NEV industries, which are most relevant to the "dual-carbon" goals, these goals are promoting—rather than hindering—economic growth. These industries have seen a growth spurt since the "dual-carbon" goals were announced. It should be particularly noted that NEVs provide a vivid example for understanding the relationship between the "dual-carbon" goals and economic growth. Such growth in the auto industry is also happening in emerging sectors such as 5G, robotics, artificial intelligence (AI), and the Internet economy.

Meanwhile, phasing out fossil fuels has a controllable impact on the economy. The impact of the "dual-carbon" goals on economic growth is most directly felt by traditional fossil fuels and related sectors. According to China's national planning, the share of fossil fuels in the country's energy consumption will drop to 75% and 20% by 2030 and 2060, respectively.

3 Synergistic Mechanism and Pathway for Carbon Neutrality and Clean Air

Climate change has become a major issue concerning the survival and sustainable development of mankind, and risks such as high temperatures and heat waves, extreme precipitation and natural hazards are growing, presenting an urgent need for collaboration among countries to achieve CO_2 reduction worldwide. China has announced the major strategic decision of peaking its carbon emissions by 2030 and achieving carbon neutrality before 2060. The "dual-carbon" goals not only point out the direction for high-quality social and economic development but also provide a basic guideline for coordinating air pollution control and GHG emissions reduction.

3.1 *Research Background*

Since the beginning of the 14th Five-Year Plan (FYP) period, China has entered a critical period of ecological conservation, during which the country will, with the reduction in carbon emissions as a major strategic goal, advance the synergistic reduction of pollution and carbon emissions, promote a comprehensive transition to green economic and social development, and bring a fundamental change to its eco-environment by accumulating small changes. Promoting concerted efforts to cut carbon emissions, reduce pollution, expand green development, and pursue economic growth is the key to the implementation of the new development philosophy in China. However, existing studies on the collaborative mechanism for pollution reduction, carbon reduction, green expansion, and growth often focus on a single area—for instance, energy mix, air quality, changes in emissions, etc.—and lack systematic, across-the-board or comprehensive assessment.

Against this backdrop, the SPS worked out 20 indicators in five areas. It established an air pollution and climate change synergistic management monitoring index system. By tracking the progress in each indicator, the SPS analyzed the achievements and obstacles in the process of China's co-governance of carbon neutrality and clean air.

3.2 *Process Evaluation of China's Co-governance of Carbon Neutrality and Clean Air*

3.2.1 Structural Transformation

The low-carbon transformation of the energy mix, the industrial and transportation structures, and the application of next-generation energy and emission reduction technologies are fundamental to a continuous reduction in CO_2 and air pollutant emissions and to synergy between air quality improvement and CO_2 reduction.

From an energy mix perspective, promoting green and low-carbon energy development and building a new electricity system with an increasing share of new energy provide important support for coordinating high-quality eco-environmental protection and high-quality economic development. From an industrial structure perspective, measures such as resolutely curbing the haphazard development of energy-intensive and high-emission projects and developing green and low-carbon industries have effectively promoted the synergistic reduction of pollution and carbon emissions. From a transportation structure perspective, it is of great significance to continuously improve the energy efficiency in the transportation sector, promote the substitution of clean energy, and gradually optimize the transportation structure, in order to propel the reduction of pollution and carbon emissions across China's society and economy.

3.2.2 Synergistic Reduction of Pollution and Carbon Emissions

Amid the overarching goals of achieving carbon peak and carbon neutrality, China is persistently intensifying the coordinated control of GHG emissions and atmospheric pollutants, as well as enhancing the synergistic control of fine particulate matter and ozone. In terms of pollution control, the potential for emissions reductions from end-of-line treatments in conventional industrial sectors and emissions-intensive sectors has been almost fully tapped. However, it is expected that deep pollution control in non-power sectors, management of volatile organic compounds, control of mobile source emissions, and promotion of clean heating in rural areas will continue to play a significant role. This is particularly pertinent for pollutants such as volatile organic compounds and ammonia, the emissions of which have yet to demonstrate a significant decline. To further the synergistic benefits of "pollution and carbon reduction" in these areas, effective emissions reduction strategies should be further implemented.

Air pollution control policies have led to the synergistic reduction of pollution and carbon emissions. On a national scale, there was a positive synergy between CO_2 emission reduction and $PM_{2.5}$ pollution improvement in China's industrial sectors between 2015 and 2020, indicating effective measures for the adjustment of the energy mix and the industrial structure during the 13th FYP period.

For transportation and residential sectors, although structural adjustment, bulk coal control, and other transformation measures taken during the 13th FYP period have begun to deliver results, CO_2 emissions were still growing on the whole, with a slight increase of 8%. In addition to structural adjustment, the implementation of end-of-pipe control measures—such as upgrading automobile emission standards and integrating the control of vehicles, fuels, and road transportation—brought $PM_{2.5}$ concentrations down by 22–23% during that period. Further structural transformation in transportation and residential sectors still enjoys great potential for the synergistic reduction of pollution and carbon emissions.

3.2.3 Air Quality

China's achievements in the reduction of pollution and carbon emissions are fully embodied in the process of air-quality improvement. In 2021, pollutant concentrations in China's 339 prefecture-level cities and five key regions—including Beijing-Tianjin-Hebei, Fenwei Plain, Yangtze River Delta, Chengdu-Chongqing, and Pearl River Delta—dropped compared to 2020 levels, with the regional annual average SO_2 and NO_2 concentrations reaching the national Grade I standards. Except for FW, the regional annual average O_3 concentration also reached the national Grade II standards. Nevertheless, many regions have not reached $PM_{2.5}$ standards.

3.2.4 Health Benefits

The goal of carbon neutrality will drive profound transformations in energy structures and technology iterations, thereby significantly improving air quality and enhancing health standards. Thanks to air-quality improvement, the levels of long- and short-term exposure to $PM_{2.5}$ in China continue to fall, and associated premature adult deaths have significantly decreased. Apart from continuous improvements in exposure, other possible causes include (1) the long-term $PM_{2.5}$ exposure–response relationship is steeper at low concentrations, indicating greater marginal benefits of exposure improvement based on low and medium concentrations than on high concentrations; (2) population aging has increased the overall population susceptibility to $PM_{2.5}$, partly amplifying the health benefits associated with $PM_{2.5}$ improvement. A comparison suggested that the number of premature deaths associated with short-term O_3 exposure was higher than or on par with (if the uncertainty is taken into account) that of $PM_{2.5}$.

3.2.5 Local Practice

Despite some successes in China's reduction of pollution and carbon emissions, there is still much room for improvement. Based on ambient air-quality monitoring data and inventories of CO_2 emissions, the report analyzed the trend in coordinated variation of $PM_{2.5}$ concentrations and CO_2 emissions in China's 335 prefecture-level cities and above between 2015 and 2020. The results showed that from 2015 to 2020, only 105 cities, or 31.3% of all the cities, witnessed a reduction in both annual average $PM_{2.5}$ concentrations and CO_2 emissions. On average, the annual average $PM_{2.5}$ concentrations in these cities decreased by 29%, and CO_2 emissions by 23%, both higher than the decreases between 2015 and 2019. On the contrary, the annual average $PM_{2.5}$ concentrations and CO_2 emissions both increased in 17 cities, accounting for 5.1%. A study found that most cities failed to achieve a synergistic decline in $PM_{2.5}$ concentrations and CO_2 emissions between 2015 and 2020, indicating that synergies to reduce pollution and carbon emissions need to be further promoted at the city level.

3.3 Policy Recommendations

Achieving synergy between pollution and carbon reduction is increasingly seen as the driving force behind the green transformation of economic and social development. Entering the 14th FYP period and under the guidance of new development concepts, strategic planning and specific policies in the field of ecological and environmental protection have gradually begun to harmonize atmospheric pollution prevention and GHG reduction efforts.

At the national level, the 14th FYP introduced in 2021, the central government's proposals on carbon peaking and carbon neutrality, and prevailing in the battle against pollution all reflect new expectations of a mutual and collaborative focus on pollution and carbon reduction. This provides guidance on how to drive a comprehensive green transformation of socio-economic development with carbon reduction at the helm, strengthen the collaborative control of multiple pollutants and GHGs, and use pollution reduction as a key measure to achieve carbon reduction goals. Building on this, MEE of the PRC is acting in areas such as environmental impact assessments, monitoring and supervision around the collaboration of pollution, and carbon reduction. These efforts aim to improve existing management systems and integrate carbon emission management requirements into ecological and environmental management systems, gradually perfecting a coordinated governance structure.

In terms of market mechanisms, the carbon emissions trading system has made significant progress. In 2021, the first compliance cycle of the national carbon emissions trading market concluded successfully. Voluntary GHG reduction trading has been further promoted.

In summary, at the societal management level, China has begun proactively establishing a management system and policy framework that promotes mutual enhancement and synergistic efficiency between pollution and carbon reduction. At the technological application level, technologies conducive to transitioning the structure of energy, industry, and transportation toward low-carbon and green models are being increasingly adopted. However, the continued increase in energy consumption due to rapid economic growth and urbanization is still the main driver of China's rising CO_2 emissions, posing the greatest challenge to achieving synergistic efficiency in reducing pollution and carbon emissions.

Overall, China has yet to fully achieve the synergistic reduction of pollutants and CO_2, and there is an urgent need for the implementation of more targeted policies. The report from the 20th National Congress of the Communist Party pointed out that "we need to synergistically promote carbon reduction, pollution reduction, green expansion and growth, and promote ecological prioritization, resource conservation, and green low-carbon development." However, as a new concept in the field of ecological civilization construction, the coordinated management of pollution reduction and carbon reduction lacks a mature theoretical framework and technical methods. To build a future mechanism for coordinating carbon neutrality and clean air, China should continue to adjust its energy mix and industrial, transportation and land-use structures, accelerate the transition from end-of-pipe control to source control of pollution, and reduce carbon emissions by addressing climate change, thus fundamentally solving the problem of environmental pollution. Specifically, China should pay attention to the following four areas.

(1) China should continue to adjust its energy mix by controlling the total consumption of fossil fuels, encourage the clean and efficient use of coal, advance the upgrading and retrofitting of coal-fired power units, and promote cogeneration (or combined heat and power, CHP) transformation projects in large coal-fired power plants. China should actively develop non-fossil energy, and vigorously promote power generation from renewable energy such as wind and solar energy. China should continue to implement projects replacing bulk coal with natural gas and electricity to strictly control bulk coal burning in rural areas.
(2) China should further adjust its industrial structure by curbing the haphazard expansion of energy-intensive and high-emission projects, shutting down outdated production facilities, scaling down overcapacity, and dynamically eliminating small, poorly managed and heavily polluting enterprises. China should speed up energy-saving upgrades and further pollutant control in key sectors such as electric power, steel, and cement. China should carry out comprehensive management of Volatile organic compounds (VOCs) and implement projects using raw and auxiliary materials and products containing zero or low levels of VOCs.
(3) China should actively adjust its transportation structure by updating the composition of the motor vehicle fleet, removing high-emission old vehicles from roads, and promoting new energy or clean energy vehicles. China should build an efficient and intensive logistics system, facilitate a shift in bulk cargo transportation and mid-long distance freight transportation from highways to railways and waterways, and make great efforts to develop multimodal transportation. China should strengthen the control over non-road mobile sources and eliminate outdated construction machinery.
(4) China should steadily adjust its land-use structure by optimizing the use of fertilizers and feed, promoting reduced yet more efficient use of fertilizers and pesticides, and stepping up pollution control and recycling of resources from livestock and poultry waste on large-scale livestock farms. China should improve the comprehensive utilization of crop straw and strengthen crop straw burning regulations. China should guide key sectors to relocate to areas with ample environmental capacity and good diffusion conditions and implement the relocation of key enterprises from cities.

4 Coal Power Reduction: Pathways to Synergistically and Efficiently Reduce Air Pollution and Carbon Emissions

Coal-fired power stations are the leading sources of air pollution and GHG emissions in the power sector globally, including in China. There are mature technology options for reducing air pollution, many of which are in place for newer coal-fired plants today: flue gas desulfurization for SO_2, selective catalytic reduction (SCR) for NO_x, baghouses for particulates, and mercury controls. Less mature technologies are in development to capture and permanently sequester fossil fuel plant carbon emissions.

Pollution standards requiring advanced pollution reduction technologies and efficiency standards for new power plants is a best practice that has resulted in remarkable pollution reduction in China, the U.S., and Europe.

Both countries have shown the effectiveness of power plant pollution controls in improving public health and air quality, but the U.S. has made larger gains in GHG emissions reductions due to three factors: economic competition from clean resources, state GHG and clean energy standards, and coal pollution standards. These have combined to dramatically reduce coal's role in the U.S. power sector, with significant pollution co-benefits.

Today, for China and the U.S., the first and most cost-effective pathway to limiting both conventional and GHG pollution more quickly is replacing coal with carbon-free renewable energy and storage. China has an opportunity to learn from the U.S. and other countries' experiences while continuing to focus on power sector reforms that promote dispatching the most efficient, cleanest resources and enhancing system energy security.

4.1 Overview of Current Coal Power and Its Phasedown Risks in China

4.1.1 Status of Coal Power in China

The power system in China still strongly relies on coal. During the 13th and 14th FYPs, total coal consumption in China gradually transitioned from a phase of rapid growth to a stable stage [2]. Against this backdrop, China's coal-fired power sector has not halted its pace of new construction. The rapid development of coal power infrastructure leads to a short average lifetime of existing coal plants in China, of which more than 75% had operated for less than 15 years by 2020 [3]. The expansion of coal power is creating higher risks of asset stranding and hindering China's progress toward building a modern power system to achieve carbon neutrality goals.

As one of the major anthropogenic emitters in China, the emissions of pollutants from coal power have been effectively controlled since clean air action begin in the country. By 2015, the ratio of coal plants equipped with FGD and de-NO_x devices reached 95.6% and 84.2%, respectively. China's Ultra Low Emissions (ULE) Standard attempts to improve coal plant efficiency while further reducing $PM_{2.5}$, SO_2, and NO_x emissions from coal plants, which will dramatically reduce air pollution and improve public health. In 2020, coal power in China contributed 1.2 Mt of SO_2, 3.0 Mt of NO_x, and 0.2 Mt of $PM_{2.5}$, which is much lower than the emissions levels in 2015 [4].

Despite this tremendous progress in reducing pollution from coal, the clean development of coal power cannot cover up the threat of coal-fired power toward climate goals. With the deepening of end-of-pipe control, the potential for emissions reductions in China's power sector is gradually narrowing. New paths must be sought to promote a synergistic reduction in carbon and pollutant emissions. Therefore, shifting the power system from relying on coal to low-carbon electricity sources is a necessary part of China's strategy to mitigate GHG emissions and further improve local air quality, due to potential challenges for climate targets and public health.

4.1.2 Necessity and Risks of the Coal Power Phasedown

One serious challenge posed by the existing coal-fired power plants is that the substantial committed emissions of GHGs hinder China's 2030 carbon peak and 2060 carbon neutrality targets [5, 6]. These negative emission technologies are not expected to satisfy the requirement of future climate targets unless most coal-fired power plants and other fossil fuel power plants are substituted by clean, renewable energies, such as wind and solar power [1].

Another serious threat is that current coal-fired power plants are expected to result in a large amount of premature death related to the emission of air contaminants, undermining the UN Sustainable Development Goal of good health and well-being, despite the widespread deployment of strong and stringent pollution controls in coming future [7, 8]. Therefore, the most feasible and cost-competitive strategy for addressing such challenges is to rapidly phase down current coal-fired power plants and to raise the penetration of clean renewables, notwithstanding the potential risks in unemployment, energy security, and stranded assets [9, 10].

The most severe risk is that a coal-fired phasedown undermines the security of energy systems. That suggests that a coal-fired power phasedown will pose great risks to the security of energy systems due to the weak stability and reliability of variable renewables regardless of their daily and seasonal cycles [11]. Moreover, a coal-fired phasedown would result in substantial stranding asset losses in China due to relatively short operation duration (the average operation duration is less than 15 years).

4.2 *U.S. Experiences: Rapid Reductions in Coal-Fired Generation Go Hand-in-Hand with an Affordable, Reliable Grid*

4.2.1 U.S. Experience Regulating Air Pollution from Coal-Fired Power Plants

In 2005, coal generation provided half of U.S. power generation. In 2022, that fell to 19%, replaced by natural gas, renewables, and greater energy productivity. This singular trend has been responsible for the majority of U.S. GHG emissions reductions since 2005, and is the result of three primary factors: stringent pollution standards for coal-fired power plants, state clean energy policies, and fuel switching to cheaper gas, wind, and solar.

The U.S. Environmental Protection Agency (EPA) is the primary regulator of coal plant pollution and GHG emissions in the United States. EPA sets health-based air-quality standards and then required technology-mandate standards to help meet these air-quality standards. The EPA also promulgates standards to reduce water pollution from coal plants, including pollution associated with the disposal of coal ash and scrubber waste. In addition to the EPA setting national standards for power plants, states have also delegated the authority to ensure air and water quality standards are met within their state boundaries. The EPA is required to conduct a cost–benefit analysis before requiring power plants to meet specific performance standards.

Over the past two decades, the EPA has been increasing the stringency of pollution standards for new and existing coal-fired power plants. These standards have internalized the pollution costs and led to hundreds of coal plants retiring or announcing their retirement this decade.

In the past decade, some coal plants have installed major new pollution controls, while others have opted instead to retire the plant and replace it with cleaner generation. The chart below shows how pollution controls for SO_x, NO_x, PM, and mercury were installed on existing coal plants at a high rate from 2005 to 2019. As a result, the average book value of regulated coal plants has doubled since 2005, meaning there is more capital asset value to pay down now compared to 15 years ago. These controls can be amortized over shorter periods than in the past, keeping the door open for retiring the plant within the next decade. Given the U.S. policy to get off of unabated coal by 2035, the better outcome for consumers would almost uniformly have been to retire rather than retrofit these plants.

Ultimately, what matters for GHG emissions and pollution is both the rate of pollution per unit of power and the total amount of fossil fuels burned.

In summary, we offer four takeaways from the U.S. experience:

1. Regulating pollution from existing coal-fired power plants can have large GHG co-benefits when markets are putting economic pressure on reducing generation, switching to cleaner fuels, or retiring the plant. Regulating GHG emissions could also come with large pollution co-benefits, but the U.S. has limited experience doing so.

2. The least-cost solution to reduce fossil power plant pollution is overwhelmingly switching to a mix of zero-carbon power plants, battery storage, demand-side management, and transmission investment.
3. The U.S. has reduced coal's share of electricity generation from 50% to less than 20% in the space of a decade while maintaining a reliable, affordable power system.
4. Utilities can manage stranded asset risks by setting long-term power sector carbon goals and ensuring all new investments in existing power plants—for pollution controls or otherwise—are amortized over a period consistent with carbon reduction targets, including President Biden's commitment to 100% clean electricity by 2035.

4.2.2 Challenges and Policy Options

Power plant cost dynamics between zero carbon and coal power are similar in China and the U.S.— recent wind, solar, and storage cost declines in China mean that new wind and solar power is cheaper than the average existing coal-fired power plant, according to IRENA. However, the Chinese market system is quite different from the U.S.'s. Economic dispatch is still evolving, so the regulation of pollution will not necessarily have the same co-benefit of GHG reduction in the U.S., where higher costs made it harder for less efficient coal plants to compete and forced some older coal plants to retire. Improving economic dispatch in China can be a GHG reduction policy because more efficient, cleaner plants will run more, and more polluting coal plants will run less. The same can be said for increasing the geographic scope of economic dispatch from provincial to regional, which can reduce costs and reduce renewable curtailment. Addressing still-growing power sector emissions will help peak and begin to meaningfully reduce Chinese emissions before 2030.

With stringent pollution standards already in place alongside robust efficiency requirements, future GHG reductions from the Chinese power sector must likely come from fuel switching, CCS retrofits, or both. Learning from the U.S.'s disjointed approach to pollutant-specific regulation, China has an opportunity to promulgate sector-wide GHG emission standards that allow subnational governments the flexibility to choose between technologies that best optimize other public policy and reliability objectives.

Setting standards now that are aligned with China's carbon goals also helps draw some boundaries around continued coal plant expansion in China that can help avoid stranded assets or higher costs in the future. Adding a coal plant likely means increasing generation—once an asset is built, it lasts for at least 20–30 years. China would be wise to avoid the looming stranded asset problem facing the uneconomic coal fleet and gas-generation infrastructure. GHG emissions standards for the power sector could help promote fuel switching now and promote the examination of clean energy resources that provide reliability value similar to new coal plants. When absolutely necessary, coal plant capacity could increase, but a GHG standard would place de facto limitations on the utilization of unabated coal and prompt inefficient plants to run less.

Reliability is key. Coal still has high reliability and energy security value. But with the advent of affordable grid-scale battery storage, clean energy resources are now able to provide significant reliability services.

China can learn from the U.S. experience by focusing on fuel switching to carbon-free sources instead of delaying the transition through more fossil investment or switching to an intermediate fossil fuel like gas. GHG emission standards that address the power system as a whole maximize flexibility and cost-effectiveness. Because competitive markets are still maturing and limiting the impact of efficiency on coal and renewable dispatch, an additional focus on GHG standards is crucial to achieving pollution and GHG reduction co-benefits.

In summary, policy will be key to reducing GHG and pollution in the power sector together. A key policy will be to create coal plant or sector-wide GHG emissions standards, allowing flexibility in which technologies plant operators and subnational governments can use to reduce emissions. Because the economics of renewables in China are so favourable, and with new cleaner options to improve reliability, it is possible to rapidly reduce coal-fired generation, grow investment in renewable energy, and more quickly achieve GHG and pollution reduction goals.

Some complementary policies will also be key to achieving these goals while maintaining reliability and enhancing affordable power, including:

- Implement economic dispatch country-wide to reward greater power plant efficiency, reduce renewable energy curtailment, and promote economic renewable and storage projects that outcompete existing coal generation.
- Further expand electric system reliability obligations from the provincial scale to the regional scale, building pilot program experience. This will improve the reliability value of renewables, improve transmission planning, and create even more opportunities for efficiency and faster renewable energy deployment.
- Set newer, more ambitious clean energy deployment standards, including battery storage deployment targets, to help continue China's leadership in growing the clean energy share.
- Change market structures to pay coal plants for reliability services without also requiring them to dispatch uneconomically. Consider developing reliability reserve products as a reliability measure supporting variable renewables until confidence in reliable system operations grows.
- Continue to gain sophistication on the reliability of renewable energy systems. Collaborate with leading grid operators in other countries and implement best-practice modelling to better compare the reliability contributions of all types of resources.
- Invest in research on dispatchable clean energy resources such as advanced geothermal, advanced nuclear, hydrogen, long-duration storage, and demand management to replace the capacity value of the coal fleet.

4.3 *Co-benefits and Pathways for the Targeted Coal Power Phasedown*

4.3.1 Incorporating Health Co-benefits in Decision-Making for the Coal Power Exit

The early retirement pathway for coal power in China is highly uncertain and requires careful design. Due to the overwhelming magnitude and the short operational lifetime of existing coal power infrastructure, the early retirement of coal power generating units would face significant risks. Current policies for a coal-fired power phase-down mainly target small-capacity, substandard generating units and auto-producers. However, a phasedown strategy with simple criteria is unlikely to seize the opportunity to maximize the benefits of decarburization policies. Policy-makers should take more factors into consideration when designing a strategy.

The coal power exit, which is designed for climate mitigation, would also reduce emissions of air pollutants, resulting in air-quality improvements and health co-benefits. The health co-benefit assessments of coal power early retirement should be integrated into decision making about climate policies. Future aging populations will further exacerbate the health burdens of coal power pollutant emissions in China [12, 13]. Protecting public health should be prioritized as a starting point for the early retirement of coal power to the synergistic governance of climate change and air pollution. Focusing on the decarbonization pathway of the power system in China, air-quality improvements due to climate measures, especially the regional health co-benefits, may offset or even exceed the mitigation costs [12, 13].

Notably, health co-benefits due to CO_2 emission reductions from the coal exit vary drastically from facility to facility [14]. When designing the coal exit pathway, the heterogeneity of co-benefits at the facility level should be considered. Early retirement strategies targeting super-polluting units could substantially reduce the health burden [8]. Therefore, incorporating a facility-level health co-benefits assessment into the policy design of the coal power exit is necessary for protecting China's public health, mitigating the risks of early retirement, and maximizing the benefits.

In addition to health, the early retirement of coal power also has positive impacts on water scarcity mitigation, ecological conservation, and other aspects [15, 16]. Except for pollutant emission removal, a coal power phasedown will reduce heat emissions contained in cooling effluents, which can lead to extended stretches of thermally polluted rivers and lakes, compromising the habitats of aquatic organisms [17].

4.3.2 Targeted Coal-Fired Power Plant Phaseout Pathway Design

A facility-level assessment of the coal power exit is in line with the current principles of refined governance in China. To explore the optimal pathway for the coal power exit, it is necessary to quantify the facility-level benefit potential and targeted phaseout of the super-polluting power plant. This can provide scientific support for minimizing the costs and maximizing the benefits of China's coal power exit.

Evaluated by unit-level technical, economic, and environmental criteria, nearly 20% of operating coal-fired power units are identified as low-hanging fruits [18]. To achieve the 1.5 °C target, it is urgent to stop coal power infrastructure construction and rapidly retire those low-hanging plants. For other existing coal-fired power plants, the average lifetime should be further shortened to 20 years or the average operating hours should be reduced from 2020s 4350 to 3750 h in 2030.

Health co-benefits of power decarbonization depend on tailored coal power retirement strategies. Pollutant emissions abatement and deaths avoided resulting from the same amount of carbon reduction vary drastically by the location and technical attributes of power plant units [6]. Under the same climate–energy and clean air pathway, early retirement strategies targeting super-polluting units could substantially reduce the health burden by avoiding millions of deaths in China [8].

To alleviate side effects caused by the coal power exit and to deepen the decarbonization process, another alternative pathway to retrofit coal-fired power units as biomass and coal co-firing power units with CCS (BCP-CCS) was evaluated based on unit-level heterogeneity information and resource spatial matching results [19].

Despite uncertainties in the above coal power retirement pathway analyses, they serve as innovative ideas and valuable references for a transformation away from a coal-based power system in China. However, designing a coal power phasedown pathway is still a complex undertaking, and current pathway analysis on coal power phaseout is still relatively weak. It is necessary to resolve the current and future drivers of coal power infrastructure construction and explore possibilities for reversing the growing trend of coal power capacity. A more comprehensive assessment framework for coal power retirement from multiple perspectives, including asset stranding, resource endowment, environmental impact, social equity, and energy demand, is urgently needed. This will provide new insights for policy-makers through a multi-dimensional quantitative analysis approach.

In addition, the early retirement of thermal power plants is not the only means to synergistically reduce pollution and carbon emissions in the power sector. The coordination of the coal phaseout with other measures, such as source control, process control and end-of-pipe control, would have substantial benefits.

4.4 Policy Recommendations for a Coal Power Phasedown

To deepen decarbonization and realize a stable transformation for China's energy systems, systematic design that integrates energy security, stranded asset, and social equity into coal power phasedown policies is urgently needed. The top priority is to maintain energy security when coal-fired power plants gradually phase down. Safeguarding the security of the energy system requires a well-designed coal power plant phasedown so that the remaining plants are capable of meeting peak load demand and preventing accidental load loss. An overly aggressive coal-fired power phasedown would likely lead to potential electricity supply shortfalls, particularly during a period of extreme weather when wind and solar power outputs drop dramatically [20]. If dispatchable energy generators and energy storage are not capable of providing enough flexible electricity, it will trigger electricity outage events and cause enormous economic losses and societal problems [20].

Moreover, coal power phasedown should combine with stranded assets, social equity, and economic development. Most of China's coal-fired power plants are no more than 15 years old, which is much younger than those in developed countries such as European Union, Australia, and the U.S. [21]. Thus, a rapid retirement policy that does not consider stranded assets will cause enormous capital losses [22]. Moreover, potential economic and social losses of coal-fired power plants may not be equally distributed across varied stakeholders, groups, and regions, particularly for areas where economic development and employment depend heavily on coal-fired power-associated industries [23, 24]. Therefore, the future strategy of the coal power plant phaseout needs to consider social equity, especially for the livelihood of stakeholders, despite the global impacts of coal-fired power plant retirement on climate mitigation [25].

Another recommendation is to accelerate the flexibility modification of remaining coal power plants, thereby realizing the safe and clean utilization of coal power and improving the ability to accommodate the high penetration of intermittent wind and solar energies in the coming future. It is projected that variable renewable energies such as solar and wind powers will dominate electricity generation after 2035, when over half of electricity demand is satisfied by solar-, wind-, and hydro-generated power, enhancing the decarbonization of the electricity system [9]. With the decarbonization of China's energy systems, the role of coal-fired power will transform from the main power source to a regulative backup that is designed to accommodate a high penetration of intermittent electricity supplies from wind and solar under the context of changing climate [18]. However, coal power plants that serve as peak load regulators for a fluctuating renewable electricity supply will result in increased operation costs and enhanced emissions of GHGs and air contaminants [26]. Therefore, the future flexibility modification of remaining coal power plants is required to satisfy the requirements of peak load regulation and reduce the emission of air contaminants.

Moreover, coal-fired power plants should be combined with renewable thermal power, such as biomass. It is urgent to promote the use of biomass liquid fuels in thermal power, which will allow enhanced agricultural and forestry waste industrialization. This effort is expected to give rise to some potential benefits. For one thing, it would reduce retired thermal power plants, thereby lowering the risk of stranded assets and the associated societal problems, such as unemployment and economic stagnation [27]. For another thing, biomass power plants are capable of clear and renewable electricity supply, which will boost the net-zero electricity system and enhance the achievement of future climate targets [1, 28]. Besides, biomass power plants are expected to provide flexible electricity generation, which will accommodate increasing peak regulation demand for variable wind and solar outputs and facilitate energy security. Moreover, bioenergy with CCS technology has been considered one of the most promising negative emission technologies to remove GHGs, despite expensive costs and potential emissions of air contaminants [1].

5 The Transportation Sector: Key Issues and Challenges in Reducing Pollution and Carbon Emissions

5.1 Introduction

As a major source of both local and global air pollutants, heavy-duty freight trucks (HDTs) should be a priority for emissions co-management. Diesel combustion in conventional HDTs produces higher levels of local and global air pollutants compared to passenger cars, which are mainly fuelled by gasoline.

Turning to GHG emissions, HDTs are responsible for about 30% of CO_2 emissions from China's transportation sector [29]. When paired with zero-emission electricity, EVs create a pathway to achieving net-zero emissions vehicles.

In addition to pollution reductions, accelerated NEV deployment would deliver two economic co-benefits: first, by increasing incentives for innovation, recommendations will boost domestic technological progress; second, by hastening technological progress and giving Chinese enterprises early advantages in scale and experience, they will improve the competitiveness of China's new energy heavy-duty truck manufacturers.

Sales of all types of NEVs are taking off globally, supported by increasingly positive economics due to an 89% decline in battery cost from 2010 to 2021, largely driven by Chinese policies and enterprises. This trend of improving performance and lowering costs will continue due to future learning curve effects. Innovation will come from advances across an increasingly diverse set of commercialized battery chemistries and continued learning by doing as production ramps up. There is also high confidence in future cost reductions due to economies of scale benefits.

Improved energy security is another reason for China to accelerate the transition to NEV HDTs, which offer three advantages. First, NEVs build on China's strong position in the battery manufacturing industry. Second, imported petroleum fuel dependency is inherently riskier than reliance on important battery minerals. Third, whereas the use of petroleum fuel involves its complete conversion to waste—unused heat and unwanted air pollutants—there exists the potential to recover and recycle battery minerals at the end of a battery's lifetime.

5.2 *China's Experience and Challenges in the Electrification of Heavy-Duty Vehicles*

5.2.1 The Status of Electric Heavy-Duty Vehicle Promotion in China

China currently has the largest number of electric heavy-duty vehicles in the world, with its ownership of such vehicles accounting for more than 90% of the global total in 2021. The last few years of HDT NEV sales in China bear similarity to the early liftoff in the NEV passenger market, which, in 2018, became the first national market to exceed 1 million EVs sold and has remained the largest market since. NEV HDT sales briefly peaked at 1.7% in 2018 [30], driven by commercial vehicle purchase incentives. In the last 2 years, NEV sales have begun taking off anew, growing to 3.5% of HDT sales in 2022. China EV 100 Vice Chairman Minggao Ouyang recently predicted NEV HDV sales would grow by at least 90% in 2023, an outlook serving as the basis for our 6.6%. So far, battery-electric trucks have captured more than 90% of HDT NEV sales, though hydrogen fuel cell vehicles also qualify, and their share of NEVs has been increasing [31].

In the process of promoting electric heavy-duty trucks, China has been attaching great importance to the role of public fleets and has vigorously promoted the electrification of taxies, buses, sanitation vehicles, and urban freight vehicles, increasing the share of electric buses nationwide from 20% in 2015 to 60% in 2020. Beijing, Shenzhen, and many other cities have set the requirements for the share of NEVs in the public sector and the deadline for full electrification.

5.2.2 Policies for Promoting Electric Heavy-Duty Vehicles in China

With respect to consumption, the Chinese government has exempted eligible electric heavy-duty vehicles from the vehicle purchase tax, in addition to subsidies and preferential loans for buyers of such vehicles. Chinese policies for supporting electric heavy-duty vehicles cover most of the life cycle of NEVs, greatly facilitating the promotion of such vehicles. Apart from national subsidies, some provinces and cities have also unveiled local policies for promoting heavy-duty NEVs.

Government initiatives supporting battery-swapping trucks have played an important role in their recent success in China. Battery-swapping technology involves driving a battery EV into the battery-swap station where the depleted battery is removed and replaced with a fully charged power pack [32].

5.2.3 The Experience of Electric Heavy-Duty Vehicles in Typical Cities

Shenzhen is a pioneer in the promotion of EVs in China. Since 2016, Shenzhen has continued to improve its system of standards for the NEV industry and has established a testing and certification system for operating vehicles and the use of power batteries, an information management system for power batteries, and a cascade utilization and recycling industry system for power batteries that provides continuous and solid support for the healthy development of NEVs. In order to further improve the road environment and control the pollution from light diesel trucks, in 2018, Shenzhen took the lead in setting up an innovative pilot "green logistics zone" program in the city to demonstrate new energy logistics vehicles.

After more than a decade of accumulation, Shenzhen has made remarkable progress in the promotion of electric heavy-duty vehicles in multiple areas, such as public transit, taxis, and logistics.

5.2.4 Challenges Facing the Promotion of Heavy-Duty EVs in China

Despite the rapid development in recent years, China's heavy-duty EVs still face real-world challenges, such as a short driving range and limited charging infrastructure. Besides, limited public charging infrastructure poses a challenge to the promotion of high-load, power-hungry, and frequently charged electric tractors.

5.3 California's Experience in New Energy HDT Policy

5.3.1 International Policy Context

The 27th United Nations Framework Convention on Climate Change Conference of the Parties (COP 27) marked the formal launch of the Global Commercial Drive to Zero initiative, which counted 27 nations as supporters by early 2023. These leading countries have committed to working together to achieve 100% sales of new zero-emission trucks and buses by 2040, with an interim target of 30% zero-emission vehicle sales by 2030. This effort aims to facilitate the attainment of net-zero carbon emissions by 2050. In addition to national commitments, an array of supporting subnational governments, automakers, truck parts suppliers, investors, and multilateral institutions has pledged their support.

5.3.2 California Policy Case Study

In 2021, California's Advanced Clean Trucks policy put the state at the forefront of efforts specifically targeting the deployment of NEV freight trucks, requiring about 60% of new heavy-duty freight trucks and buses sold in 2035 in California to be NEVs [33]. As with China's dual credit policy, Advanced Clean Trucks is a technologically neutral policy, meaning either electric trucks or hydrogen fuel cell trucks qualify.

NEV HDT technology and model availability have progressed substantially since 2020, when California first established the Advanced Clean Trucks policy. On April 28, 2023, the California Air Resources Board's (CARB's) approval of the Advanced

Clean Fleets rule opened a new chapter in commercial vehicle regulation, ramping up the state's NEV truck sales requirement to 100% by 2036 for all commercial vehicles above 3.86 tonnes. Another noteworthy feature of the Advanced Clean Fleets rule is a policy innovation that offers to reduce reliance on publicly funded consumer incentives, phasing in an increasing NEV purchase requirement for commercial vehicle fleets. This novel way to support the demand side of the market's transition to NEVs frees up government revenue for other investments [34].

California's embrace of NEV sales standards to drive commercial deployment builds on the state's successful experience with its analogous policy for passenger cars (i.e., light-duty passenger vehicles).

5.4 *Assessment of the Pollution and Carbon Reduction Benefits of Electrifying Heavy-Duty Vehicles in China*

5.4.1 An Analysis of the Potential of a Synergistic Sustainable Power System for Pollution and Carbon Reduction Throughout the Life Cycle of NEVs

The potential of NEVs for reducing pollutants and CO_2 emissions needs to be systematically evaluated through a life-cycle assessment (LCA). The Greenhouse Gasses, Regulated Emissions, and Energy Use in Transportation (GREET) model developed by the Argonne National Laboratory (ANL) is an LCA model widely used in the transportation sector, including Well-to-Wheels (WTW) and Vehicle Cycle evaluation, with the former focusing on the application during energy production and driving and the latter on the entire process of vehicle materials/parts from raw material exploitation to scrapping and recycling.

However, considering the high reliance on coal-based electricity generation, there remains ongoing public debate on the actual impacts of Battery Electric Vehicles (BEVs) on mitigating CO_2 emissions. The GREET model-based analysis results suggest that the full life-cycle CO_2 intensity of EVs varies from region to region due to different electricity mixes in regional power grids.

In the future, with a series of improvements in conditions for EV application, such as more electricity generation from clean energy, progress in battery technologies, and higher vehicle energy efficiency, the life-cycle CO_2 intensity of EVs in China will be significantly reduced.

Under the scenario of coordinated charging, further reductions in CO_2 emissions from EVs and sustainable electricity generation can be achieved in a synergistic manner if EVs maximize electricity from renewable energy.

5.4.2 Modelling and Evaluation of Different Policy Pathways to Accelerate the Electrification of Heavy-Duty Vehicles

To quantitatively evaluate emission reduction potential from NEV HDT deployment, we use the China Energy Policy Simulator (China EPS), a model developed by Energy Innovation: Policy & Technology, LLC and the innovative Green Development Program (iGDP).We use the China EPS to analyze two accelerated NEV deployment scenarios, as Table 1 details.

The first set of results focuses exclusively on Recommended Scenario impact analysis and disaggregates economy-wide effects using three categories: transportation, electricity, and all other sectors.

1. Transportation. The transportation sector impacts capture changes in pollutant emissions with a narrow focus on emissions for which vehicles themselves are the source, i.e., narrowly considering emissions emanating from vehicle tailpipes, which excludes emissions associated with petroleum refining to produce gasoline and diesel transportation fuels, as well as emissions from new energy sources. The China EPS separately tracks emissions associated with electricity generation, petroleum refining, and hydrogen production under industrial emissions.
2. Electricity. The electricity sector impacts account for added emissions from the increased use of electricity as a transportation fuel. Additional electricity-related emissions in the results that follow are based on pollutant emissions intensities calculated for the 14th Five-Year Plan Scenario in which the share of clean energy—including renewable, hydroelectric, and nuclear technologies—reaches 40% in 2030 and 52% in 2040.
3. All other sectors. The aggregation of effects outside of transportation and electricity: changes in petroleum refining emissions are the most significant single contributor.
4. Net emissions reductions. The net effect calculates the sum of economy-wide impacts across transportation, electricity, and all other sectors.

Results illustrate that net benefits from NEV deployment depend on the pollutant emissions intensity of electricity generation, especially in the case of $PM_{2.5}$. The public health impact is contingent on how exposed people are to air pollutants at the time of initial release, when they are most concentrated.

The next set of results, compares net emissions reductions in the Recommended and Mid Scenarios. Comparison of net emissions reductions in the Recommended

Table 1 China EPS scenarios analyzed (% NEVs as a share of all new HDT sales) [35]

	2025	2030	2035	2040	2045	2050	2055	2060
Recommended scenario (%)	13	45	75	100	100	100	100	100
Mid scenario (%)	9	30	55	75	90	100	100	100
14th five-year plan scenario (%)	6	19	36	44	45	46	47	47

and the Mid Scenarios shows that the Recommended Scenario yields significantly greater emissions reductions in the next two decades.

- Technical Feasibility

The respected International Council on Clean Transportation concludes: "Commercial availability and cost of ownership projections demonstrate that 45% zero-emission HDV sales in 2030 and 100% sales in 2040 are feasible goals." [36] This concluding section explores technological, investment, and market trends enabling a rapid shift to NEV HDT sales.

One indicator of recommended ambitious NEV HDT deployment feasibility is the fact that success would require a slower increase in NEV HDT sales than has occurred historically in China's NEV light-duty passenger market. Continued market momentum is expected due to the trend of increasing industry investment, considering historical outlays and forward-looking commitments.

Turning to technological trends, high confidence in the continuation of battery innovation in coming years exists because of the range of innovations with the potential to reach commercial success. Another promising techno-economic development concerns the rising commercial success of lithium-ion phosphate batteries, which require no cobalt and are offering an affordable alternative, albeit at the cost of less energy density. The increasingly diverse array of battery storage technologies boosts the feasibility of rapid NEV HDT deployment through related innovation and economic benefits.

Because of these trends and others, rapid NEV HDT deployment in line with recommendations is feasible, which is not to say simple or easy. The recommended NEV transition will present challenges, but efforts to overcome them will be well worth the effort, delivering a potent combination of air pollutant emissions, energy security, and economic benefits.

5.5 *Economic Benefits: Innovation Stimulus and Enhanced Economic Growth*

Recommended policies would deliver economic co-benefits in technology innovation and economic growth, in addition to the local and global air pollutants that are their primary target. By accelerating the transition to NEV HDTs, China's policymakers can spur additional movement up the learning curve, delivering innovation, better performance, and lower cost.

Domestic innovation stimulus due to policies supporting accelerated NEV deployment will, in turn, enhance the competitiveness of China's NEV HDT producers. The significance of this benefit will grow more obvious over time, considering the growing consensus that NEVs are on track to become the preferred transportation technology.

China has considered NEVs an industry of economic importance since their designation as such in the 12th Five-Year Plan, and export data indicate that China's supportive NEV policies are already paying economic dividends. In the first quarter of 2023, China claimed the mantle of the globe's leading motor vehicle exporter, surpassing Japan for the first time. In 2020, surging NEV exports enabled China to pass Germany for the first time to become the second-largest motor vehicle exporter globally. In 2022, heavy-duty NEV exports climbed 131% over 2021, while passenger car NEV exports increased 120% [34]. Buses represent most of China's heavy-duty NEV exports so far, but the freight truck segment is growing.

5.6 Recommendations

We recommend a three-part strategy: (1) establish clear, ambitious long-term targets for NEV HDT deployment and help meet them by (2) establishing a NEV HDT sales standard and (3) continuing to refine and expand China's existing portfolio approach, recognizing the necessity of a multitude of instruments to optimally manage the transition. Increased certainty and clarity on future market conditions will help unlock greater investment, supporting supply chain development and stimulating additional innovation.

Regarding long-term targets for NEV HDT deployment, we suggest setting quantitative NEV HDT sales targets of 45% by 2030, 75% by 2035, and 100% by 2040. The Ministry of Industry and Information Technology (MIIT) is reported to be considering NEV HDT deployment targets and supportive policies, but these have yet to be announced.

The recommended long-term schedule aligns with International Council on Clean Transportation (ICCT) research, which, using the term heavy-duty vehicle to refer to both freight trucks and passenger buses over 3.5 tonnes, concludes: "Commercial availability and cost of ownership projections demonstrate that 45% zero-emission HDV sales in 2030 and 100% sales in 2040 are feasible goals" [36]. Maximum feasible ambition is needed due to inertia in the transportation energy system due to long vehicle life. Continuation of conventional combustion vehicle sales locks in pollution levels for years to come. A delayed transition to NEV HDTs would sacrifice local clean air, global climate, and economic development benefits, in addition to increasing the difficulty of reaching net-zero goals.

To reach ambitious targets, a second recommendation is to establish a NEV sales requirement for HDTs as soon as is practical. Such a policy can build on China's successful Dual Credit Policy for light-duty passenger vehicles, which the ICCT credits as "the main driver for the market growth in China." With respect to NEV HDT sales standard design, in its first phase, the Dual Credit Policy gave enterprises the choice to use surplus NEV credits for compliance with fuel efficiency regulations.

China's successful light-duty passenger NEV deployment strategy has used multiple, complementary policy instruments, and such a "portfolio approach" is important for NEV HDT deployment too. It should include measures encouraging vehicle energy efficiency, developing NEV industry standards, expanding infrastructure serving new energy vehicles, and continuing both fiscal incentives and non-fiscal inducements. Regarding fiscal incentives, the sales tax exemption for new energy commercial vehicles is scheduled to expire at the end of 2023, and we recommend its extension through the end of 2025. A promising non-fiscal policy option is to allow NEV HDTs preferential roadway access, alleviating significant restrictions that conventional HDTs currently face.

We highlight one new policy meriting addition to China's NEV HDT portfolio: commercial fleet vehicle purchasing requirements, which are a novel policy recently adopted in California, as described further in our California Case Study, below.

6 Regulatory and Enforcement Mechanisms for Coordinated Control

6.1 California's Experience in Coordinated Control

6.1.1 Introduction

The objective of this section is to describe strategies for effective action to address traditional pollution (e.g., ozone-forming pollutants, particulate matter, toxic pollutants) and climate change by holistically focusing on emission reduction goals, strategy design, and implementation. Specifically, as described below, many of the key emission sources for traditional pollution and GHGs are the same (e.g., transportation, combustion-based electricity generation, and industrial sources). Thus, properly chosen strategies can deliver reductions for each of the pollutants of interest, resulting in reduced costs, greater efficiencies, and better outcomes.

This section also explores the principles necessary to realize emission reductions anticipated from measures emphasizing that well-designed measures must be coupled with effective implementation to successfully deliver on their ambition.

6.1.2 The History of Air Pollution Control in California

CARB was created by state law in 1967 with the goal of establishing an approach to address the state's severe air-quality problems, particularly in southern California and the central valley. As remains the case today, there was broad public concern about air pollution, its impact on public health, and the need for action that led to the creation of CARB.

Since its formation, CARB has worked with the public, communities, the business sector, and local governments in its effort to identify solutions to California's air quality and climate problems. The federal Clean Air Act provided CARB unique authority to establish motor vehicle standards given California's extreme air-quality problems. Over the past several decades, the authority provided to CARB has set the stage for some of the most creative emission reduction strategies in the world. Many of the measures established by CARB have become the U.S. standard and have been adopted by international jurisdictions.

Over the past few decades, California's cars and trucks, along with the fuels they use (primarily gasoline and diesel), have become the cleanest in the world. CARB, which had already eliminated lead in gasoline, adopted standards for cleaner-burning gasoline, as well as standards for cleaner diesel fuel for trucks, buses, and other on/off-road equipment. CARB also began work to reduce smog-forming emissions from thousands of common household products (e.g., adhesive remover, air freshener, automotive brake cleaner, electrical cleaner, general purpose degreasers, and hair care products).

6.1.3 Principles of Effective Emission Reduction Programs

Based on the last several decades of air pollution control, we can apply learnings to the development and implementation of plans and measures that simultaneously focus on reducing emissions of criteria pollutants and toxic pollutants, as well as GHGs. Doing so will afford the opportunity to design more effective measures to deliver the needed reductions at reduced costs. These principles are foundational to prioritizing, developing, and implementing emission control measures.

- Clear and Measurable Targets

Establishing clear, trackable, and enforceable emission reduction targets that are subject to regular reporting, enabling stakeholders to easily assess progress, is the cornerstone of an effective program/measure.

- Support for Action/Leadership

Progress in reducing emissions of GHGs, criteria pollutants, and toxic pollutants is predicated on actions that include establishing regulations, incentive programs, enforcement, and industry investment in new, cleaner technologies.

- Authority to Act

Taking effective action to address air pollution and climate change requires a suite of measures, including the adoption of enforceable regulations. Regulatory organizations must have clear and unambiguous authority to develop, implement, and enforce the necessary requirements to provide the market certainty that it will prevail if challenged.

- Data/Analysis (i.e., robust technical foundation for proposed measures)

Rigorous analysis informs the development of effective plans and mitigation measures.

- Partnerships (e.g., academic, governmental, industry, community, environmental)

Effective control plans and measures rely on partnerships. This includes reaching out to academic institutions to identify the availability of useful studies as well as potential new work to address gaps.

- Establish Comprehensive Plans to Prioritize/Guide Measure Development

Achievement of emission reduction targets requires careful planning to inform a package of interrelated measures to understand how measures interact with one another.

- Capacity and Expertise to Develop and Implement Measures

Developing and implementing effective plans and measures requires a network of resources, including in-house experts; contractors, including those associated with academic institutions; and funding.

- Transparency (i.e., open public process/broad-based engagement)

The most effective plans and measures are developed through a public process that affords easy engagement for interested stakeholders in a variety of forms (e.g., web-based meetings, in-person workshops, one-on-one meetings, web postings outlining the development process with related resources) with proposals posted well in advance of meetings. Also, public comments should be posted and easily accessible, with written responses to stakeholder proposals that provide the rationale for why and how recommendations were or were not incorporated.

- Take Action (adopt/implement measures)

Ambitious and achievable emission reduction targets are important, but emission reductions result from taking action (e.g., adopting and enforcing regulations, providing incentive funding, etc.).

- Ongoing Measurement/Monitoring/Reporting

Once adopted emission control programs must be subjected to careful monitoring and reporting to assess their effectiveness. The reports should be updated frequently and be broadly available for independent analysis by academics and other stakeholders.

- Vigorous Enforcement

The vast majority of regulated parties comply with emission reduction program requirements. That may include investing in newer/cleaner control technologies and fuels as well as meeting reporting requirements.

- Adjustments to Strengthen Measures

Throughout program implementation, staff must carefully evaluate program data as well as engage stakeholders to assess any elements of the program that are not working as intended. This ongoing assessment can be used to support issuing guidance or inform about where regulatory amendments may be needed.

- Results

The metric for assessing the effectiveness of programs is whether they meaningfully reduce emissions, as intended, in support of reaching targets. And, that the reductions and associated benefits (e.g., reduced premature mortality, reduced asthma cases, reductions in hospitalizations, and reductions in lost work and school days) are commensurate with the costs and associated resources.

6.1.4 Results: Summary of Emission Reductions Achieved

Emission reduction measures without the elements described in this paper are unlikely to be successfully implemente. Over the same period, California's GDP grew substantially while per capita GHG emissions continued to decrease.

6.1.5 Applying the Principles: Identifying and Developing Multipollutant-Focused Measures

The objective of this section is to provide a few illustrations of multipollutant measures that have been successfully implemented in California. The measures deliver multipollutant benefits and are built on the principles described in this paper. But first, an abbreviated roadmap for prioritizing, selecting, and implementing measures is provided.

Prioritize Measures

The emission inventory helps to inform the most significant sectors of emissions as well as the potential opportunities for the greatest emission reductions. A theme that emerges from holistically considering the emissions inventory is that transportation is the main source of GHG, NO_x, and diesel PM emissions in California, as well as in many regions across the globe. Guided by the principles discussed throughout this paper, multipollutant measures can be identified, developed, and implemented, thus delivering air quality, community, and climate benefits while more efficiently using limited resources and investments.

Measure Development

Measure development should be informed by data and analysis of the emission reduction opportunities, feasibility of the measures, costs, benefits, and implementation timetable. The analysis should be developed in close coordination with a broad spectrum of stakeholders, published/posted, and broadly distributed allowing for public engagement, questions, and the crafting of specific recommendations. A transparent public process is as important as the analysis and supports the development of more effective measures.

Implementation

As noted earlier, establishing a process for the careful monitoring of measures throughout the implementation process supports the early identification of problems that may arise.

Illustrations

Below are summaries of three examples, including the underlying rationale for their selection: (1) zero-emission vehicles; (2) a low-carbon fuel standard; and (3) building standards.

Zero-Emission Vehicles (Cars, Trucks, and More)

California developed several programs that support the transition to zero-emission transportation. This includes zero-emission requirements for manufacturing cars, trucks, buses, and other equipment, as well as programs that support "market pull," such as fleet purchase requirements. Incentives also played a significant role in supporting the transition to zero-emission transportation, and, as previously noted, cleaner transportation results in reduced emissions of GHGs, criteria pollutants, and toxic. The efforts also reduce the reliance on petroleum-based fuels, saving consumers money at the pump while delivering billions in benefits due to reductions in premature mortality, asthma cases, hospitalizations, and lost work and school days.

Low-Carbon Fuel Standard (LCFS)

As jurisdictions around the globe transition to electrifying the transportation sector, it is clear that liquid and gaseous fuels will continue to play a role throughout the transition for several decades, particularly for the most challenging sectors to electrify (e.g., aviation). Thus, strategies that facilitate cleaner traditional fuels, as well as investments in next-generation fuels are integral parts of the solution.

The Low-Carbon Fuel Standard (LCFS) requires a progressive reduction in the carbon intensity of transportation fuels and, in doing so, is catalyzing unprecedented investments in fuels that reduce GHG, ozone, and particulate-forming pollutant emissions, as well as a myriad of toxic pollutants, including diesel particulates. The LCFS developed and successfully implemented in California is being replicated by other states/jurisdictions due to its effectiveness.

Building/Appliance Standards

Commercial and residential buildings rely on a broad spectrum of appliances to provide space heating and cooling, hot water, and to support cooking. The fuel source for hundreds of thousands of buildings is natural gas. The natural gas supplied to these appliances is burned in water heaters, furnaces, boilers, and stoves, resulting in emissions of ozone and particulate-forming pollutants, as well as various toxic compounds.

6.1.6 Overcoming Barriers

The section identifies elements necessary for effectively developing and implementing co-pollutant mitigation efforts. However, several barriers must be identified and overcome to successfully implement co-pollutant control programs. The barriers may differ between jurisdictions, thus requiring a focused assessment of the problem, as well as the solutions. Two common barriers that must be overcome are administrative and legal institutional norms.

As traditional air quality programs have typically been operational in many jurisdictions for decades, there is an established institutional structure with staff, managers, organization reporting structure, communications, budget, oversight, etc. As a result, climate programs are typically established as separate and distinct units from traditional air-quality programs. A key to breaking through these common institutional barriers includes clear and consistent expectations from the highest levels of leadership that well-coordinated co-pollutant programs are a priority. These priorities must be reinforced with structural adjustments (e.g., a common director overseeing traditional pollutant and climate programs, budgets that reward co-pollutant programs, and promotions of team members with a track record of cross-department collaboration).

Another institutional barrier is related to enabling legislation and authority. Traditional air-quality programs were typically established over decades with a series of laws and directives, as well as a track record on the rulemaking process that has evolved and been optimized. The optimization process may include modifications in response to legal challenges impacting how programs are designed and documented. Climate programs are typically more recent and established with new directives, priorities, and authorities. In addition, identifying conflicts, inconsistencies, and collaborative opportunities with existing law can inform opportunities for legislative alignment to better deliver on efficiencies with programs that fully consider the opportunities with co-pollutant control efforts.

6.1.7 Recommendations

The body of evidence underscoring the adverse impacts of air pollutants and climate change on public health and the economy is overwhelming. The commitment to action at the international, national, regional, and local levels is increasing at an unprecedented rate, recognizing both the science and growing pressure to respond; this is placing strains on government and industry resources. The need to focus limited resources more efficiently is paramount for organizations interested in targeting opportunities that deliver on multiple objectives. Emission reduction strategies that address air quality, toxic pollutants, and climate change will rise in importance, as will the focus on efficient and effective identification and implementation of holistic emission reduction measures. This paper proposes principles and processes foundational to successfully identifying, developing, and implementing multipollutant emission control measures and is intended to support a range of stakeholder forums focused on developing effective, comprehensive action plans.

6.2 The Extension of the Environmental Public Interest Litigation Mechanism to GHGs Control

6.2.1 Background

Environmental public interest litigation (EPIL) is a legal tool that has been used in the United States and some European countries through public interest groups (e.g., non-governmental organizations [NGOs]) to raise litigation cases against polluters for existing or potential environmental damages.

Starting from January 1, 2016, when the Amendment of the Environmental Protection Law came into effect, EPIL became available in China as a new environmental enforcement tool. In the past seven years, EPIL practice has played a positive role in China's environmental governance.

Thousands of EPIL cases have been raised and adjudicated or settled in recent years, signaling the emergence and rapid development of EPIL in China.

- In addition to China-based NGOs, the Procuratorates have also raised many EPIL cases since 2017. In many EPIL cases raised by the Procuratorates, NGOs provided the information regarding the litigation cases. Collaboration between the Procuratorates and NGOs is a special feature of EPIL in China that is not seen in other countries.
- Chinese court and environmental authorities have been working together on EPIL cases. Based on EPIL procedure, after receiving an EPIL case, the court should refer the case to the local environmental authority.
- Some EPIL cases are very significant. For example, an EPIL case in Ningxia Autonomous Region resulted in the highest penalty in the history of environmental pollution cases in China: RMB 569,000,000 for soil restoration and pollution prevention, and RMB 6,000,000 for environmental public interest funds.
- Aiming to avoid potential EPIL claims against them, many companies—including state-owned enterprises, Chinese private companies, and multinational corporations—have become more vigilant in their environmental law compliance.

6.2.2 Key Challenges

Despite the positive developments and results of EPIL in China, there are still many challenges to the implementation of EPIL.

- Under the current legal system in China, the scope of EPIL is narrow, and there is no clear guideline that states that GHG emissions can be targeted by EPIL. Under the current legal system in China, EPIL cases can be raised against polluters based on environmental damages that have already happened, e.g., air pollution and soil contamination. Such EPIL cases are "damage-based" causes of action. China has not established a legal mechanism or practice through which NGOs or prosecutors can raise EPIL cases against future construction projects that may be harmful to the environment and climate (e.g., the planned coal power plants).
- The grounding capacity of EPIL and environmental governance is still weak. Only a very limited number of NGOs in China have raised EPIL cases since EPIL became available in China. Generally speaking, these NGOs do not have sufficient legal or environmental expertise to raise and handle EPIL cases in the most professional manner. In addition, EPIL is new for most prosecutors, particularly at the local level.

6.2.3 Policy Recommendations

With the above background and challenges established, CCICED has analyzed international experiences on EPIL and the local needs to improve EPIL and would like to provide the following policy recommendations to the Chinese government.

Expand the Scope of EPIL into GHG Enforcement

To address the urgent climate challenges and improve GHG enforcement, the scope of EPIL should be expanded from environmental damage to GHG enforcement.

- The National People's Congress (NPC), the Supreme Procuratorate of China, and the Supreme Court of China should issue a regulation or interpretation document that allows NGOs and procurators to raise EPIL cases targeting GHG emissions and climate damages.
- A practical guideline to raise, adjudicate, and settle EPIL cases targeting GHG emissions and climate damages should be provided by the Supreme Procuratorate of China, and the Supreme Court of China. It is a positive signal that the Supreme Procuratorate of China issued an opinion in February 2023 that calls for better justice service on carbon peaking and neutralization.
- Training programs should be provided to local NGOs and prosecutors so that they can obtain better knowledge and experience to raise EPIL cases on GHG enforcement.
- Case studies on EPIL targeting GHG emissions and climate damages should be collected, compiled, and distributed.
- National authorities should encourage NGOs and prosecutors to raise EPIL cases related to GHG emissions and climate damages, including cases involving coal power plants that were approved by provincial authorities in recent years but have not yet started construction.

Provide Further Political Support to EPIL Mechanisms and Implementation

Generally speaking, the Chinese government has supported EPIL in China. EPIL language has been included in the overall National Economic and Social Development Five-Year Plan approved by the NPC and other high-level government documents. That said, it is still important for the Chinese government and top political leaders to further support EPIL mechanisms and implementation.

- In high-level documents and speeches, including those issued by the Party and/or the State Counsel, EPIL should be repeatedly emphasized.
- Chinese government agencies —particularly those at the provincial and local levels—should recognize that environmental enforcement by administrative agencies has been historically weak; therefore, China needs a new and additional approach (i.e., EPIL) for better environmental and climate enforcement. While government officials at the national level generally support EPIL as an alternative tool of environmental enforcement, some local environmental officials may be reluctant to support EPIL cases. Internal education should be conducted to build stronger and more consistent support of EPIL by local officials.

Establish a Preventative EPIL Mechanism

To avoid potential environmental and climate damages and reduce stranded costs, EPIL should be expanded to prevent potential environmental and climate damages. It is important for China to introduce "preventative EPIL" into China's legal system and practice so that certain future construction projects can be sued by NGOs or prosecutors and therefore potentially be stopped or delayed.

- National authorities—namely the NPC, the Supreme Procuratorate of China, and the Supreme Court of China—should issue a regulation or interpretation document through which the owner or proponent of a construction project may be sued for its potential environmental or climate damages.
- If a construction project is sued as an EPIL case for potential environmental and climate damages, such a construction project should be placed on hold unless a court decision is made.
- A clear definition of "potential environmental and climate damages" should be provided so that preventive EPIL is not abused.
- International experiences, best practices, and leading cases may be introduced in designing China's preventive EPIL mechanism and operational procedures.

Promote Provincial EPIL Regulations

Currently, most EPIL regulations are promulgated by national authorities. Generally speaking, provincial authorities are not active in setting up their own EPIL regulations and implementation details. Provincial regulations of EPIL should be promoted with the following actions.

- Provincial authorities should be allowed and encouraged to set up their own EPIL implementation details in areas and subjects where no national implementation details have been provided already.
- National authorities may organize 3–5 provincial governments as pilots to draft and promulgate provincial EPIL implementation details.
- National authorities may provide a "template" of EPIL implementation details that provincial and local governments may use to improve their environmental and climate governance.

Build Better Capacity for EPIL Operation

The Chinese government may make efforts to build better capacity for EPIL operations and encourage other interested stakeholders to take such capacity- building efforts.

- Chinese government authorities, such as the NPC and MEE, should organize or encourage EPIL training programs targeting stakeholders that include environmental officials, lawyers, NGO representatives, and industry managers.
- The Supreme Procuratorate of China and the Supreme Court of China should establish internal training systems through which the procurators and judges are trained from time to time. EPIL should be added to such internal training programs.
- Publication and information sharing on EPIL cases and experiences should be encouraged by all government agencies.
- Leading Chinese social organizations, such as the China Environmental Protection Foundation (CEPF), should expand their existing programs to mentor and provide grants to grass NGOs at a more meaningful scope.
- International climate foundations and environmental NGOs should be encouraged to offer grants for EPIL training, publication, and other capacity-building actions.

7 Gender Analysis

This chapter is devoted to analyzing and identifying gender-related issues within the coordinated management of carbon reduction, pollution reduction, green expansion and growth. We aim to understand the roles, rights, opportunities, needs, and contributions of different gender groups in the green transformation. Through gender analysis, we can better identify the origins of inequality and address these issues by designing and implementing more equitable and effective policies, mechanisms, and services. The specific research questions proposed in this study are:

1. What gender-related issues exist in the coordinated mechanism of carbon reduction, pollution reduction, green expansion, and growth?
2. In the key areas of carbon/pollution mitigation and green transition, how can gender equality be ensured through institutional design?

Given the current lack of research on gender equality in green, low-carbon development domestically, this study plans to use a literature review to identify specific gender issues. By combining the background of domestic carbon/pollution mitigation and the status of industry transformation, we provide preliminary policy suggestions, laying the foundation for subsequent research.

7.1 Gender Issues in the Coordinated Mechanism of Carbon Reduction, Pollution Reduction, Green Expansion, and Growth

The impacts of climate change and air pollution are not uniformly distributed worldwide; certain social groups, such as women and children [37], are commonly and often more severely affected [38–40]. On one hand, natural disasters, extreme weather events, climate change, and exposure to air pollution particularly impact the health of pregnant women and children [41], increasing the risk of malnutrition, acute respiratory infections, diarrheal diseases, low birth weight, and premature death [42]. On the other hand, traditional roles and socio-economic statuses influence how different gender groups access and manage resources, leaving women and other gender minorities often in weaker positions in public domains, such as pollution control.

The issue of gender equality is often overlooked in the transition to a green economy. There is a risk of adopting a "gender-blind" approach in the economic transformation. [43] Consequently, the new policies established during the green revolution could potentially generate long-term negative effects on women's employment, participation, work environment, and opportunities for education and training.

Globally, many international initiatives, agreements, and policies have specifically highlighted issues of gender equality and social justice. Among these, the United Nations' Sustainable Development Goals (SDGs) are perhaps the most well-known framework. On the matter of climate change, the United Nations Framework Convention on Climate Change, the Paris Agreement, and the Kyoto Protocol have all firmly consolidated the importance of gender equality at multiple levels, including policy framework, execution strategies, and financial mechanisms [44]. At the same time, many countries have actively integrated gender equality into their policies and practices while promoting green growth.

Introducing gender issues and actively practicing gender mainstreaming in the process of promoting green growth are not only key strategies for improving gender equality, but also important foundations for promoting sustainable development. China has shown unique advantages in the competition for green development. Not only do we possess a firm understanding of the new development concept, but we also have strong government coordination capabilities. Moreover, we have achieved some remarkable successes in promoting social fairness, as evidenced by our strong execution and coordination skills demonstrated in the fight against poverty. Now, as we shift our focus to the coordinated advancement of carbon reduction, pollution control, green expansion and growth, it is crucial to stress the importance of gender mainstreaming. We should also take this as an opportunity to further promote and practice the concept of gender equality.

7.2 *Gender Equality in the Green and Low-Carbon Transition of the Power and Transportation Sectors*

7.2.1 Power Sector

The green and low-carbon transition in the power sector contributes to eliminating gender inequalities caused by air pollution and climate change, but the process could potentially lead to new inequalities. Phasing out coal-fired power and widely adopting clean energy are major steps in the low-carbon transition of the power sector. The environmental and health co-benefits these steps generate often outweigh the policy costs. [41, 45–47] The coal phaseout process can also provide new job opportunities for women, increasing their confidence, self-esteem, and financial independence [46, 47].

Given that coal-fired electricity still accounts for around 60% of China's total power generation, decarbonizing the power sector is of utmost importance among all economic sectors undergoing a transition. Hence, diverse strategies are needed to mainstream gender, especially in regions heavily invested in the coal-fired power industry. First, female workers in the coal industry should be given due attention. Second, we need to promote gender equality in the renewable energy industry. Last, it is crucial to strengthen policy research and development from a gender perspective. By integrating a gender perspective, we can more comprehensively consider and address various issues arising in the low-carbon transition process, thereby achieving a more equitable and inclusive energy transition.

7.2.2 Transportation Sector

In the green and low-carbon transition process within the transportation sector, potential gender inequality issues might arise from four areas: travel behavior, job positions, traffic safety, and traffic pollution exposure. Regarding travel behavior, studies find that women tend to travel less frequently and less distance, [48] and they are more likely to use walking facilities [48] and public transportation, [49] hence producing less environmental impact [50]. Additionally, as the green and low-carbon transition progresses, certain traditionally male-dominated jobs, such as fossil fuel mining and heavy industry, may decrease, and new employment opportunities, such as the development of the renewable energy industry, may not benefit all genders equally due to educational and skill barriers. Furthermore, a study suggests improving the "walkability" of cities by creating excellent walking conditions and facilities to promote inclusiveness and equality. [51] Last, women are more likely to choose slow modes of transportation, such as walking, cycling, e-bikes, and public transportation, exposing them more to exhaust pollution from motor vehicles, which poses higher potential risks to respiratory and cardiovascular health. Pregnant women exposed to exhaust pollution may risk adverse health effects for both the mother and the fetus.

To mainstream gender in the low-carbon transition of the transportation sector, we need to take action at multiple levels, including policy, planning, and practice. First, at the policy level, we need to ensure that all transportation and environmental policies take gender differences into account and develop specific measures based on this foundation. Second, in terms of green employment, we should enhance women's skills and knowledge through education and vocational training, enabling them to find work in the emerging green transportation industry. Furthermore, at the planning level, we must ensure that city and transportation planning adequately consider the needs and interests of all genders, particularly in terms of public transportation and non-motorized travel facilities. In addition, we need to enhance women's representation and influence in transportation planning and decision making through public participation and social dialogue. Finally, at the practical level, we must boost women's leadership and influence in various ways, shattering gender stereotypes and elevating women's roles and status in the green, low-carbon transition. Specific measures could include providing leadership training, promoting professional networks and mentoring systems, and increasing the proportion of women in critical decision-making positions.

7.3 Gender Strategies in Collaborative Management

In summary, as China works to reduce carbon, decrease pollution, expand green development, and stimulate growth, gender mainstreaming is indispensable due to the varied impacts policy formulation and implementation often have on different genders. Gender equality in the green transition can be promoted from the following three perspectives.

- Establish scientific research plans, conduct continuous research, and provide periodic feedback

A medium- to long-term research plan should be developed to track and thoroughly investigate the development of gender mainstreaming in our country's green transition process, providing a scientific basis for policy-making and improvement. The first step of the research plan should start with a literature review and setting research objectives to establish a solid theoretical foundation and clarify the direction of the study. The second step involves a deep analysis of the impacts of green low-carbon transition policies on both men and women, furthering our understanding of the specific roles of gender factors in the green transition process through case studies. The third step is to try to quantify the impacts of green transition policies on men and women through quantitative analysis, and based on existing analysis results, propose preliminary gender-sensitive policy suggestions. The fourth step could involve expert interviews and focus group discussions to gather more firsthand data and in-depth insights, further optimizing our policy recommendations. The fifth and final step is to compile the final research report and disseminate the research findings and policy recommendations to relevant decision-makers and stakeholders, serving as a reference for advancing gender mainstreaming in China's green low-carbon transition. This research process should continuously track specific industry situations and form a periodic feedback mechanism, providing timely advice to policy-makers and promoting gender equality in the green transition.

- Addressing universal gender issues while considering unique industry challenges

In the green and low-carbon transition of various industries, there exist both universal gender issues and unique sector-specific challenges. Therefore, it is important to distinguish between these in problem identification, research methodology determination, and policy design. Universal measures should include the implementation of gender-responsive budgeting, where financial resource allocation takes gender impacts into account, thus reflecting fair treatment of the needs of both men and women. Gender indicators should be incorporated in the execution and evaluation of policies. In employment and skills training within relevant industries, there should be an increase in support and encouragement for women. Additionally, women's participation in decision-making should be amplified.

- Learn from international practices, summarize our own experiences, and promote international communication

In the early stage of promoting gender mainstreaming, it is essential to continuously learn from exemplary international practices and successful cases, setting a positive trajectory for promoting gender equality in China. At the same time, China can share its own experiences and achievements in advancing gender equality and social justice on a global scale.

8 Policy Recommendations

8.1 Green Growth

- Carbon neutrality is a major opportunity for China, and the dual-carbon goal is becoming a new driving force for China's economic growth. China must set visionary goals and develop strong programs, especially in the energy sector, to see that the goals are met.

- The "1 + N framework provides an excellent basis for building a comprehensive strategy. It must be followed by detailed plans, organized around a healthy race to the top, and supported with detail at the provincial and municipal level.
- Controlling both conventional pollution (NO_x, $PM_{2.5}$, SO_x, and other particulates at the same time as the GHGs of CO_2, CH_4, and N_2O) can save vast sums of money and steer investments to far better choices.
- In its great progress in renewable energy investment and NEV development, China benefits the world by providing supply and lowering technology costs. The work to push important technologies down the learning curve offers broad and rapid returns. China should identify ever more realms where it can lower the cost of key technologies (and practices).
- Besides advances in the economy and standards of living, co-control of conventional pollutants and GHGs will spur better health and livability.
- Almost all of this can be wrapped up in a carefully designed mix of performance standards, sector-specific goals, and economic signals that can drive rapid environmental improvements while making new markets for advanced goods and services.

8.2 *Cross-sectoral Opportunities*

- The essence of our recommendations in this SPS is that China can reap vast health, economic, and environmental benefits from smart advances in policy. It is far less expensive to solve conventional pollutants together than apart: new buildings, for example, should entail quality construction (circular economy), be ultra-efficient, be heated and cooled and supplied with hot water from heat pumps, and use clean electricity for their power.
- In the early days, financial incentives can drive this four-part (economy, environment, health, and climate change) set of benefits. As prices drop and markets grow, high-performance buildings should get mandated by an advanced building code. Build once, build right, and reap forever.
- A synchronized administrative system requires the following:
  - A robust technical foundation of data and analysis
  - Clear and measurable targets
  - Ongoing monitoring and vigorous enforcement
  - Comprehensive plans that prioritize and measure deployment
  - Increased capacity and expertise to develop and implement measures
  - Thorough and accurate measurement and reporting.

  Important steps include:

- Developing a regulatory system and plans that simultaneously drive higher reliability and lower use of coal.

- Phasing down coal consumption and steadily and rapidly increasing the proportion of clean energy.
- Restricting enterprises with serious pollution and energy consumption, reducing excessive production capacity, and paying attention to source control.
- Promoting the pollution control of agricultural surface and dust sources.
- Expanding the scope of EPIL into GHG enforcement, providing further political support to EPIL mechanism and implementation, establishing preventive EPIL mechanism, promoting provincial regulations of EPIL, and building better capacity for EPIL operation.

8.3 Power

Clean electricity is at the heart of co-control. This requires leaders to:

- Ensure that there are no electric power shortages: reliability, affordability, and low- to zero-carbon must be achieved together. There are many examples of how to achieve all three.
- Design a comprehensive coal phasedown strategy that takes account of technical attributes, profitability, stranded assets, health impacts, and environmental equality to realize the cost-optimal energy system transition from coal power to renewables.
- Accelerate the flexibility modification of remaining coal power plants to improve the ability to accommodate the high penetration of variable wind and solar energies and to satisfy the requirement of peak load regulation.
- Continue to focus on implementing economic dispatch, focusing on competitive exposure for less efficient coal plants.
- Continue expanding electric system reliability obligations from the provincial scale to the regional scale, reducing the incumbent coal advantage and creating more opportunities for efficiency and faster renewable energy deployment.
- Set newer, more ambitious clean energy deployment standards and storage deployment targets.
- Change market structures to maintain much of the existing coal fleet as backup for reliability services, without also requiring them to dispatch a certain amount. Consider developing reliability reserve products if necessary.
- Collaborate with other leading grid operators in other countries and implement best-practice modelling to better compare the reliability contributions of all types of resources.

8.4 Transportation

- Keep up the transition to EVs. Make sure that vehicle manufacturers have a clear public and market signal to continue to switch to electrics; set charging station goals and assign responsibility; invest in R&D to develop ever-better battery chemistries.

- Heavy-duty vehicles will require a smart mix of incentives, fleet requirements, and technological development. This pursuit is a top priority.
- With fleet electrification, decarbonization in the transportation sector ties to the carbon intensity of the electricity grid. To further reduce emissions in the future, it is necessary to efficiently use sustainable power through measurements like coordinated charging.
- Establish clear, ambitious long-term targets for NEV HDT deployment (% NEVs in HDT sales: 45% by 2030, 75% by 2035, 100% by 2040), and help meet them by.
- Establishing a NEV HDT sales standard.
- Continuing to refine and expand China's existing portfolio approach, recognizing the necessity of a multitude of instruments to optimally manage the transition. Defining these NEV policies and the specific long-term NEV targets for heavy-duty vehicles will increase certainty and clarity on future market conditions, spurring greater investment, supporting supply chain development, and stimulating additional innovation.
- Update the vehicle fleet and optimize transportation methods
- Ensure cities are designed to:
- Support fast, reliable public transit
- Have a rich network of physically protected bikeways
- Have ample room for pedestrians on shaded walkways
- Make a "15-min city" a reality for most Chinese
- New urban developments should follow the "Emerald Cities" principles, building a safe, pleasant, and efficient lifestyle.

References

1. Fuss, S., Lamb, W. F., Callaghan, M. W., Hilaire, J., Creutzig, F., Amann, T., Beringer, T., de Oliveira Garcia, W., Hartmann, J., Khanna, T., Luderer, G., Nemet, G. F., Rogelj, J., Smith, P., Vicente, J. L. V., Wilcox, J., & del Mar Zamora Dominguez, M. (2018). *Negative emissions—Part 2: Costs, potentials and side effects*. https://doi.org/10.1088/1748-9326/aabf9f
2. State Statistical Bureau. (2022). *Statistical bulletin of the People's Republic of the People's Republic of China 2021*.
3. Qiang, Z., & Dan, T. Global Energy Infrastructure Emissions and Their Lock-in Effect. https://www.efchina.org/Reports-en/report-lceg-20220303-en?set_language=en
4. Liu, F., Zhang, Q., Tong, D., Zheng, B., Li, M., Huo, H., & He, K. B. (2015). High-resolution inventory of technologies, activities, and emissions of coal-fired power plants in China from 1990 to 2010. *Atmospheric Chemistry and Physics, 15*(23). https://doi.org/10.5194/acp-15-13299-2015
5. Liu, Z., Deng, Z., He, G., Wang, H., Zhang, X., Lin, J., Qi, Y., & Liang, X. (2022). *Challenges and opportunities for carbon neutrality in China*. https://doi.org/10.1038/s43017-021-00244-x
6. Tong, D., Zhang, Q., Davis, S. J., Liu, F., Zheng, B., Geng, G., Xue, T., Li, M., Hong, C., Lu, Z., Streets, D.G., Guan, D., & He, K. (2018). Targeted emission reductions from global

super-polluting power plant units. *Nature Sustainability, 1*(1). https://doi.org/10.1038/s41893-017-0003-y

7. Geng, G., Zheng, Y., Zhang, Q., Xue, T., Zhao, H., Tong, D., Zheng, B., Li, M., Liu, F., Hong, C., He, K., & Davis, S. J. (2021). Drivers of $PM_{2.5}$ air pollution deaths in China 2002–2017. *Nature Geoscience, 14*(9). https://doi.org/10.1038/s41561-021-00792-3
8. Tong, D., Geng, G., Zhang, Q., Cheng, J., Qin, X., Hong, C., He, K., & Davis, S. J. (2021). Health co-benefits of climate change mitigation depend on strategic power plant retirements and pollution controls. *Nature Climate Change, 11*(12). https://doi.org/10.1038/s41558-021-01216-1
9. Cui, R. Y., Hultman, N., Cui, D., McJeon, H., Yu, S., Edwards, M. R., Sen, A., Song, K., Bowman, C., Clarke, L., Kang, J., Lou, J., Yang, F., Yuan, J., Zhang, W., Zhu, M. (2021). A plant-by-plant strategy for high-ambition coal power phaseout in China. *Nature Communications, 12*(1). https://doi.org/10.1038/s41467-021-21786-0
10. Goforth, T., & Nock, D. (2022). Air pollution disparities and equality assessments of US national decarbonization strategies. *Nature Communications, 13*(1). https://doi.org/10.1038/s41467-022-35098-4
11. Tong, D., Farnham, D. J., Duan, L., Zhang, Q., Lewis, N. S., Caldeira, K., & Davis, S. J. (2021). Geophysical constraints on the reliability of solar and wind power worldwide. *Nature Communications, 12*(1), 1–12. https://doi.org/10.1038/s41467-021-26355-z
12. Liu, Y., Tong, D., Cheng, J., Davis, S. J., Yu, S., Yarlagadda, B., Clarke, L. E., Brauer, M., Cohen, A. J., Kan, H., Xue, T., & Zhang, Q. (2022). Role of climate goals and clean-air policies on reducing future air pollution deaths in China: A modelling study. *The Lancet Planetary Health, 6*(2). https://doi.org/10.1016/S2542-5196(21)00326-0
13. Chang, S., Yang, X., Zheng, H., Wang, S., & Zhang, X. (2020). Air quality and health co-benefits of China's national emission trading system. *Applied Energy, 261*. https://doi.org/10.1016/j.apenergy.2019.114226
14. Li, J., Cai, W., & Li, H. (2020). Incorporating health cobenefits in decision-making for the decommissioning of coal-fired power plants in China. *Environmental Science and Technology, 54*(21). https://doi.org/10.1021/acs.est.0c03310
15. Zhang, X., Liu, J., Tang, Y., Zhao, X., Yang, H., Gerbens-Leenes, P. W., van Vliet, M. T. H., & Yan, J. (2017). China's coal-fired power plants impose pressure on water resources. *Journal of Cleaner Production*, 161. https://doi.org/10.1016/j.jclepro.2017.04.040
16. Raptis, C. E., Van Vliet, M. T. H., & Pfister, S. (2016). Global thermal pollution of rivers from thermoelectric power plants. *Environmental Research Letters, 11*(10). https://doi.org/10.1088/1748-9326/11/10/104011
17. Raptis, C. E., Oberschelp, C., & Pfister S. (2020). The greenhouse gas emissions, water consumption, and heat emissions of global steam-electric power production: A generating unit level analysis and database. *Environmental Research Letters, 15*(10). https://doi.org/10.1088/1748-9326/aba6ac
18. Lu, X., Chen, S., Nielsen, C. P., Zhang, C., Li, J., Xu, H., Wu, Y., Wang, S., Song, F., Wei, C., He, K., McElroy, M. B., & Hao, J. (2021). Combined solar power and storage as cost-competitive and grid-compatible supply for China's future carbon-neutral electricity system. *Proceedings of the National Academy of Sciences of the United States of America, 118*(42). https://doi.org/10.1073/pnas.2103471118
19. Wang, R., Li, H., Cai, W., Cui, X., Zhang, S., Li, J., Weng, Y., Song, X., Cao, B., Zhu, L., Yu, L., Li, W., Huang, L., Qi, B., Ma, W., Bian, J., Zhang, J., Nie, Y., Fu, J., Zhang, J., & Wang, C. (2022). Alternative pathway to phase down coal power and achieve negative emission in China. *Environmental Science and Technology, 56*(22). https://doi.org/10.1021/acs.est.2c06004
20. Raynaud, D., Hingray, B., François, B., & Creutin, J. D. (2018). Energy droughts from variable renewable energy sources in European climates. *Renewable Energy*, 125. https://doi.org/10.1016/j.renene.2018.02.130
21. Tong, D., Zhang, Q., Liu, F., Geng, G., Zheng, Y., Xue, T., Hong, C., Wu, R., Qin, Y., Zhao, H., Yan, L., & He, K. (2018). Current emissions and future mitigation pathways of coal-fired power plants in China from 2010 to 2030. *Environmental Science and Technology, 52*(21). https://doi.org/10.1021/acs.est.8b02919

22. Lu, Y., Cohen, F., Smith, S. M., & Pfeiffer, A. (2022). Plant conversions and abatement technologies cannot prevent stranding of power plant assets in 2 °C scenarios. *Nature Communications, 13*(1). https://doi.org/10.1038/s41467-022-28458-7
23. Mercure, J. F., Pollitt, H., Viñuales, J. E., Edwards, N. R., Holden, P. B., Chewpreecha, U., Salas, P., Sognnaes, I., Lam, A., & Knobloch, F. (2018). Macroeconomic impact of stranded fossil fuel assets. *Nature Climate Change, 8*(7). https://doi.org/10.1038/s41558-018-0182-1
24. Svobodova, K., Owen, J. R., Kemp, D., Moudrý, V., Lèbre, É., Stringer, M., & Sovacool, B. K. (2022). Decarbonization, population disruption and resource inventories in the global energy transition. *Nature Communications, 13*(1). https://doi.org/10.1038/s41467-022-35391-2
25. Li, L., Zhang, Y., Zhou, T., Wang, K., Wang, C., Wang, T., Yuan, L., An, K., Zhou, C., & Lü, G. (2022). Mitigation of China's carbon neutrality to global warming. *Nature Communications, 13*(1). https://doi.org/10.1038/s41467-022-33047-9
26. Zhao, Y., Wang, C., Liu, M., Chong, D., & Yan, J. (2018). Improving operational flexibility by regulating extraction steam of high-pressure heaters on a 660 MW supercritical coal-fired power plant: A dynamic simulation. *Applied Energy*, 212. https://doi.org/10.1016/j.apenergy.2018.01.017
27. Ahlström, J. M., Walter, V., Göransson, L., & Papadokonstantakis, S. (2022). The role of biomass gasification in the future flexible power system—BECCS or CCU? *Renewable Energy*, 190. https://doi.org/10.1016/j.renene.2022.03.100
28. Meckling, J., & Biber, E. (2021). A policy roadmap for negative emissions using direct air capture. *Nature Communications, 12*(1). https://doi.org/10.1038/s41467-021-22347-1
29. IEA. *World Energy Investment 2022.* https://www.iea.org/reports/world-energy-investment-2022.
30. IEA. *Global EV Outlook 2022.* International Energy Agency. https://www.iea.org/reports/global-ev-outlook-2022.
31. Mckerracher, C. (2022). *China's electric trucks may well pull forward peak oil demand.* Bloomberg.
32. Bernard, M. R., Tankou, A., Cui, H., & Ragon, P. L. Charging solutions for battery electric trucks. https://theicct.org/wp-content/uploads/2022/12/charging-infrastructure-trucks-zeva-dec22
33. CARB. (2021). *Advanced clean trucks fact sheet.* https://ww2.arb.ca.gov/resources/fact-sheets/advanced-clean-trucks-fact-sheet
34. CARB. (2022). *Proposed advanced clean fleets regulation preliminary language revisions workshop—Staff presentation.* https://ww2.arb.ca.gov/sites/default/files/2023-02/acfpres230213_ADA
35. EI. (2022). *China energy policy simulator.*
36. Xie, Y., & Tim Dallmann, R. M. (2022). Heavy-duty zero-emission vehicles: Pace and opportunities for a rapid global transition. *ZEV Transition Council Briefing Paper* (2022). https://theicct.org/wp-content/uploads/2022/05/globalhvsZEV-hdzev-pace-transition-may22
37. COP26. (2021). *Women bear the brunt of the climate crisis, COP26 highlights*, May 26, 2023. https://news.un.org/en/story/2021/11/1105322
38. Sugden, F., de Silva, S., Clement, F., Maskey-Amatya, N., Ramesh, V., & Philip, A. A framework to understand gender and structural vulnerability to climate change in the Ganges River Basin: Lessons from Bangladesh, India and Nepal. In *IWMI working papers.* https://doi.org/10.5337/2014.230
39. ADB. *Integrated flood risk management sector project*. Philippines.
40. EIGE. (2016). *Gender in environment and climate change.* http://eige.europa.eu/rdc/eige-publications/gender-environment-and-climate-change
41. Vandyck, T., Ebi, K. L., Green, D., Cai, W., & Vardoulakis, S. (2022). *Climate change, air pollution and human health.* https://doi.org/10.1088/1748-9326/ac948e
42. IPCC. *Climate change 2022, impacts, adaptation and vulnerability.*
43. Dupar, M., & Odi E. T. (2022). *Women's economic empowerment—The missing piece in low-carbon plans and actions* (Vol. 19, pp. 1–16).

44. Liu, L. (2021). Mainstreaming gender in actions on climate change. *Climate Change Research, 17*(5), 548–558. https://doi.org/10.12006/j.issn.1673-1719.2020.298
45. Chanthamith, B., Wu, M., Yusufzada, S. & Rasel, M. (2019). Interdisciplinary relationship between sociology, politics and public administration: Perspective of theory and practice. *Sociology International Journal, 3*(4): 353–357. https://doi.org/10.15406/sij.2019.03.00198
46. Rauner, S., Bauer, N., Dirnaichner, A., Van Dingenen, R., Mutel, C., & Luderer, G. Coal-exit health and environmental damage reductions outweigh economic impacts. *Nature Climate Change*, 10.
47. Zhang, S., Wu, Y., Liu, X., Qian, J., Chen, J., Han, L., & Dai, H. (2021). Co-benefits of deep carbon reduction on air quality and health improvement in Sichuan Province of China. *Environmental Research Letters, 16*(9). https://doi.org/10.1088/1748-9326/ac1133
48. Mahadevia, D., & Advani, D. (2016). Gender differentials in travel pattern—The case of a mid-sized city, Rajkot, India. *Transportation Research Part D: Transport and Environment*, 44. https://doi.org/10.1016/j.trd.2016.01.002
49. Spitzner, M., Hummel, D., Stiess, I., Alber, G., & Roehr, U. *Interdependente Genderaspekte der Klimapolitik.* http://www.umweltbundesamt.de/publikationen
50. Kronsell, A., Smidfelt Rosqvist, L., & Winslott Hiselius, L. Achieving climate objectives in transport policy by including women and challenging gender norms: The Swedish case. *International Journal of Sustainable Transportation, 10*(8). https://doi.org/10.1080/15568318.2015.1129653
51. Harumain, Y. A. S., Azmi, N. F., & Yusoff, S. M. (2021). Assessing elements of walkability in women's mobility. *Journal of the Society of Automotive Engineers Malaysia, 1*(3). https://doi.org/10.56381/jsaem.v1i3.61

Chapter 5
High-Quality Development of River Basins and Adaptation to Climate Change

1 Introduction

A watershed is a complex system characterized by changes at different temporal and spatial scales. There are uncertainties associated with predicting and managing these changes. Some countries and regions have developed relatively mature governance practices and experience in many aspects of watershed management, including the identification of watershed problems and pressures, the promotion of a conceptual consensus to undertake holistic planning of watersheds, and some specific working methods, such as nature-based solutions, encouraging diverse stakeholder participation, including that of women and other marginalized groups, and integrating long-term needs and short-term actions. At the same time, however, there is still a long way to go to integrate watershed management as a system. Building on previous NCC studies on integrated watershed management (2004) and ecological compensation (2018 and 2019), as well as several CCICED studies of the Yangtze River, the 2023 joint report of China and The Netherlands, a CCICED event at the 2023 UN Water Summit, and other developments, this report focuses on identifying effective regional cooperation mechanisms in the increasingly urgent context of climate change impacts across the Yangtze River basin, from extreme flooding to drought. The report promotes the use of relevant research recommendations in China as well as globally.

Green Empowerment and High Quality Development,
https://doi.org/10.1007/978-981-96-4218-2_5

2 Governance History, Issues, and Challenges

2.1 *The Evolution of the Yangtze River Basin Governance*

Based on the results of the analysis of the change in the proportion of topics in Chinese government policy documents on the Yangtze River (based on the semantic analysis method) and the case study of the impact of major events on river basin governance, the study finds gradual and intermittent change in the governance policy of the Yangtze River basin. Based on the identification of key time points, the governance history of the Yangtze River basin since 1949 can be divided into four phases.

2.2 *The Initial Stage (1949–1978)*

In the context of the planned economy at that time, various management activities in the Yangtze River basin were mainly arranged and deployed at the national level, with the focus on two key areas of management: flood control and water conservation, and power generation. Under the unified arrangement of the state, a number of embankments, water conservancy, and hydropower hubs and other projects were built in the Yangtze River basin during this period.

Economic Development-LED Phase (1978–1998)

After 1978, with the implementation of China's reform and opening-up policy, the Yangtze River basin governance entered a stage with economic development as the core. In addition to flood control and power generation, water resources development and deployment, integrated development of shipping and industrial development along the Yangtze River basin also became the focus of river basin management. The structure of governmental bodies involved in river basin governance has become more diversified. In addition to the water conservancy department, the importance of development and reform, ecological and environmental protection, urban and rural construction, land and resources, and transportation departments at the national level has increased in the governance of the Yangtze River basin, but the coordination of different basin governance bodies is insufficient.

Initial Transformation Phase (1998–2012)

After the 1998 Yangtze River floods, the Chinese government and public opinion paid more attention to the risk of flooding in the Yangtze River, and the policy of watershed management changed to integrate development and protection and strengthen systemic management, but the change in the concept of watershed management was not complete due to the limitations of the development stage. In addition to the central government, local governments at the provincial, municipal, and county levels within the river basin have significantly increased their role in the governance of the river basin from the goal of developing the local economy, and have carried out a large number of economic development-type governance activities, such as urban construction, industrial development zone construction, port development, and mining exploitation, but the regional coordination mechanism across administrative regions at the local level is not perfect.

High-Quality Development and Protection Phase (2012 to Present)

Since the Chinese government incorporated the construction of ecological civilization into the overall national strategy in 2012, the governance policy of the Yangtze River basin has gradually changed; especially after the release of the Outline of the Yangtze River Economic Belt Development Plan in 2016, the governance orientation of the basin has clearly changed from "prioritizing economic development" to "prioritizing ecology and green development."

Although coordinated regional governance has been strengthened since 2012, collaborative governance in the Yangtze River basin is still in its infancy in terms of biodiversity conservation, coordination between hydropower development and environmental protection in upstream areas, and efficient development of shorelines in the middle and lower reaches. Flooding and drought events, like the ones that have occurred in recent years, may significantly impact hydropower output and lead to an increase of investments in coal-fired power stations to counter the decline of hydropower output, when renewable energy capacity by wind and solar is not yet sufficient to cover this decline.

2.3 *The Evolution of the Governance of the Rhine River Basin*

The history of Rhine basin governance shows that systematic shifts in basin interventions, and the accompanying emergence of new governance structures, are often triggered by natural disasters, i.e., devastating events with significant social impacts. In most cases, however, interventions were proposed before the corresponding major destructive events occurred. Natural disaster events only drive the consensus on the urgency of implementing these measures and programs at the social and political levels. The following major events are particularly relevant regarding the evolution of Rhine basin governance.

French Invasion of Holland (1672)

In 1672, the Dutch dug an artificial channel in the lower Rhine to defend against the French invasion. This channel later formed a new tributary of the Rhine and now marks the division of the Rhine into the Waal River and the Pannerden Canal.

Mannheim Act (1868)

In 1868, the countries of the Rhine basin signed the Mannheim Convention, which guaranteed a regime of freedom of navigation on the Rhine. Since then, it has been an important treaty for facilitating trade relations between European countries.

Zuiderzee ('Southern Sea') Flood (1916)

In 1916, a major coastal flood occurred of the Zuiderzee ("Southern Sea"), an inland sea that was connected to the North Sea and reached from the north to the middle of The Netherlands.

Establishment of the International Commission for the Protection of the Rhine (ICPR, 1950)

In 1950, the International Commission for the Protection of the Rhine (ICPR) was established jointly by The Netherlands, Switzerland, France, Germany, and Luxembourg in Basel, Switzerland, seeking to address the increasing river pollution of the Rhine through cooperation.

North Sea Flood (1953)

After the North Sea flood of 1953, the Dutch government initiated the construction of the Delta Project. The construction of the project was directed by the Special Committee on the Delta. Although the Dutch government made a clear decision to build the Delta Project, the project was also hindered by social opposition in its implementation.

Sandoz Chemical Leak Disaster (1986)

In 1986, a fire at the Sandoz Chemical Company in Basel, Switzerland, spilled a large quantity of pesticides into the Rhine River, causing an environmental disaster throughout the river's lower reaches. In response to this disaster, the Rhine riparian countries initiated ecological restoration of the Rhine, including the implementation of the ICPR Rhine Action Plan (1987).

Extremely Large River Discharges (1993, 1995)

The Rhine experienced extreme river discharges in 1993 and 1995, which put the dikes along the river to the test. In response, the Dutch government launched the "Room for the River" project: a series of measures along the Rhine to increase the river's water discharge capacity. A planning tool was developed to visualize the effect of all sorts of interventions and to facilitate the participation of different stakeholders and government levels in the decision-making process.

Development of the EU Water Framework Directive (2000) and the Floods Directive (2007)

The development of the EU Water Framework Directive in 2000 meant that member states had to work together to develop river basin management plans. In the context of water resources management, the Directive focuses on controlling the water ecological quality of the basin. Similarly, for flood risk management, flood risk management plans for each basin must be developed within the framework of the European Floods Directive.

2.4 *The Evolution of the Mississippi River Basin Governance*

The Mississippi River basin covers 41% of the contiguous United States. The Mississippi River and its tributaries have approximately 10,000 miles of government-maintained navigable waterways. Navigation in the Upper Mississippi, Ohio and Red River basins is secured through a series of locks and dams. Navigation on the Missouri, Middle Mississippi and Lower Mississippi rivers is maintained through levee construction, bank improvement, and channel dredging.

The history of Mississippi River governance illustrates the complexity of the division of responsibilities in watershed management between the U.S. federal government and state governments. The complexity of watershed governance has been exacerbated and regional governance has become less efficient by the large number of governance responsibilities vested in the states and the lack of cooperation among the states.

Division of Responsibilities Between the Federal Government and the State Governments Since the Early Nineteenth Century

Development of the Mississippi River basin began in the early 1800s, after which the U.S. federal government assumed responsibility for maintaining navigation on America's major rivers. As the settled population in the basin grew, local governments assumed responsibility for providing flood protection for the basin's flood-risk residents. Under the U.S. Constitution, all powers not expressly granted to the federal government in the Constitution were left to the states. Therefore, the power to manage water resources (excluding navigation) belongs to the state governments.

Mississippi River Commission (1879)

In 1879, the U.S. Congress approved the creation of the Mississippi River Commission to oversee activities involving the Mississippi River and assigned the Commission specific responsibilities for maintaining river navigation in the basin.

Great Floods (1927, 1936)

In 1927, a catastrophic flood swept through the Lower Mississippi River basin, disrupting navigation and causing millions of dollars in damage. In 1928, the federal government assigned the Mississippi River Commission the responsibility of managing navigation on the Lower Mississippi River and working with the states to adopt a systematic approach to flood control. In 1936, after another major flood, the U.S. federal government began to strengthen flood control programs. The federal government would assume responsibility for the construction of major flood control projects when the benefits of these projects exceeded their costs and proposed that the implementation of major flood control projects be carried out on a case-by-case basis in each locality. But the United States still has not assigned responsibility for overall watershed-level oversight to any federal agency. Meanwhile, at the request of the Missouri River Basin states, the U.S. federal government has assumed responsibility for the construction of six major dams on the mainstem of the Missouri River for flood control, irrigation, and maintenance of the Missouri River navigation system, but not for watershed oversight.

Establishment (1965) and Abolition (1981) of the Basin Commission

In 1965, recognizing the real challenges to the operation of water systems by state and local agencies, the U.S. Congress enacted a water resources planning law requiring the establishment of basin commissions and water resources councils in major basins to coordinate federal and state activities in the area of water resources. In 1981, largely because the states complained that the basin commissions were interfering with their constitutional authority, the President of the United States ordered The River Basin Commission abolished and support for the Water Resources Board was withdrawn. These rulings essentially told the states that the federal government would operate navigation and flood control dams on the Missouri River until the states worked out a cooperative solution.

Current Challenges

Today, the Mississippi Basin faces significant challenges in achieving cooperation and collaboration among the relevant governance actors in water resources development, flood control, navigation, and ecological protection. At the top of the list is the federal system in place in the United States, which has transferred responsibility for management of most water resource activities to the state governments. The states also face challenges in internal cooperation, depending on the nature of their constitutions and the provisions in those constitutions regarding the powers of lower levels of self-governance within the states. The second challenge is the upstream and downstream coordination issues faced worldwide. On tributaries located entirely within a state, that state can manage most of the watershed activities. In other cases, too little or too much water in a river can quickly lead to disagreement or even conflict if states share a river segment.

2.5 Relevant Experience to Learn from

Although the situation varies from basin to basin, both in China and in other countries, basin stakeholders are facing increasing challenges from climate change and need to continuously improve and strengthen regional coordination mechanisms to address them. China is strengthening regional coordination of major river basins through a more systematic and comprehensive approach, such as the successive enactment of the Yangtze River Protection Law and the Yellow River Protection Law.

When designing policies and interventions for watershed governance, they must always be evaluated at appropriate spatial and temporal scales, and all consequences must be assessed in a multi-objective dimension, including the costs and effectiveness, co-benefits, and negative side effects of the interventions. The assessment should consider not only the short-term impacts on river flow and ecology, but also the long-term impacts; the impacts on the surrounding areas need to be considered within the watershed system, not just the local geographical impacts; and the trade-offs between the multi-dimensional objectives of achieving the benefits of engineering measures such as navigation and flood control and maintaining the long-term stability of the river ecosystem should also be taken into account.

Changes in basin governance policies are often in response to major events and are abrupt in nature but require advance preparation of policies and plans to promote timely implementation of policies when the window of opportunity comes. The IPCC Sixth Assessment Report states that watersheds that are poorly adapted to climate change will be at great risk, and as with watersheds around the world, advance integrated watershed resilience planning is not only necessary to address the multiple and acute impacts of climate shocks on the Yangtze River Basin, but is one of the key government tasks with a great sense of urgency.

3 Upstream River Section: Analysis and Synergistic Governance Thinking in the Middle and Lower Jialing River Region

3.1 Regional Overview

The Jialing River is an important tributary on the left bank of the upper reaches of the Yangtze River. The mainstream of the middle and lower reaches of the Jialing River is 740 km long with a natural drop of 303 m. The main tributaries in the middle and lower reaches include the Bailong River, Dong River, Xi River, Dui River and Ful River, and the watershed is dominated by low mountains and hills. The major cities in the region are Guangyuan, Nanchong and Guang'an in Sichuan Province and Hechuan, Beibei, Yubei, Jiangbei, Shapingba and Yuzhong in Chongqing.

The region is a less-developed area and is in a rapid development stage with an increasing demand for energy, water and other resources. The region belongs to two provincial-level units, placing higher demands on the synergistic management of the basin. Chongqing became a municipality directly under the central government in 1997, after which the middle and lower reaches of the Jialing River were divided between Sichuan Province and Chongqing Municipality.

3.2 Risks and Challenges Faced

The complexity of the problems brought about by climate change and accelerated development has posed new challenges to the existing integrated watershed management approach. In recent years, the middle and lower reaches of the Jialing River have faced both more frequent heavy rainfall floods and high temperature droughts due to climate change, and accelerated development as a late-developing region has brought more energy and resource demands and environmental pressures. The risks of water security and water shortage in the non-riverine rural areas of the basin are more prominent, the single energy structure dominated by hydropower is challenged by the increased number of droughts, and the organization of water transportation is also facing difficulties in the coordination of multi-stage locks, and the problem of surface pollution caused by agricultural development is further aggravated.

Rural areas are lagging in disaster prevention and mitigation facilities, with high impacts of flood and drought disasters and agricultural pollution. Weak flood-prevention facilities in rural areas lead to rising agricultural losses every year. Drought caused a reduction in rural food production and difficulties in drinking water for humans and livestock. Water quality is generally improving, but agricultural pollution has become the main source of pollution.

The hydropower industry's ability to cope with high temperatures and droughts is relatively weak, and there is a need to improve the energy mix and increase the resilience of energy supply.

3.3 International Case Study

The Rhine River originates in Switzerland at the northern foot of the Alps and flows through Austria, France, Germany, and The Netherlands before emptying into the North Sea near Rotterdam, The Netherlands. Major interventions were implemented in the nineteenth and twentieth centuries to optimize the economy (with the main aims of flood control and improving shipping conditions). These interventions have—partly unforeseen—long-term, adverse impacts that threaten the river's functions. To mitigate these negative impacts, and because of the growing concern for sustainable functioning of watersheds (hydrodynamics, morphology, and ecology), several watershed management plans and governance mechanisms have been established, including the return of land to the river's floodplain, the introduction of integrated river basin management, and the establishment of the International Committee for the Protection of the Rhine (ICPR).

Compared to most large river systems, the Rhine is relatively well-coordinated at the basin scale, as all Rhine riparian countries are committed to the EU directives on ecology, biodiversity, and flood protection and appreciate the functioning of the International Commission for the Protection of the Rhine. Since 2000, the European Water Framework Directive (WFD) gives new impetus to the river basin approach and to international cooperation in European catchments. It aims at transforming mainly water quality-oriented management into the more integrated approach of ecosystem management. It applies to inland, transitional, and coastal surface waters as well as groundwaters. It ensures an integrated approach to water management, respecting the integrity of whole ecosystems, including by regulating individual pollutants and setting corresponding regulatory standards. It is based on a river basin district approach to make sure that neighbouring countries cooperate to manage the rivers and other bodies of water they share.

The Rhine River has the following lessons learned in its management: When intervening in a river system, there is a need to assess the impact on the appropriate scale in time and space (i.e., a scale of expected effects); the need to seek alternatives to infrastructure measures, again, using appropriate scopes and system boundaries; and the need to seek collaboration on appropriate scales (corresponding to the scale of the physical system).

3.4 Collaborative Governance Mechanisms for Improving Basin Resilience

Past Interventions and Their Effectiveness

In 1986, the Jialing River started the construction of the first terrace Mahui Power Station, and in 1999, the Sichuan Provincial People's Government approved the implementation of the "Jialing River Drainage Planning Report," which considered flood control, water conservancy, hydropower, and water transportation factors, and planned the construction of interconnected terrace navigation and power projects from Guangyuan to Chongqing.

The water conservancy projects have improved the flood control conditions of cities along the river and promoted the development of hydropower and shipping. The hydroelectric project has provided a lower cost and sufficient supply of hydropower to cities along the river, stimulating the development of high energy-consuming industries. The entire middle and lower reaches of the Jialing River have been canalized, and the navigability has been significantly improved.

The hydroelectric projects bring certain ecological and environmental impacts, and the low price of hydropower leads to structural risks of high energy-consuming industries and insufficient improvement of non-riverine rural areas in the basin. As a result of the implementation of the stepped navigation and power project and human activities, the watershed environment of the Jialing River has been damaged to a certain extent. The ecological flow in the downstream section of the river is insufficient, the aquatic plant system and biological population structure have been affected, and the migration and spawning paths of rare fish such as coelacanth have been blocked, resulting in a decline in the number and diversity of fish. The construction of water conservancy projects has improved the flood control and water supply conditions of towns along the river, but the water supply coverage of non-riverine areas, especially rural areas, is insufficient, and the resistance to climatic disasters, such as high temperature and drought, is more vulnerable.

The irrigated area and proportion of newly cultivated land irrigated in the three cities of Sichuan province from 2010 to 2018 are shown in Table 1.

The middle and lower reaches of the Jialing River have initially established a coordination mechanism for flood and drought prevention and control. Sichuan and Chongqing have developed mechanisms such as "Sichuan Jialing River Basin Flood Control Consultation System," "Sichuan Jialing River Basin Flood Information Transmission System," and "Memorandum on Information Sharing and Notification System for Flood and Drought Disaster Prevention in Chengdu-Chongqing Economic Zone." Sichuan and Chongqing share rainfall and water information from nearly 100 stations and conduct joint reservoir scheduling in response to floods.

The basin governance faces synergy challenges in multiple fields such as shipping, flood control, drought relief, and hydropower generation. At present, the Jialing River obeys flood control scheduling during flooding periods and mainly obeys power scheduling during non-flooding periods. Influenced by the security of the power grid system and fluctuations in peak and valley power demand, the flow rate in the reservoir area may rise and fall steeply, and the water level may rise and fall steeply, posing certain risks to navigable vessels. The increase in extreme weather due to climate change makes it more difficult to synergize multiple areas such as shipping, flood control, drought relief, and hydropower generation.

The water conservancy hub in the middle and lower reaches involves six owners, the government involves two provinces (cities), six cities (districts), and several county-level units, and the departments of water conservancy, energy, transportation, flood control, and environment are involved in the management of the basin, making the synergy between the two provinces, between departments, and between the government and enterprises increasingly difficult. After the construction of water

Table 1 Irrigation area and proportion of newly cultivated land irrigated in the three cities of Sichuan province

	2018			2010–2018		
	Cultivated land	Irrigated arable land	Proportion of irrigated area to cultivated land (%)	New arable land	New area of arable land irrigated	New irrigated area (%)
	Thousand hectares			Thousand hectares		
Sichuan Province	6723	2992	45	2712	439	16
Among which						
Nanchong	534	235	44	233	28	12
Gwangan	308	108	35	134	13	9.5
Guangyuan	353	95	27	187	7.2	3.8

Source Sichuan Statistical Yearbook

conservation projects, flood control, water supply, and navigation conditions in the near-shore areas of the mainstream have been improved, and the socio-economic development gap between the vast countryside of the inland hinterland has increased, requiring further strengthening of regional coordination.

Importance of Collaborative Governance Mechanisms for Resilience

The future development of the middle and lower reaches of the Jialing River faces extremely complex challenges, both in terms of regional security, environmental protection, and control of energy consumption and carbon emissions due to climate change and in terms of meeting the resource and energy needs resulting from rapid development. Europe has made strategic environmental impact assessment a prerequisite for synergistic watershed management, with good results. In China, the synergistic management of watersheds is more complex and requires synergies in flood, drought, flood control, power generation, urban and rural water security, navigation, and biodiversity, etc. The following specific measures are needed to develop "multi-objective integrated solutions."

Transform the economic and social development mode and explore the path of regional economic and social development with low energy consumption and low water consumption. Moderate adjustment of industrial structure and gradually limit the development of metal smelting, chemical, non-metallic minerals and other high-energy-consuming industries in the basin.

Establish a synergy mechanism between governments, departments and enterprises, and between the river and the hinterland, considering multiple objectives and multiple interests.

Focusing on non-riverine rural areas, localized, natural, low-impact rural governance approaches need to be explored. Protect and restore the weir and pond system to build a green and resilient ecological base for the countryside. Develop a three-dimensional, composite rural industry or "agriculture-wet-fruit-raising-tourism." development model.

The water conservation hub projects in the middle and lower reaches of the Jialing River that have not yet been constructed should be reassessed. One reason is to assess the changes in the construction conditions of water conservancy projects brought about by the expansion of urban construction land, the second is to assess the impact of water conservancy projects on urban safety and accident risks, the third is to assess whether it is conducive to the improvement of the regional energy structure, and the fourth is to assess the ecological and environmental risks brought about by the construction.

4 Great Lakes Basin: Analysis and Collaborative Governance Thinking in the Taihu Lake Basin

4.1 Regional Overview

Taihu Lake Basin is located at the mouth of the lower reaches of the Yangtze River, involving three provinces and one city, Jiangsu, Zhejiang, Shanghai, and Anhui, with a basin area of about 36,900 km^2, a resident population of about 68.11 million in 2021, and a population density of 1846 people/km^2, equivalent to 3.5 times that of The Netherlands; a regional GDP of about 112,736 billion yuan, and a per capita GDP of 165,000 yuan, equivalent to the national average. It is a typical delta region with high population density, high economic vitality, and high intensity of land development.

The topography of Taihu Lake Basin is a disc-shaped terrain with a high periphery and low middle. The general topography is high in the west and low in the east. In 80% of the plain area, the elevation is generally below 5 m, and more than half of the ground elevation is lower than the flood level. The current water surface area of the basin is about 5551 km^2, the water surface rate reaches 15.0%, equivalent to 1.5 times The Netherlands. In the period of ancient agricultural society, the water problems in Taihu Lake Basin were mainly floods and droughts. Nowadays, there has been a tight balance between human activities, land development and water management, resulting in a comprehensive and complex water problem where water disasters, water environment, and water ecology coexist and intertwine. Therefore, this report focuses on synergistic management of watershed flood safety and synergistic management of agricultural surface source pollution.

History and Effectiveness of Collaborative Flood Safety Management

The watershed flood control engineering system has been gradually established, relying mainly on the backbone river flooding. In 1991, a huge flood occurred in Taihu Lake Basin. In order to promote the comprehensive management of Taihu Lake Basin, in September 1991, the State Council held a meeting on the governance of Huaihuai and Taihu, made a decision on the comprehensive management of Taihu Lake, approved the implementation of the "Overall Plan for the Comprehensive Management of Taihu Lake Basin," 11 backbone projects for the comprehensive management of the basin (a round of treatment of Taihu), and 21 backbone projects for the comprehensive management of the water environment in the basin (the second round of treatment of Taihu) were implemented one after another, laying a three-way drainage pattern in the Taihu Lake Basin and initially the basin flood control and water resource control engineering system has been formed. Among them, four projects, including Tai Pu River, Wang Yu River, Hangjia Lake South Drainage and Lake Ring Dike, are World Bank flood control projects for Taihu Lake, which is the first cross-province (municipality directly under the central government) flood control project in China with World Bank loans.

Urban flood control to large polder areas as a unit and consantly expand the scope of the polder and raise the defense standards. The construction of the Taihu Lake

Urban Flood Control Polder has to some extent improved the standard of flood control in the urban centre and low-lying areas.

Preliminary completion of the basin scheduling system and implementation of graded scheduling in different time periods. In the face of excessive floods in Taihu Lake, the Taihu Bureau, together with the provinces (municipalities directly under the central government) in the basin, based on the "Taihu Lake Basin Flood and Water Scheduling Plan," "Taihu Lake Basin Water Allocation Plan," and other scheduling documents, focused on improving infrastructure, integrating system resources, promoting information sharing and deepening business applications, dispatched backbone water conservancy projects in different time periods and built the "Smart Taihu Lake "system. At present, the system has made partial achievements in basin information sharing, integrated monitoring and early warning, scheduling decision making and comprehensive supervision, realizing a certain degree of wisdom.

History and Effectiveness of Agricultural Surface Pollution Management

After the outbreak of cyanobacteria, the government began to implement a comprehensive management program for the water environment in the Taihu Lake Basin. In 2007, a large-scale cyanobacteria outbreak in Taihu Lake led to a water supply crisis in the region, which seriously affected the normal life of the local people and aroused widespread concern in the whole society. In 2008, the State Council approved the "Overall Plan for the Comprehensive Management of the Water Environment in Taihu Lake Basin," which clearly put forward the two goals of ensuring the safety of drinking water and ensuring that no large area of black water odour occurs.

The joint meeting system has been established with multi-party collaborative comprehensive management, improving legislative protection. In 2008, the National Development and Reform Commission (NDRC) took the lead, and relevant departments and three provinces and municipalities participated in the joint provincial-ministerial meeting system for the comprehensive management of the water environment in the Taihu Lake Basin. In 2011, the Ministry of Water Resources, the Ministry of Environmental Protection, and the State Council Legislative Office jointly issued China's first comprehensive administrative regulation, "Regulations on the Management of Taihu Lake Basin," and took the lead in establishing and improving the five-level river and lake governor system at provincial, municipal, county, and village levels, building a platform for a collaboration mechanism between the governors of Taihu Lake and Dianshan Lake, and vigorously promoting pesticide and fertilizer use reduction, and rectifying illegal farming near the Taihu Lake, Ge Lake, and other rivers and lakes.

Comprehensive management of the water environment has achieved initial results, including a significant improvement in the basin water quality. In recent decades, with the joint efforts of all parties, the comprehensive management of the water environment in the basin has achieved phased results, drinking water safety, pollution prevention and control, cyanobacteria prevention, and ecological restoration and other key tasks continue to advance; the total amount of pollutants into the lake has been significantly reduced, and the quality of the water environment has been gradually improved.

4.2 Risks and Challenges Faced

Flood Safety

Overall, the possibility of extreme events in the basin causing disasters is on the rise. The sub-basin conditions of the basin have changed significantly, and the storage capacity has weakened. The lack of coordination in the construction of urban polders has led to increased difficulty in flood scheduling.

Extreme rainfall events and sea level rise will further test the basin's ability to discharge and refine the level of prevention and control. The problem of insufficient pumping capacity along the river and Hangzhou Bay has become more prominent. In addition, sea level rise and ground subsidence will increase the difficulties of drainage in coastal cities, so that the head of drainage pumping stations will increase and drainage capacity will decrease.

Agricultural Surface Source Pollution

As Taihu Lake is a typical dish-shaped lake with limited water environment capacity and insufficient self-purification ability, and considering the developed agriculture in the basin, the input load of nitrogen and phosphorus and other nutrients is large. In the context of global warming, the risk of cyanobacterial blooms is high.

4.3 International Case Study

In response to the flooding event of the Southern Sea (Zuiderzee), Lake IJssel in the Dutch Rhine river system was created in 1932 by damming this inland sea (connected to the North Sea) with a 32-km long closure dike. In the face of climate change, Lake IJssel is becoming increasingly important as a freshwater buffer (in times of drought) and as a water storage area (in times of flooding). In terms of flood protection, climate change is expected to lead to an increase in peak discharge from the Rhine, while higher sea levels will eventually mean that the water of the Rhine can no longer be drained through this lake into the North Sea by gravity alone.

Therefore, due to climate change and spatial scarcity, the pressure on the system from different users in and around Lake IJssel is increasing, and stakeholders at the local, national, and even international scales need to work together at these different spatial scales to balance these interests.

In terms of agricultural nitrogen and phosphorus pollution, the creation of Lake IJssel has had some negative ecological consequences, with algal blooms and stratified/anoxic zones when tidal seawater becomes a stagnant freshwater lake. In response to water quality issues and adaptation to climate change, EU regulations require a new agricultural transition toward less intensive and more regenerative methods, with less environmental impact in terms of emissions and allowing more biodiversity, now addressed by reducing phosphate emissions.

4.4 Collaborative Governance Mechanisms for Improving Basin Resilience

Transforming urban-regional development, carrying out regional subsurface control, and improving the overall safety and resilience of the watershed.

Taihu Lake Basin, should properly handle the relationship between people and nature. From the perspective of watershed security and resilience, priority should be given to the protection of non-construction space such as arable land, garden land, forest land, and water area and "returning land to water" (restrictions are identified based on areas of ecological importance, current levels of urban congestion, pollution runoff, etc.).

Strengthen the integrated management of flood risks in the basin, the implementation of polder refinement linkage management, strengthen upstream and downstream cooperation in flood control.

Optimize the polder construction mode, the implementation of polder refinement linkage management. In terms of polder construction level, it is recommended to restrict large-scale joint polder and polder projects in the basin, strict control of new polders, and avoid building polders in the semi-upland. Before adding polder areas, carry out flood risk analysis along with moderate reduction of low-risk polder construction standards. Polder management level should be coordinated with the relationship between water levels inside and outside the polder to ensure the safe operation of urban polder, moderate reduction in pumping power, slowing the rate of rise of water levels outside the polder.

Strengthen flood control cooperation in the basin and improve the upstream and downstream synergy-feedback mechanism. In the face of severe challenges brought by climate change, the model of relying on each administrative region alone for flood control and management is unsustainable. Combining with the experience of cross-industry cross-regional coordination mechanism of Lake IJssel in The Netherlands, it is suggested to accelerate the establishment of a basin-regional coordination mechanism in the Great Lakes region, cooperate in the construction of flood control coordination zones, integrated management of flood control and drainage, etc. For major projects in dispute, the scheduling authority can be handed over to the basin management agency for unified scheduling.

Collaborate to build a smart water network in the basin and improve the level of engineering scheduling and risk warning. Promote the deep integration of advanced information technology and water business, realize the change from water and rain forecasting to flood impact and disaster risk forecasting, and improve the level of information openness; strengthen the cooperation and exchange of monitoring and early warning between basin agencies and related cities in the field of water safety and security by opening up the smart water platform at different levels of the basin, and improve the flood control decision-making command system.

Protect and utilize the great heritage of agricultural water conservancy and natural wetland resources, shape the urban and rural ecological space of Tangpu Polder, and promote natural solutions.

Establish an ecological buffer zone around the lake to enhance the ecological value and social value of natural resources. Strengthen the control of transboundary rivers and lakes, and gradually restore the basin water storage space. Learn from history and promote the protection and utilization of water cultural heritage with high quality. Respect site characteristics and carry out low-impact development and construction with the polder as a unit.

Promote the integration of modern agriculture, collaborate to carry out joint treatment of pollution and carbon reduction, and manage agricultural surface pollution through scientific and technological innovation.

Promote green production methods in agriculture and collaborate to promote pollution and carbon reduction. Promote regenerative agriculture and other natural farming methods, implement chemical fertilizer and pesticide reduction and efficiency actions, and improve the efficiency of nitrogen and phosphorus fertilizer utilization; improve the level of comprehensive utilization of straw, and strengthen the control of straw burning; promote the resource utilization of livestock and poultry manure, and the construction of ecological breeding of rice and fishery, and reduce carbon emissions in key breeding links.

Improve access to green and low-carbon agricultural industries and strengthen the leadership of science and technology innovation in agricultural surface source pollution control.

Establish ecological compensation mechanisms for watersheds in economically dense areas and diversify funds to protect watershed management.

Establish a watershed water fund to share ecological benefits. The downstream gets clean water and pays the upstream to support the pollution remediation action in the upstream area. Over and above this, establishing a "water fund" trust can guide multiple parties to participate in the protection of water sources and share the benefits, such as developing green industry, building nature education, ecological agriculture, culture and tourism projects.

Build an integrated accurate prediction and monitoring system for the watershed as a basis for ecological compensation. Increase the research and application of total phosphorus control, cyanobacterial water bloom mechanism and prediction in the lake area, accelerate the pilot integrated prevention and control of agricultural and rural pollution into the lake river flux monitoring, and identify areas with serious agricultural surface source pollution. Conduct background surveys on key rivers and lakes and build a surface source monitoring system that integrates remote sensing monitoring, ground monitoring and model accounting in the watershed.

Water and land together, the establishment of ecological agriculture special compensation mechanism. To protect water first, strengthen the synergistic governance of ecological space, such as farmland forestry and inland river pollution in the watershed, and special compensation covers important components of ecological agriculture, such as rice, ecological public welfare forests, and wetlands. Through compensation, further mobilize to improve the enthusiasm about the development of ecological agriculture in the watershed, to reduce surface pollution from the source, taking into account food security and ecological safety.

5 Estuaries: Analysis and Synergistic Governance Considerations for the Coastal Zone Area of the Pearl River Estuary

5.1 *Regional Overview*

The coastal region of the Pearl River Estuary has one of the highest population and town densities and fastest urbanization processes in the world, and the harmonious development of town development and ecological environment is crucial. In 2020, the resident population reached 54 million, and the urbanization rate exceeded 80%. Over the past 10 years, the region's population growth rate (35%) is about 7 times the national average, and the average annual GDP growth rate (8.9%) is about 1.35 times the national average. Shenzhen, Hong Kong, and other cities have population densities approaching 7000 people/km^2, with a high overlap between areas of intense human activity and important biological habitats. It is also a hotspot for industrial development and infrastructure construction, with numerous large infrastructures, such as ports and river crossings, putting great pressure on the ecological environment.

The coastal area of the Pearl River Estuary is in the subtropical zone, with rich ecological resources and diverse types. It is also home to many national animal and plant sanctuaries. It is a migratory resting place and wintering ground for internationally important migratory birds (including black-faced spoonbills, black-billed gulls, yellow-breasted flounder, and other national-level protected birds).

The coastal area of the Pearl River Estuary is one of the areas in the world with the closest interaction between sea and land and the most concentrated conflicts between urban development and ecological protection, and the development trend of towns toward the sea is obvious.

The coastal zone of the Pearl River Estuary includes the first county (district) administrative unit of the five major cities of the Pearl River Estuary (Guangzhou, Shenzhen, Dongguan, Zhongshan, and Zhuhai), as well as the whole area of Hong Kong and Macau. The cities in the Pearl River Estuary are facing common challenges brought by climate change, and strengthening regional collaboration in coastal zone protection and governance is the only way to meet the challenges, but the cities in the Pearl River Estuary coastal zone involve "one country," "two systems," three customs zones, and three legal systems. This also brings unique opportunities and challenges for regional cooperation in watershed governance [1].

5.2 *Risks and Challenges Faced*

The environmental problems of surface water in the Pearl River Estuary affect the water quality of the near-shore sea. With the accelerated urbanization of the coastal areas of the Pearl River Estuary, the environmental pollution of surface water has become increasingly prominent.

The coastal ecosystem is in a subhealthy state, and biodiversity is threatened. Water quality in the coastal waters of the Pearl River Estuary has deteriorated, leading to frequent eutrophication and red tides [2]. Currently, the density and biomass of zooplankton and macrobenthos, as well as the density of fish eggs and larvae, are low in the Pearl River Estuary, and the entire marine and terrestrial ecosystem is in a subhealthy state. The risk of some disasters is exacerbated by rising sea levels caused by climate change.

5.3 Domestic and Foreign Experience and Case Study

Inspiration from Hong Kong and Macau

Hong Kong and Macau have been in line with international standards earlier than cities in Guangdong Province, faced the conflict between ecological protection and urban construction earlier, and have a longer history of governance.

Hong Kong and Macau have a higher degree of public participation in reclamation decision making. Protests against reclamation projects, such as the formation of the Hong Kong Association for the Protection of the Harbour, forced the Hong Kong government to establish the principle of reclamation that there is an urgent and overriding immediate need (including the economic, environmental, and social needs of the community) and that reclamation should only be considered when there is no other feasible alternative and when it will cause the least damage to the harbour.

Hong Kong's seawater quality monitoring standards are more stringent, and monitoring and control methods are also more complete. In contrast, municipalities in Guangdong Province carry out marine monitoring work under the supervision of the province and the city, and district-level units carry out their own monitoring work according to actual needs. For better disaster prevention and mitigation, environmental protection and sustainable development, the Greater Bay Area is vigorously developing and improving the regional ocean monitoring and forecasting system, which greatly enhances the capability and level of monitoring.

Based on the existing operational platforms, the Greater Bay Area ocean monitoring system should focus on the development of new intelligent platforms such as unmanned vessels, unmanned aircraft, underwater submersibles, micro and nano-satellite constellations, ground wave radar, smart fibre optic cables and their networking applications; while the Greater Bay Area forecasting system should focus on the development of new technologies such as ultra-high-resolution regional models, deep learning, and visualization digital twins under the framework of the Earth simulation system, to provide user-friendly public services and management support [3].

Mangrove conservation and restoration in Hong Kong started earlier and has a more mature governance mechanism. Mangrove protection and restoration in Hong

Kong has been tracked for more than 20 years, including continuous monitoring of various core indicators that started earlier in the operation of governmental and non-governmental organizations. In recent years, mainland provinces and cities have been paying more and more attention to mangrove protection, especially through large-scale mangrove restoration operations, which have led to a sustained increase in the area of mangroves. At the same time, through learning from international experience and cooperation with Hong Kong, cities such as Shenzhen have begun to actively introduce NGOs that have also begun to join in the protection and management of mangroves. For example, the Mangrove Forest Foundation (MCF), developed in Shenzhen, is the first privately initiated environmental foundation in China and the only non-profit social organization in mainland China that operates and maintains mangrove nature reserves and mangrove parks.

Hong Kong pays more attention to the eco-friendly and low-impact seawall construction mode. In seawall construction, Hong Kong pays more attention to the low impact on the natural environment and biodiversity and adopted the eco-friendly seawall construction mode earlier. Most of the seawall construction in Guangdong Province is based on the traditional physical seawall construction, which focuses more on engineering safety and lacks the eco-friendly construction concept. In recent years, eco-friendly seawall construction in Nansha New Area of Guangzhou, Shenzhen, Maozhou River of Dongguan, and Hengqin Guangdong-Macao Deep Cooperation Zone has shown initial results.

Revelation of the southwest Rhine Estuary

Starting in the 1970s, there was growing concern about the poor condition and further deterioration of the geological ecosystem. Delta construction included a series of dams, sluices, and storm surge barriers to protect the coastline from flooding. These constructions were based on a simple cost–benefit analysis: to maximize flood protection while minimizing costs and maintenance efforts.

The Delta Project had a huge negative impact on the ecology of the region. Since the Haringvliet estuary is isolated from the sea but maintains an open connection with the Rhine and Meuse rivers, it was affected by severe pollution of the Rhine and Meuse rivers in 1970. River sediments contaminated with heavy metals and organic compounds settled in the freshwater lake, leading to poor water quality and ecosystem deterioration. The transformation of the Grevelingen from an estuary to a stagnant brackish lake led to changes in species composition;

Social influences were considered in the decision, and conservationists, fishers, and other interest groups effectively lobbied to change the plan to close the Eastern Scheldt. The decision was made to build a storm surge barrier consisting of a series of gates that would only be closed during storms rather than permanently. In this way, the brackish environment of the Eastern Scheldt and its unique ecosystem and associated species (e.g., mussels) were protected.

Spatial planning requires a new vision, especially for strategies such as agriculture. As ecological conditions deteriorate and sea levels rise, many large infrastructures in estuaries are approaching the end of their safe lifespan, and these challenges require that the flood risk management strategies adopted in the future should be jointly determined by the relevant subjects in the context of regional cooperation. It will also require combining different interests while protecting/enhancing ecological values, such as freshwater supply, flood risk management, fishing industry, renewable energy, etc.

5.4 Collaborative Governance Mechanisms for Improving Basin Resilience

The protection and management of the coastal zone area of the Pearl River Estuary as an important issue of regional collaboration.

Facing the multiple challenges of river basin management in the estuary, a stronger mechanism is needed to cross regions and sectors and strengthen regional collaboration. It is recommended to draw on the international experience of coastal zone scope delineation and unify both sides of the Taiwan Strait and Hong Kong to delineate the coastal zone area of the Pearl River Estuary. The government should learn from Hong Kong's public participation in coastal zone governance, encourage collaboration among non-governmental organizations in various cities or regions, and guide the participation of multiple subjects in the protection and governance of the coastal zone of the Pearl River Estuary.

Gradually harmonize standards and approaches to environmental monitoring and large-scale infrastructure development through regional coordination.

Develop a collaborative environmental impact assessment mechanism among the regions in the Guangdong, Hong Kong, and Macau Bay Areas as a pilot in the Yangtze River Basin. Strengthen regional cooperation in the construction of large-scale infrastructure in the Pearl River Estuary, collaborate on environmental impact assessment, gradually coordinate and unify standards and approaches in the construction of large infrastructure such as seawalls, ports, airports, bridges, etc., improve the operation and service efficiency of facilities, reduce the scale of construction and the occupation of shorelines, and reduce the impact of large-scale infrastructure construction on the impact of the ecological environment of the Pearl River Estuary.

Modern developments in strategic environment impact assessment (EIA) should be taken advantage of.[1] Modern EIA is much more than a tick-box mechanism to obtain the formal go-ahead of a law or development program. The latter will likely be necessary given the longevity of many projects, innovation, and uncertainties in climate change, especially at the regional level. Such assessment practices are emerging in various parts of the world under varying names. Participants in events related to this SPS in 2022 and 2023 recalled these modern developments, quoting examples of the Mississippi Delta and Gulf of Mexico and the EIA development assistance program sponsored by the government of The Netherlands.

[1] See, for instance, https://www.eia.nl/en/countries/china, specifically the December 2014 publication.

Establish a collaborative mechanism for environmental management in the Pearl River Estuary coastal zone area by linking rivers and seas, unifying standards for pollutant monitoring and control, cooperating in monitoring and early warning of marine environmental disasters, and gradually unifying standards and specifications for mangrove protection, shoreline restoration, and seawall construction.

Delineate unified ecological reserves and promote transboundary cooperation and public participation in biodiversity conservation.

Suggest GBA promote the practices such as Shenzen mangrove that identify potential nature-based solutions for win–win outcomes, for example in terms of nature and climate resilience. Optimize the integration of various existing ecological reserves and develop cross-border cooperation in biodiversity conservation. It is recommended to break the administrative division of cities, unify the designation of ecological reserves in geographical proximity, and establish a conservation and operation platform for regional cooperation.

Suggest regional joint advocacy for public education and public participation in biodiversity conservation. Fully mobilize Shenzhen's Mangrove Forest Foundation, the Cross-border Environmental Concern Association (CECA) and other NGOs to promote public participation, expand public participation channels in various policy areas and decision-making processes, such as environmental impact assessment, territorial spatial planning, and engineering construction decisions, use online and offline multiple media and platforms to enrich public articipation methods, promote regional cooperation in public science popularization and education, and jointly enhance the attention and participation of the whole society to biodiversity conservation in the coastal zone.

6 Cross-Cutting Issues: Energy Transition and Agricultural Modernization

6.1 Experience in Energy Transition

China is currently faced with the challenge of having limited land available in densely populated areas and the need to meet multiple demands. As China continues to build wind and solar projects, one challenge is the amount of land area these projects take up. Currently, China has established a system of ecological protection red lines as a policy tool to guide integrated spatial planning across watersheds. Further development of remote sensing technology has made such forward planning more feasible.

6.2 Thoughts and Discussions on Energy Transition

A Comprehensive Approach

Given the challenges of maintaining the resilience of the basin in the coming years, a comprehensive application of the eco-redlining system throughout the basin would be beneficial. Likewise, an integrated approach to hydropower use, flood control, floodplain management, and integrated water resources management should be applied to achieve a more optimal allocation among these resources, requiring comprehensive, basin-wide planning. Such an approach should.

- Incorporate requirements for regional ecological conservation planning into hydropower development planning to ensure that the natural ecology of biologically valuable and representative portions of rivers in the basin are protected and that the free flow of rivers is maintained.
- Assess multiple resources and industry needs, including hydropower, environmental resources and ecosystem services such as biodiversity and fisheries, reservoir flood management, and flood risk in the mid- and downstream floodplains.
- Use China's considerable spatial planning tools to identify the best places for siting wind and solar assets, e.g., with consideration going first to contaminated lands where other uses are less feasible, possibly with consideration of planting of biofuels among such RE assets.

All of this represents an area for both immediate implementation and further study, including looking at how different jurisdictions are handling climate impacts on hydropower as well as the development and siting of RE assets.

Sustainable Hydropower Fund

China could consider the concept of a sustainable hydropower fund, particularly in the Yangtze River basin. The benefits of such an approach include:

- Fully utilizing the power generation potential of the lower Jinsha River terrace reservoirs to increase power supply, especially to address the demand of peak electricity consumption in summer.
- Providing sustainable financial support for the Yangtze River flood risk management system, including revegetation and management activities in the forestry sector.
- Improving ecological flow, promoting the protection and restoration of important lakes and wetlands in the middle reaches of the Yangtze River, and avoiding the risk of flooding, as well as the risk of insufficient reservoir recharge after the season.
- Providing sustainable financing mechanisms for projects, such as ecological compensation triggered by hydropower projects, systematic protection of freshwater ecosystems, ecosystem monitoring, and adaptive management.

6.3 Comments on Agriculture and Basins Management Modernization

CCICED studies have addressed agriculture extensively in recent years. An SPS report in 2014 on "Institutional Innovation of Eco-Environmental Redlining" looked at the need to ensure adequate productive agricultural lands. Another report in 2015, "Soil Pollution Management," emphasized the reduced use of fertilizers to protect and restore China's soils. Last year's report on "Sustainable Agrifood Systems—Meeting China's Food and Climate Security Goals," went into more detail on practices that can improve soil resilience and long-term productivity while reducing GHG emissions related to food production. The report identified cover and rotational crops among those regenerative practices that can lead to benefits on both mitigation and adaptation. While modern agricultural methods have greatly increased food production in China and elsewhere, the heavy use of chemical fertilizers and other chemical additives has polluted surface runoff while affecting soil health. China is moving toward conservation agriculture, which saves money and increases soil biodiversity, with benefits for soil and water conservation, resilience, and long-term productivity. A recent demonstration program of conservation tillage ("low-till") across part of the Yellow River Basin produced dramatic results:

Outcomes of Two-year Scientific Monitoring:

- Soil fertility improved by around 10%
- Soil water storage improved by around 7%
- Underground biodiversity increased.
- Fossil fuel usage decreased by around 58% due to simplified machinery operations
- Costs decreased by around 17%.

The cases of the Yangtze, Mississippi, and Rhine river basins illustrate the negative consequences of the excessive pursuit of agricultural yield growth, especially the large amount of nutrients being discharged into rivers and coastal waters. Therefore, modernizing agriculture so that food production can keep up with population growth while dealing with the adverse consequences of climate change (droughts, heat, floods) and reducing stress on water systems is one of the major issues facing us all. A growing body of research shows the economic benefits and risk reduction benefits of conservation practices like cover crops and conservation tillage.

6.4 Thoughts and Discussions on Agricultural Modernization

Regenerative agriculture can be a nature-based solution to balance food security and farm income to ensure long-term food production. This approach can also contribute to the goals of the CKPA.

To reduce stress on watersheds and increase human resilience, China can focus on the excessive intake of modern dietary habits that have led to health problems and overconsumption in developed countries.

7 Gender Equality and Social Inclusion Considerations in Watershed Governance

7.1 *Situation Analysis and Problem Identification*

The previous report of the Special Policy Study found that gender equality is a key issue in global watershed governance today. Gender inequality in watershed governance is most prominent in rural areas of developing countries, which is caused by the disaster resilience gap between rural and urban areas and regional areas on the one hand, and the concentration of women and poverty in rural areas due to population outflow on the other. This section aims to further investigate the gender-differentiated impacts of climate change in three typical case areas in the upper and middle reaches of the Yangtze River and the corresponding strategies to address them.

Women in the Yangtze River basin are more vulnerable to the negative effects of climate change than men, which is particularly evident in the upper Jialing River region.

First, it is more difficult for women to secure water and food in shortage situations after a disaster than men. Women, who bear disproportionate responsibility for household work due to rigid gender roles that are based on and perpetuate gender inequality, require water to cook, wash clothes, and feed livestock, amongst other tasks. Since their husbands engage in income-generating work outside the home, it is usually the women who are responsible for carrying water. When experiencing a drought and water shortage, women are forced to spend additional time and travel further to obtain water for domestic use, exacerbating their time poverty.

Second, the reduction in food and cash crop production caused by climate change has a significant impact on women's livelihoods and incomes. Female workers in developing countries are primarily engaged in informal agricultural work, with low income and few protections. Women, as the mainstay of agriculture, have to work many times harder to cope with the reduced income and livelihood difficulties brought about by drought and other climate change impacts.

Third, due to multiple layers of gender discrimination, rural women typically have lower incomes, lower education levels, and less access to training to acquire skills than men, reducing their ability to cope with climate change. The Study on Climate Change Vulnerability from a Gender Perspective in China shows that women have less income, land resources, access to credit, and access to non-farm employment than men, reducing their opportunities to improve their adaptive capacity.

Despite the direct and disproportionate impacts of climate change on Chinese women, they have low levels of participation and leadership within watershed governance.

First, women's participation in decision making on safe environments in the basin is still insufficient.

Second, there is a relatively serious gender bias in the workplace in watershed management-related industries. At the same time, men are also more involved in marine research and education, leading and implementing more scientific research activities. In addition, women face serious barriers to participation in marine conservation, such as a lack of economic and technical support and policies that fail to take their perspectives into account. Unequal opportunities and rights ultimately show up as unequal pay.

Third, most watershed management policies fail to integrate a gender perspective. Among the regulations and programmatic policy documents, only the China Climate Change Adaptation Strategy and the China National Program for Implementing the 2030 Agenda for Sustainable Development focus on marginalized groups such as women, while most of the watershed regulations and programmatic policy documents generally lack a gender perspective.

7.2 *Social Equity and Gender Strategies in Watershed Governance*

Improve drinking water, irrigation, and other measures to cope with climate disasters and encourage the use of solar energy and other low-carbon energy facilities.

To make up for the shortcomings in rural drinking water and irrigation facilities in rural areas of the Jialing River Basin, and to reduce the risks women face in times of drought and other disasters such as long-distance access to water resources, water shortages, and reduced food production and income.

Promote women's participation and leadership in watershed governance through watershed collaboration.

First, we need to revise watershed management regulations and institutional documents to ensure that impacts on women are considered and mitigation measures incorporated in watershed governance practices, such as strengthening research on women's perspectives on ecological compensation mechanisms. Second, we need to develop a unified public participation system and regulations for watershed governance based on gender equity, and guarantee channels for women to participate in decision-making, such as establishing programs to enhance women's leadership and valuing women's local knowledge and experience. Chinese women have unique advantages in marine biodiversity conservation.

Implement gender-specific statistics in multiple statistical indicators on watersheds and climate change.

Currently there is a lack of gender-disaggregated statistics for disaster loss hazards and other climate change impacts. This is one of the many factors that contribute to the invisibility of women and girls and their needs and priorities to policymakers. Because there are no statistics on women, women's needs are often overlooked. Therefore, government and educational institutions should work together to advocate for gender-disaggregated statistics on disaster loss hazards to provide an evidence base for informing the development of gender-responsive policies, programs, and budgets.

Drawing on the experience and foundation of Hong Kong and Macau, support the enhancement of women's capacity to participate in watershed environmental protection and management through training and education.

This includes: providing training for rural women in agricultural technology, disaster response, and off-farm employment; guiding women to engage in scientific research and technology development related to the field of engineering science and technology; implementing employment equity policies and practices to eliminate gender discrimination in the workplace; and implementing an education policy that addresses gender bias in education and incorporates gender-sensitive curricula and teaching methodologies on watershed governance and environmental protection for adolescent girls and boys.

8 Main Policy Recommendations

Recommendation 1: It should be fully recognized that enhancing river basin resilience and climate resilience has a high comprehensive value of safety and security, economic investment, low-carbon development, and ecological protection, and it should be taken into consideration in the overall economic and social development of the watershed. Enhancing basin climate resilience not only enhances the level of resilience of urban and rural settlements in the watershed and safeguards life and property, but it also expands effective investment, promotes green and low-carbon construction, and reduces the comprehensive economic and social risks brought about by climate disasters. The collaborative governance strategy in the basin should not only adopt "defensive" measures to cope with the urgent pressure of climate risks but also "innovative" measures in terms of technology and governance models. Reaching resilience to climate change and implementing the "double carbon" target is not just an urgent pressure for basin development but an important driving force for green transformation development.

Recommendation 2: The urgency of climate risk and the necessity of climate adaptation cannot be ignored. Drawing on the experience of the EU Framework Directive, and under the guidance of the applicable laws of the Yangtze and Yellow rivers, we should establish as soon as possible a comprehensive cooperation mechanism in the basin with the participation of multiple subjects, such as cross-sector and cross-administrative regions, government, and enterprises, as well as the public and NGOs, and formulate an immediate action plan to strengthen cross-spatial, cross-sectoral and cross-time regional synergy to achieve sustainable development and resilience of the basin.

- Strengthen spatial synergy, coordinate trunk and tributaries and upstream and downstream, form specialized coordination mechanisms based on spatial units of sub-basins and sub-regions, and specific synergistic matters and issues to avoid transferring problems to other areas.
- Strengthen sectoral collaboration, relevant sectors such as economic development, agriculture, ecology, transportation, water conservancy, disaster prevention, energy, etc. should develop common basin development strategies to improve cross-sectoral effectiveness and not transfer negative impacts to other sectors.
- Strengthen temporal synergy, should learn from history (evolutionary best practices) and look ahead to future climate change impacts, with foresight studies spanning at least 100 years to prevent shifting the problem into the future.

Recommendation 3: Referring to the experience of international water funds, explore the "Sustainable Basin Synergy Fund," comprehensively coordinate the synergy of interests among hydropower, flood control, flood storage areas, fisheries and aquatic biodiversity protection in the basin. Promote the further deepening of the existing ecological protection compensation system and improve the horizontal transfer payment mechanism between upstream and downstream areas. Develop modern agriculture (renewable agriculture), promote the transformation of the energy structure, promote high-quality and green development in the basin, and implement the "double carbon" strategic goal.

Recommendation 4: Drawing on the experience of the European "Strategic Environmental Impact Assessment," a more comprehensive assessment mechanism should be established to systematically assess the long-term pressures and short-term impacts of climate change at the basin level, analyze the adaptability of current policy measures, construction practices and disaster prevention standards, and develop systematic emergency scenarios at the regional level. In line with new practices worldwide and in China, this renewed assessment system for large programs and legislation should map responsibilities throughout the basin, enable broad engagement and establish a frame for monitoring progress and periodic re-evaluation in view of new knowledge. The latter will likely be necessary given the longevity of many projects, innovation, and uncertainties in climate change, especially at the regional level. This will be a prerequisite for the formulation of resilience policies and major construction actions in the upstream and downstream basins. In response to the uncertainty of climate change risk, systemic safety thinking should be advocated, and engineering measures should be combined with nature-based solutions in watershed management to form comprehensive management measures. Avoid engineering construction based on a single safety standard for disaster probability analysis, carry out systematic emergency scenario construction work, and study and establish a technical standard specification system for adapting to future climate change.

Recommendation 5: Special attention should be paid to gender and social equity issues, and the benefits and participation of disadvantaged groups, such as women, in watershed governance should be increased through various means. Strengthen infrastructure development in less-developed areas of the basin, promote statistical analysis based on population characteristics, and reduce the risks faced by rural women and low-income people under climate change; through policy and mechanism improvement, continuously promote education equity and workplace equity for women, and increase opportunities for women, the elderly, and other vulnerable groups to participate in decision making in basin governance.

Recommendation 6: With the increasing risk of extreme weather events such as floods and droughts, resilience is essential for economic and energy security and shared human prosperity, and immediate action should be taken to accelerate the development of the Yangtze River Basin Development Plan and the Territorial Spatial Plan. We need to accelerate the development of the "Yangtze River Basin Development Plan" and "Territorial Spatial Plan" to form a "strategic and comprehensive solution" with a basin-wide, multi-sectoral and multi-objective approach, to promote the green and sustainable development of the river basin through regional synergy, and to provide a model for the management of rivers in China and the world.

- The Jialing River and other secondary tributaries should actively explore "integrated resilience solutions" to provide a demonstration of green and low-carbon models for the modern development of less-developed inland areas. We should adopt a negative list to promote the sustainable development of the industrial structure of the river basin and the diversification of the energy structure; explore and improve the synergy mechanism between cities, governments, and enterprises in the river basin to promote the integration of flood control, drought relief, power generation, and shipping; and establish a comprehensive assessment mechanism to evaluate and mitigate the impact of large water conservation facilities on biodiversity conservation, power generation, navigation, water supply, etc.
- Taihu Lake and other important lake basins should explore the transformation of the regional spatial development model, establish a synergistic mechanism between flood control in the basin and drainage in the polder area, and form a "nature-based solution" with the collaboration of multiple parties. Optimize regional-urban flood control and drainage, adjust the "big polder" flood control and drainage model, and improve the overall safety and resilience of the basin; control the development intensity of the urban-regional sub-base, and strengthen the protection and control of non-construction land; protect the Taihu Lake agricultural and water conservancy heritage, and reshape the "Tangpu polder field" urban and rural ecological space, explore regenerative agriculture practices, protect water quality and maintain the ecological health of the lake.
- Pearl River Estuary and other large river estuaries should strengthen the coastal zone area governance of land and sea integration, forming a "comprehensive solution for cross-regional cross-system cooperation." Promote the unified delineation of coastal zone areas and ecological protection zones in Guangdong, Hong Kong, and Macao, and establish cooperative institutions and mechanisms; carry

out an environmental impact assessment of large-scale infrastructure construction to reduce disturbance of the natural environment of land and sea and biological habitat migration. Draw on the experience of Hong Kong's NGO and public participation models to promote the participation of multiple subjects in the protection and governance of coastal zone areas.

References

1. Fan, Z., Fang, Y., & Zhao, Y. (2020). *The future and consensus of the Greater Bay Area of Guangdong, Hong Kong, and Macau* (pp. 17–64). China Architecture & Building Press.
2. Guangdong Provincial Department of Ecology and Environment. (2020). *Guangdong provincial ecological and environmental status report. 2021* (pp. 15–18).
3. Chen, D. (2023). Greater Bay Area marine environmental monitoring and forecasting system. In *Sustainable blue economy for carbon neutrality in the Greater Bay Area Seminar.*
4. Wang, Y., Wang, J., Wu, M., et al. (2019). Impact of land use and climate change on hydrological characteristics in the Jialing River Basin. *Water and Soil Conservation Research, 26*(01), 135–142.
5. Wang, S., Liu, K., & Meng, C. (2022). Analysis of spatial and temporal evolution of drought and flood in the Jialing River basin based on SPEI. *Water Resources and Hydropower Fast Journal, 43*(05), 12–19.
6. Feng, B., Qiu, H., & Ji, G. (2022). Preliminary study on characteristics and causes of meteorological drought in the Yangtze River Basin in the summer of 2022. *Yangtze River, 53*(12), 6–15.
7. Li, X., Li, W., Guan, Y., et al. (2019). Discussion on flood control status and dispatching strategy in the middle and lower reaches of the Jialing River. *China Flood and Drought Management, 29*(12), 27–32.
8. Liu, M., & Li, H. (2021). Suggestions for promoting the navigation of the Jialing River and connecting Northwestern China. *China Water Transport, 681*(02), 38–39.
9. Zeng, Z. (2021). Analysis of navigation development strategies in the Jialing River Basin. *China Water Transport, 699*(08), 21–23.
10. Wu, H., & Lu, Z. (2021). Review and reflection on water control practice in the Taihu Basin. *Journal of Hydraulic Engineering, 52*(03), 277–290.
11. Yang, J., & Zhang, R. (2022). Land use change in the taihu basin and its impact on non-point source pollution. *Journal of Jiangsu Ocean University (Natural Science Edition), 31*(01), 37–44.
12. Zhou, H., Li, M., Wang, T., et al. (2015). Analysis of the impact of drainage in the Embankment Area of the Taihu Basin on regional flood control. *Water Resources Planning and Design, 11*, 1–2+21.
13. Shan, Y., Cai, W., Xue, X., et al. (2018). Impact and countermeasures of flood control encirclement construction in the Taihu Urban agglomeration. *China Flood and Drought Management, 28*(02), 56–59+65.
14. Li, B. (2016). Preliminary exploration of the impact of sea level rise and ground subsidence on water security in the Taihu Basin. *People's Pearl River, 37*(01), 38–41.
15. Ding, L., & Yang, K. (2014). Inspiration of comprehensive management of Lake IJssel in the Netherlands for Taihu Lake Governance. *Environmental Protection of Water Resources, 30*(06), 87–93.
16. Yang, J., Zhang, R., Zhang, Y., et al. (2022). Study on changes in non-point source nitrogen and phosphorus load in the Taihu Basin from 1980 to 2018. *Environmental Protection Science, 48*(06), 93–101.

17. Zhang, J. (2021). Spatial distribution and source identification of water quality in the Taihu Basin. *Jiangsu Science and Technology Information, 38*(10), 48–54.
18. Wu, H., Qin, H., He, B., et al. (2022). Discussion on the development trend of agricultural non-point source pollution control model based on carbon neutrality. *Journal of Ecology and Environmental Sciences, 31*(09), 1919–1926.
19. Lu, S., Yao, J., & Cao, X. (2020). Analysis of current situation, causes, and countermeasures of agricultural non-point source pollution in the Taihu Basin. *Water Resources Development Research, 20*(02), 40–44+53.
20. Liu, X., Deng, R., Xu, J., et al. (2017). Analysis of spatio-temporal changes and driving forces of coastline in the Pearl River Estuary Area in recent 40 years. *Journal of Earth Information Science, 19*(10), 1336–1345.
21. Yu, L., Lin, S., Jiao, X., et al. (2019). Ecological problems and protection strategies faced by mangrove wetlands in the Guangdong-Hong Kong-Macao Greater Bay Area. *Journal of Peking University (Natural Science Edition), 55*(04).
22. Guangdong Provincial Meteorological Bureau. (2022). Macao geophysical and meteorological Bureau, Hong Kong Observatory. In *2021 Greater Bay Area climate monitoring report.*

Chapter 6
Reshaping Land Use Toward Synergy Among Biodiversity, Climate Change, Food, and Water

1 Foreword

Currently, the world is facing significant risks resulting from human activities. These risks, in turn, threaten the stability and development of human society. The "Global Risks Report 2023" released by the World Economic Forum in January 2023 highlights five risks: the failure to mitigate and adapt to climate change, natural disasters and extreme weather events, biodiversity loss and ecosystem collapse, and environmental damage incidents, making it the fourth consecutive year they have been ranked among the top 10 global risks since 2020. Biodiversity loss is particularly considered one of the most rapidly deteriorating global risks for the next decade. Climate change and biodiversity loss directly threaten global food security.

Issues such as climate change, biodiversity loss, and water and food security are directly related to land use. As for terrestrial and freshwater ecosystems, changes in land use have been the most significant direct drivers for biodiversity loss since 1970 [1]. As the world accelerating energy transition, it's easily neglected that solar panels and wind tribunes will need a lot of land. Utility-scale solar and wind farms require at least ten times as much space per unit of power as coal fired power plants, including the land used to produce and transport the fossil fuels.

The transformation of land use practices to address issues such as biodiversity loss, climate change, and food security has become a major and urgent topic. However, the transformation of land use is not merely a significant land planning issue; it is fundamentally a question of transforming development patterns. Traditional industrialization models, centred around large-scale production and consumption of material wealth, are based on a foundation of "high carbon emissions, high ecological damage, and high resource consumption," leading to inherent conflicts between the environment and development.

Land use serves as the primary locus of interaction between human economic activities and the natural world. Different economic activities require different land use practices, which in turn have varying impacts on nature. In China, the past changes

Green Empowerment and High Quality Development,
https://doi.org/10.1007/978-981-96-4218-2_6

in land use and their ecological and environmental consequences are largely products of the traditional industrialization model.

Land use function change due to extraction of natural resources, and releasing waste back into the environment were types of consequences closely associated with rapid industrialization and development-centric model. This process inevitably brings about significant pollution, resulting in the loss of land functionality and environmental and biodiversity destruction. Urbanization has been a key driver of China's rapid economic growth over the past four decades. China's urbanization rate[1] increased from 17.9% in 1978 to 65.22% in 2022, leading to changes in living and consumption patterns of residents, as well as in significant changes in land use. Under the driving force of the traditional industrialization model, land use practices continue to change. On the one hand, in addition to land for industrialization and urbanization, the demand for agricultural land also continuously increases.

If the transition to green development from traditional industrialization is not made, changing the land use efficiency alone cannot fundamentally solve the issues. Biodiversity loss and climate change are clear examples. On December 19, 2022, under the presidency of China, 196 parties adopted the "Kunming-Montreal Global Biodiversity Framework," a landmark for global biodiversity conservation. However, making the ambitious goals proposed in this framework self-enforcing presents a significant challenge. The failure to achieve the "Aichi Biodiversity Targets" is fundamentally due to seeking biodiversity protection within the framework of traditional industrialization, which ironically contributed to biodiversity destruction.

Therefore, the effective realization of biodiversity conservation goals relies on transforming the traditional development model, shifting the relationship between development and conservation from a trade-off to a synergy, and creating an environment where they are mutually reinforcing. Numerous studies demonstrate that environmental protection presents substantial economic opportunities [2–4].

China's ecological civilization construction provides a fundamental direction and assurance for green transformation. The 20th National Congress of the Communist Party of China regards Chinese-style modernization as its central task. The fundamental and strategic position of ecological civilization is reflected in the essential characteristics, nature, and goals of the Chinese path to modernization. China's "14th Five-Year Plan" and China Vision 2035 have also outlined specific plans for the construction of ecological civilization.

Therefore, reshaping land use to address biodiversity loss, climate change, food security, environmental pollution, and other issues must transcend the traditional industrialization mindset. Through paradigm shifts in development, the conflicting relationships among these goals under the traditional industrialization model can transform into mutual reinforcement under the framework of ecological civilization, ultimately forming a nature-positive economy.

[1] The National Bureau of Statistics (NBS) reflects the urbanisation rate mainly in terms of the urbanisation rate of the resident population, which refers to "the proportion of the resident population in the urban territory of a region to the total resident population of the region, reflecting the urban-rural distribution of the resident population." Refer to NBS website: http://www.stats.gov.cn/zs/tjws/tjzb/202301/t20230101_1903783.html.

In this scoping study, under the requirements of ecological civilization, the main problems in transforming China's land use practices to achieve biodiversity targets, "dual carbon" goals, water security, and food security goals were identified. The domestic and international research status and policy progress regarding these issues were evaluated, exposing existing shortcomings, and innovative policy research directions were proposed, laying the groundwork for subsequent Special Policy Studies (SPS) to identify innovative policy approaches.

This project primarily investigates five key topics, each of which includes the following four main components:

- Firstly, identifying problems. Identifying significant issues in these five areas.
- Secondly, analyzing problems. Analyzing the identified key issues, evaluating the governance structure, policy, and research status to reveal the key, challenging aspects, and reasons behind these issues.
- Thirdly, solving problems. Building on the previous analysis, providing conceptual policy recommendations for these issues.
- Lastly, based on this foundation, proposing ideas for the focus of future five-year Special Policy Studies (SPS).

2 Topic Study

2.1 Pursuing Development from the Height of Harmonious Coexistence Between Humans and Nature

Research Question: Thinking beyond the traditional industrial civilization, under the concept of ecological civilization, research how to promote a paradigm shift in development and establish a mutually reinforcing relationship between ecological environment protection and economic development. To address environmental and developmental challenges, it is necessary to "plan development from the perspective of harmonious coexistence between humanity and nature" (Report of the 20th National Congress of the Communist Party of China) and fundamentally transform the development approach. Therefore, it is essential to comprehensively incorporate international conventions such as the "Kunming-Montreal Global Biodiversity Framework" and the "Paris Agreement", as well as sustainable development goals, into the overall layout of ecological civilization construction, to promote a modernization characterized by harmonious coexistence between humanity and nature.

2.1.1 Rethinking Modernization

The 20th National Congress of the Communist Party of China sets the central task of the party in the new era as "uniting and leading the people of all ethnic groups in the country to comprehensively in building a strong modern socialist country, realize the second centenary goal, and promote the great rejuvenation of the Chinese nation through the Chinese path to modernization." The development of the Chinese path to modernization, guided by a harmonious coexistence between humanity and nature, breaks away from the unsustainable modernization model rooted in anthropocentrism established after the industrial revolution. It represents a redefinition of the unsustainable concept of modernization that emerged post the industrial revolution [5].

Following the Industrial Revolution, social productivity made unprecedented advancements, and a few industrialized nations led the way in achieving what is commonly referred to as modernization. The prevailing global conception of modernization largely equates it with adopting the standards of developed countries as the default norm. If we divide modernization into two dimensions—"What kind of modernization to achieve" (What) and "How to achieve modernization" (How)—the modernization endeavors of developing nations have primarily focused on emulating the developmental path of developed countries. However, there has been relatively limited reflection on the actual substance of modernization.

Undoubtedly, following the Industrial Revolution, developed countries established a modernization model rooted in traditional industrial civilization, significantly propelling the progress of human civilization. China, too, has been among the greatest beneficiaries of this modernization concept. However, this form of modernization based on the traditional industrialization model possesses inherent limitations: firstly, it struggles to avoid the divergence between developmental goals and means; secondly, as it relies on high resource consumption and environmental degradation, it inevitably leads to unsustainable ecological environments; thirdly, due to the high resource and environmental costs associated with this model that in turn affects the long-term productivity of the traditional modernization development model, it allows only a minority of the global population to enjoy a modern lifestyle, and expanding it further could lead to a global sustainability crisis.

Therefore, merely considering "how to achieve modernization" is insufficient; a deeper reflection and redefinition of "what kind of modernization to achieve" is necessary, establishing a forward-looking and globally applicable discourse on the Chinese path to modernization. This modernization fundamentally involves profound reflection and reconstruction of the modernization concept formed after the Industrial Revolution [6].

2.1.2 Changes and Existing Issues in China's Land Use

The industrialization, urbanization, and modernization of agriculture have brought about rapid economic development in China, but concurrently, they have exerted significant impacts on land use patterns. Land use includes arable land, construction land, and unused land, categorized into 12 primary classes and 73 secondary classes (GB/T 21010-2017). According to data from the Third National Land Survey (referred to as the "Third Survey"), the respective areas of these three types of land in

China are as follows: China's arable land area reaches 101.72 billion mu, accounting for 70.64% of the total land area, making it the dominant land use type. The area of unused land is 3.614 billion mu, constituting 25.10% of the total land area. Among these, areas such as saline-alkali land, sandy land, bare land, and rocky gravel land account for 2.512 billion mu. Construction land covers an area of 613 million mu, representing a mere 4.26% of the total land area. This signifies that Nature-Based Solutions (NBS) hold significant potential as nature-centred approaches, as discussed in detail below:

(1) Farmland resources. The overall quantity of farmland resources has declined. This change is largely influenced by agricultural structural adjustments and land greening initiatives, all the while strictly adhering to a balanced approach when non-agricultural construction occupies farmland. The substantial production of animal-based and processed food products in the agri-food industry directly or indirectly increases the demand for land for such products, leading to the conversion of farmland into grasslands and orchards. According to the data from the "Third Survey," the area of farmland (1.918 billion mu) decreased by 113 million over a decade relative to the "Second Survey." Under the strictest ecological and environmental protection regime, around 12 million mu (8 billion square meters) of farmland have been converted into ecological land such as forests and wetlands.
(2) Grassland resources. The overall quantity of grassland resources has declined. This change is also influenced by agricultural production structural adjustments and land greening initiatives. The extensive production of animal-based food products in agriculture has led to a sharp increase in livestock and poultry farming, resulting in overgrazing on limited grassland. The average livestock overload rate on key natural grasslands in China exceeds 10% [6]. According to the "Third Survey" data, China's grassland area is 3.968 billion mu, ranking second in the world, but it has still decreased by 342 million mu (228 billion square meters) compared to the Second National Land Survey ("Second Survey").
(3) Forest land and wetland. The overall area of forest land and wetlands shows an increasing trend, mainly driven by government policies. According to data from the "Third Survey," the forest land area in China is 4.262 billion mu, increasing by 453 million mu (302 billion square meters) compared to the "Second Survey," with a growth rate of 11.88%. This contributes a quarter of the world's newly added forest area. China's wetland area is 352 million mu, ranking first in Asia, including 42 types of wetlands classified under the Convention on Wetlands. The expansion of forest and wetland ecological land relies largely on government policy support, essentially forming a national forest (wetland) policy pathway. In wetlands, projects such as conversion of marginal farmland to wetland, cessation of fishing to restore wetlands, and wetland water replenishment have been implemented, establishing a wetland conservation system primarily centred around national parks, natural wetland reserves, and wetland parks.

(4) Construction land. There has been rapid expansion in the total area of construction land, resulting not only in significant reductions in other land types but also in the problems of idle and inefficient use of a large amount of construction land. The results from the "Third Survey" indicate that the total area of construction land in China is 613 million mu (about 409 billion square meters), an increase of 128 million mu (about 85 billion square meters) compared to the "Second Survey," with a growth rate of 26.5%.The scale of village land in the country reaches 329 million mu, with idle land in rural residential areas accounting for around 10–15%.
(5) Desertification and land degradation. The area of desertification and land degradation has continuously decreased, yet the primary trends of desertification and land degradation remain persistent. According to data from the Sixth National Desertification and Land Degradation Investigation, as of 2019, the area of desertification and land degradation reached 42.615 million hectares, accounting for 44.4% of the total land area. However, compared to 2014, there was a net reduction of 7.1232 million hectares in the area of desertification and land degradation over 5 years, indicating significant achievements in China's desertification control efforts. Nevertheless, the primary trends of desertification and land degradation have not weakened. The areas with evident trends of desertification include grassland, farmland, and forest land, totaling 26.894 million hectares. Hence, if the land use problems contributing to desertification and land degradation are not addressed, the expansion of desertification and sandy land areas will continue in arid and low-rainfall climates.

2.1.3 Development in Harmony with Nature

The above-mentioned issues are not merely about land re-planning; they also signify a profound shift in the development paradigm. Only by strategically planning development in harmony with nature and fundamentally altering land utilization methods can we transform the conflicting relationships between biodiversity, food security, and environmental protection into mutually synergistic and even mutually reinforcing relationships.

Based on the ecological civilization, the Chinese path to modernization offers the possibility for this transformation. The fundamental and strategic position of ecological civilization in the Chinese path to modernization is reflected in what this kind of modernization is, how to build it, and its goals.

Firstly, it is reflected in "what the Chinese path to modernization is." The modernization that China aims to achieve on its path includes "harmonious coexistence between humans and nature" as one of the five fundamental characteristics of Chinese-style modernization, as well as its essential requirement.[2]

---

[2] The five fundamental characteristics of Chinese-style modernization are as follows: it is a modernization characterized by a large population scale; it is a modernization where the entire population achieves shared prosperity; it is a modernization that harmonizes material and spiritual civilization; it

Second, it is reflected in "how to achieve the Chinese path to modernization." The 19th National Congress Report of the Communist Party of China (CPC) pointed out that "achieving high-quality development is the Party's primary task in comprehensively building a socialist modernization country," which requires the "comprehensive, accurate, and thorough implementation of the new development concept."

Third, it is embodied in the "Goals of the Chinese path to Modernization." The 19th National Congress Report of the Communist Party of China clearly lays out a strategic arrangement for the comprehensive construction of a prosperous socialist modernized country in two stages: from 2020 to 2035, the basic realization of socialist modernization; from 2035 to the middle of the century, building China into a prosperous, strong, democratic, culturally advanced, and harmonious socialist modernized country.

The 20th National Congress of the CPC has laid out a new strategic deployment for ecological civilization construction, fully opening a new chapter in ecological civilization construction. At the strategic level, the Chinese-style modernization is the central task proposed in the Party Congress report for China's future. The essential characteristics, intrinsic requirements, and objectives of Chinese-style modernization will be comprehensively embodied in China's economic and social development strategies and actions.

The 20th National Congress Report of the Communist Party of China dedicates its 10th section specifically to the theme of "Promoting Green Development and Achieving Harmonious Coexistence Between Humans and Nature," emphasizing the importance of ecological civilization construction and making corresponding strategic arrangements.

In conclusion, the Chinese-style modernization is a rethinking and redefinition of the modernization model established after the Industrial Revolution. Among these, the modernization characterized by harmonious coexistence between humans and nature serves as the foundation for Chinese-style modernization. The transformation of the modernization model implies changes in both development content and methods.

2.1.4 Future Key Research Directions

Planning development from a perspective of harmonious coexistence between humanity and nature essentially signifies a profound transformation of the development concepts and paradigms formed after the Industrial Revolution. When the traditional industrialization model needs to undergo transformation due to its unsustainability, the underlying development theories, industrialization patterns, urbanization models, agricultural modernization models, infrastructure, and more, all require changes.

---

is a modernization of harmonious coexistence between humans and nature; and it is a modernization pursued through the path of peaceful development.

Firstly, there are major theoretical issues. The profound transformation of development paradigms involves a reconsideration of fundamental development questions, including why development is pursued, what content should be developed, how to achieve development, and the global applicability of development models.

Secondly, there's the transformation of the traditional industrialization model and its impact on land use. Disrupt specific mechanisms that are causing ecological environment degradation through transformations in land use patterns.

Thirdly, there's the transformation of the traditional urbanization model and its impact on land use. Changes in urbanization methods and their associated content have implications for the ecological environment.

Fourthly, there's the transformation of green agriculture and its impact on land use. The current "modernization" of agriculture in various countries largely occurs within the framework of traditional industrialization thinking, encompassing both the content and methods of agricultural production.

Fifthly, there's the impact of changes in development methods on the economic geographical pattern. Different development models carry different spatial implications.

2.2 The Green Transformation in Agriculture

The Green Revolution that emerged in the mid-twentieth century greatly boosted agricultural productivity and led to significant transformations in the content and methods of agricultural supply and production. However, simultaneously, monoculture farming, chemical-intensive agriculture, and industrialized agriculture have posed substantial challenges to agricultural biodiversity. Globally, carbon emissions from agriculture, forestry, and land use account for nearly 20% of the total. Therefore, China's agriculture urgently requires an upgraded version of the Green Revolution to effectively address issues such as food security, increased income for farmers, and ecological environmental protection.

2.2.1 The Urgency of China's Agricultural Green Transformation

Over the past four decades of reform and opening up, China's agriculture has achieved remarkable accomplishments. The Per capital grain output has reached 486 kg, and the per capital disposable income of rural residents has exceeded 10,000 yuan. However, due to the development of industrialized and chemical-intensive agriculture, which is based on high resource consumption and severe environmental degradation, there are sustainability issues arising from the excessive use of arable land and water resources, heavy reliance on fertilizers and pesticides, resulting in land degradation, environmental pollution, climate change, and loss of biodiversity.

Industrialized agriculture and its consequences largely stem from the traditional industrialization model's transformation of agriculture into "modernization." This transition involves a shift in agricultural production content (what) from plant-based to animal-based products and a change in agricultural production methods (how) from diverse ecological agriculture to single-focused industrial and chemical agriculture [7]. The proportion of agriculture in crop production continues to decline, accounting for only about half of the total agricultural output in 2021,[3] while livestock and fisheries have grown significantly, with their output in 2021 more than doubling that of 1978, reaching 37%. Consequently, the production of animal products such as meat, eggs, and milk has sharply increased, reaching a staggering 90.74 million tons in 2021, significantly surpassing other countries. Currently, the proportion of overweight and obese adults in China has reached 50.7% [8], leading to a death rate of 41 people per 100,000. Moreover, China's per capita protein consumption rapidly increased from 1999 to 2019, going from far below the global average to surpassing the average of OECD countries and only slightly lower than the protein consumption of the highest-ranking countries such as the United States and France, nearly on par with Australia's, which ranks third [9].

At the same time, the rapid growth of livestock and fisheries implies more consumption of feed grains and forage, driving land use change, increased consumption of agricultural inputs such as fertilizers and pesticides, and increased emissions of greenhouse gases like methane. Due to the land consumption and greenhouse gas emissions of animal-based food production being much higher than that of plant-based products, extensive production of animal-based food leads to greater land resource consumption and greenhouse gas emissions, exacerbating chemical pollution and climate change, which in turn pose a significant threat to agricultural production and food security.

If China's agricultural development were to converge toward the industrialized agriculture (chemical agriculture) based on the traditional industrialization model, it would undoubtedly result in significant harm to people's health and the ecological system. Therefore, a pressing need exists for China's agricultural development model to undergo transformation, returning to the fundamental purpose of agricultural development—its original intention—in order to promote food security, water security, human health, curb land degradation, reduce environmental pollution, address climate change, and safeguard biodiversity, thereby establishing a synergistic relationship among multiple objectives.

2.2.2 Existing Issues and Prominent Issues in Practice

Agricultural Green Transformation: An Upgraded Version of the Agricultural Green Revolution of the 1960s (Agriculture 3.0 Era). FAO recommended agroecological approaches to conserve the basis for production, which includes 10 elements.[4]

---

[3] The proportion of output value of plantation industry decreased from 80.0% in 1978 to 53.29% in 2021.

[4] FAO developed the 10 Elements of Agroecology framework to assist countries in fostering transformative change, including: Diversity, Co-creation and sharing of knowledge, Synergies, Efficiency,

Existing literature on research related to agricultural green transformation primarily focuses on discussing the greening of agricultural production methods ("how"), with limited attention given to how agricultural development content ("what") can be transformed into environmentally friendly practices. Green technological innovation is regarded as a pivotal means to achieve agricultural green development.

Regarding the key aspects of China's agricultural green development system and policy framework, there is also a greater emphasis on how to make agricultural production methods environmentally friendly in order to reduce the negative impact of agriculture on the environment. However, there is a lack of attention given to how agricultural development content should be transformed to achieve agricultural green development.

The Chinese government has consistently attached great importance to green agricultural development and has implemented a series of measures that have yielded significant results. After carrying out actions to reduce chemical fertilizer usage and promote the substitution of organic fertilizers for chemical ones, the total fertilizer application in 2021 decreased by 13.8% compared to 2015, with a fertilizer use efficiency exceeding 40%. Initiatives aimed at straw disposal and agricultural film recycling have resulted in a comprehensive straw utilization rate of 88.1% and an agricultural film recycling rate exceeding 88% [10]. However, China's practice of agricultural green development primarily focuses on reducing the negative impact of agricultural production on the environment.

It is evident that if the focus remains solely on changing agricultural production methods without altering the underlying agricultural development content, it will not fundamentally resolve the conflicting relationship between agricultural development and environmental protection.

2.2.3 Approaches to Achieving Green Agricultural Transformation

The direction of agricultural green transformation is rooted in meeting the fundamental objective of providing healthy and nutritious food for people. Therefore, it is not only essential to emphasize the transformation of agricultural production methods but also to underscore the transformation of agricultural production content. In accordance with the principles of ecological civilization, agricultural production activities need to gradually shift from being carbon sources to becoming carbon sinks, reducing excessive production and consumption of animal products under traditional consumption patterns and increasing the production of plant-based meat and dairy substitutes. In terms of agricultural production methods, the shift is toward transitioning from chemical-intensive and monoculture farming to an agriculture that harnesses natural fertilizers and employs climate-smart technologies to cultivate

---

Recycling, Resilience, Human and social values, Culture and food traditions, Responsible governance, Circular and solidarity economy. Reference: https://www.fao.org/agroecology/overview/overview10elements/en/.

crop diversity. Agricultural green transformation changes the conflicting relationship between agricultural development and environmental protection into a mutually supportive one, thereby achieving a win–win situation for both economic and ecological benefits [7].

The fundamental approaches to achieving green agricultural transformation are as follows:

Firstly, it is essential to adopt a broader perspective of harmonious coexistence between humans and nature, re-examining the content of the agricultural green development system. This includes exploring the essence of agricultural green development, evaluation frameworks, policies, and implementation mechanisms, all aimed at fostering a mutually reinforcing relationship between agricultural development and environmental protection. This approach involves demonstrating and promoting regenerative agriculture that ensures high and stable yields, thereby achieving synergies across multiple goals such as health, biodiversity conservation, carbon neutrality, and food security.

Secondly, there's a need to re-evaluate the costs and benefits of agricultural development, encompassing both non-monetary and monetary aspects. Drawing inspiration from international carbon-labelling systems, it is important to establish a comprehensive mechanism reflecting the carbon, water, and resource intensities of different food products. This approach aims to internalize the social costs of agricultural production to the greatest extent possible and transform agricultural development by altering relative product prices.

Thirdly, optimizing and adjusting agricultural support policies to facilitate the transition of agricultural production toward green and healthy agricultural products and ecosystem services is crucial. This involves discontinuing subsidies for agriculturally toxic and harmful practices while strengthening fiscal support for the production of environmentally friendly products that promote human physiological health with lower environmental impact. This support includes subsidies for eco-friendly inputs during production and bolstering the information system for pricing at the sales end, all of which contribute to boosting the supply of such products.

Lastly, the establishment of a robust system for agricultural technological innovation and dissemination is vital. This entails increasing support for revolutionary and integrated green technologies, breaking through technological barriers within the agricultural sector, enhancing the efficiency of agricultural resource utilization, and mitigating negative impacts such as chemical pollution, climate change, and environmental degradation.

2.2.4 Future Focus Areas of Research

Building upon existing research, future studies on the issue of agricultural green transformation should focus on several key areas:

Firstly, it's essential to re-evaluate the cost and benefit assessment of China's agricultural development from the perspectives of environmental and health objectives. This entails a systematic analysis to assess the health, resource, and environmental costs associated with China's agricultural development, along with the benefits gained from factors such as boosting farmers' income and rural development. The goal is to uncover the benefits of China's agricultural green transformation.

Secondly, there is a need to evaluate the effectiveness of China's agricultural policies and optimize them to support agricultural green transformation. This involves constructing economic models to analyze the impacts of different agricultural policies on aspects such as agricultural output, health, resources, environment, and biodiversity. The aim is to enhance China's agricultural policies in a way that promotes agricultural green transformation.

Thirdly, research should be conducted on China's agricultural green innovation system, focusing on overcoming barriers to green technology and driving the transition of chemical-intensive agricultural production to environmentally friendly methods. This research should take an economic perspective, systematically analyzing the key and challenging aspects of agricultural green innovation and the underlying mechanisms.

2.3 Pursing Food Security in the Context of Ecological Security

Food security is the foundation of the national economy. Existing definitions of food security mostly focus on how food supply meets the demand for food but lack emphasis on whether food demand is reasonable. Currently, a prominent issue is that the market demand for food is primarily driven by commercial forces, failing to adequately reflect the requirement for a healthy life as defined in food security ("healthy life") and deviating significantly from health-related dietary needs (both in terms of quantity and food structure). Rethinking food security calls for consideration not only of market stability but also health and environmental requirements. This reevaluation brings new implications for agricultural development direction, food security, and the impact on resources and the environment.

2.3.1 Achievements and Challenges of China's Food System

Since the opening-up and reform, China's grain system has achieved remarkable success. By 2022, grain production in China had increased by 125.3% compared to 1978, far surpassing the population growth rate of 46.7%. This has contributed greatly to reducing hunger, extending life expectancy, reducing infant mortality, and alleviating poverty. However, along with these achievements, numerous health and environmental issues related to grain have also emerged. Currently, the overweight and obesity rate among Chinese adults has reached 50.7% [7], and obesity-related mortality stands at 6.4%.[5]

Food security, health, climate change, and related problems are all closely connected to shifts in food demand. In terms of food security, China's continuously increasing demand for grains has led to a persistent tight balance between supply and demand, despite the growth in grain production and escalating net grain imports. Loss rates throughout the entire chain of China's three main grain crops account for approximately 20.7% of total production. Reducing these losses by 40% could save 110 billion kilograms of grain.[6] Concurrently, the transition in residents' dietary patterns has led to a deviation from nutritional and health requirements, resulting in numerous health issues.

Consequently, achieving the synergistic realization of goals related to food security, health, climate change, and biodiversity protection urgently requires a transformation in food demand patterns. This transformation should align with the essence of nutritional health needs and reflect a people-centred development philosophy.

2.3.2 Prominent Issues in Existing Research

Currently, the literature on food security mainly discusses how to increase supply to ensure a balance between food market demand and supply. However, there is limited attention to whether the food demand is reasonable and its impacts. For instance, discussions of food demand primarily focus on market demand for food and lack an analysis of the demand for food based on health considerations [11, 12]. Regarding the nutritional perspective on food demand, these works mainly use national dietary guideline to perform simple calculations of food demand [13] yet lack an analysis of the reasons and mechanisms behind the deviation between actual demand and healthy dietary demand. Regarding the research on the impact of the food system on health and the environment, the literature focuses more on discussing the benefits of healthy eating and reducing food waste losses [14], lacking analysis of the underlying mechanisms behind the evolution of current dietary structure.

The Chinese government attaches great importance to food security and has formulated a series of policies and regulations to ensure it. In general, China's food security policies have consistently centered around expanding production and increasing supply. If the food system is solely considered from the perspective of supply, overlooking the demand side that is central to the development of the food system, achieving food security goals becomes challenging.

---

[5] https://ourworldindata.org/obesity/.

[6] China Agricultural Industry Development Report 2023.

2.3.3 Food Security Approach Under Environmental and Health Goals

If we aim only to meet the market's demand for food by increasing supply under the traditional definition of food security, it's difficult to truly achieve food security and may exacerbate health and environmental issues. Only by recognizing its essential health requirements and embodying a people-centred development philosophy can the food system effectively promote the simultaneous realization of food security and ecological safety goals. The basic approaches are as follows:

The first approach involves optimizing the definition of food security to provide scientific guidance for the sustainable development of the food system. Under the conditions of synergistically achieving multiple objectives, such as food security, health, and ecological safety, it is necessary to reconsider the framework of the current definition of food security and assess the limitations of these aspects in achieving multiple goals.

The second approach entails adjusting food security policies to optimize the supply of the food system. Shift the direction of food fiscal support policies. Enhance financial support for nutritious and sustainable food to increase the supply of high-quality staple foods and reduce the production of inferior-quality grains.

The third approach involves optimizing China's dietary guidelines to facilitate the simultaneous realization of multiple objectives such as food security, health, and environmental protection. Evaluate China's current dietary structure and the health and environmental effects of dietary guidelines.

2.3.4 Future Key Research Directions

Building upon the existing research, addressing the issue of food security should encompass the following directions:

- Conducting research on dual-benefit health and environment dietary guidelines to guide the transformation of residents' dietary structures toward healthier patterns, and the shift of agricultural and food system production toward healthier and sustainable food production.
- Undertaking research on the cost and benefit assessment of food consumption demands to promote the transformation of the food system. The aim is to propose a multi-objective synergistic development model for the food system.
- Conducting research on China's food security policies under the objectives of health and environmental (water, air, soil) protection to promote the simultaneous realization of multiple objectives, such as food security, dual-carbon goals, and biodiversity conservation.

2.4 National Spatial Governance and Policies

Land resources, including soil, water, and biodiversity, provide fundamental products and services for humanity, such as food, water, fibre, energy, raw materials, and places for living and working. Land is a finite resource, and its functions are crucial for our economic, environmental, and socio-cultural well-being. However, under the current land use system, the functions of these lands are not always compatible or can even conflict, resulting in dysfunctional trade-offs for sustainability under conventional land use systems. To seek systematic solutions from a more holistic and synergistic perspective, national spatial planning could play a significant role at the highest level.

Traditional land use, particularly unsustainable expansion of arable land, hinders the progress of sustainable development. The IPCC report further highlights [1] that land degradation exacerbates climate change, while climate change, in turn, intensifies land degradation and desertification, leading to food security issues. While food security is a global goal that requires consideration of multiple factors, soil health, especially its fertility status is the fundamental building block on which all agricultural production systems are built.[7] Therefore, in formulating the integrated national spatial governance system, decision-makers must address the challenges of rapid urbanization while also coordinating grand objectives such as biodiversity conservation "30 × 30",[8] "dual-carbon" goals, water, and food security, which can be highly demanding.

China's spatial planning system is unique worldwide and might hold important lessons for other countries. It is extremely relevant for operationalizing the GBF globally, especially for Target 1.

2.4.1 Challenges and Current Situation of Multi-objective Integrated Governance in China's National Spatial Planning

China's vast population of over 1.4 billion people has surpassed the total population of all developed countries combined [15]. At the same time, suitable space for production and living is limited and unevenly distributed. Although overall natural resources are abundant, the per capita share of land, energy, minerals, and other major resources is far below the global average [16]. Rapid urbanization has resulted in significant reductions in farmland [17], and by 2030, China's urbanization rate is

---

[7] Referred to the insights from Ronald Vargas, the Secretary, Global Soil Partnership, Food and Agriculture Organization of the United Nations, https://www.un.org/en/un-chronicle/soils-where-food-begins.

[8] Hereby referring to the targets of the Kunming-Montreal Global Biodiversity Framework on conservation and restoration, including TARGET 2: Ensure that by 2030 at least 30 per cent of areas of degraded terrestrial, inland water, and marine and coastal ecosystems are under effective restoration; TARGET 3: Ensure and enable that by 2030 at least 30 per cent of terrestrial and inland water areas, and of marine and coastal areas, especially areas of particular importance for biodiversity and ecosystem functions and services, are effectively conserved and managed, so global communities call it "30 × 30".

projected to reach 70%, potentially leading to a loss of around 20 million acres of high-quality arable land [18], posing a potential threat to food security.

The "Outline of National Overall Land and Spatial Planning (2016–2030)" issued by the State Council highlighted four key points: [19] (1) Since the reform and opening-up, China's industries and employment have continuously concentrated in the eastern coastal areas, leading to a spatial mismatch between market consumption and resource-rich regions. (2) Structural contradictions between urban, agricultural, and ecological spaces are becoming more pronounced. (3) The intensity of land development in some regions does not match their resource and environmental carrying capacity. (4) The land development in coastal areas does not align with the marine resources and environmental conditions.

Before the 18th National Congress, various departments in China had a variety of planning systems, each with its own framework, and various spatial constraint plans lacked sufficient strength. A relatively common phenomenon was that the same piece of land was categorized as basic farmland in national land planning and as forest land in forestry planning. The specific reasons hindering the capacity for national spatial governance include: (1) different planning timelines, and (2) different technical standards and information platforms, especially the use of different technology platforms, different basic maps, inconsistent statistical criteria, and non-uniform land classification.

In 2019, the country issued the "Several Opinions on Establishing the National Spatial Planning System and Supervising Its Implementation", requiring the integration of main functional zone planning, land use planning, urban and rural planning, and other spatial planning into a unified national spatial planning, implementing a "multiple plans integration" approach [20], and calling for the establishment of a "Five-level, Three-category" national spatial planning system. In the same year, the Chinese government issued the "Notice on Comprehensive National Territory Spatial Planning (Ministry of Natural Resources [2019] No. 87)", officially launching the compilation of national spatial planning at all levels. Currently, "The Outline of National Overall Land and Spatial Planning (2021–2035)" (hereinafter referred to as the "Outline") has been approved by the State Council but has not yet been publicly released. The "Outline" encompasses the overall arrangement of the national spatial planning, including policies for the protection, development, utilization, and restoration of land and serves as the fundamental basis for local spatial planning.

However, as pointed out in the "Master Plan of National Important Ecosystem Protection and Restoration (2021–2035)" published by the National Development and Reform Commission in 2020 [21], there is still a considerable gap in understanding the intrinsic mechanisms and laws of mountains, rivers, forests, farmland, lakes, and grasslands as a community of life, which hampers the implementation of the concept and requirements of integrated protection, systematic restoration, and comprehensive governance. Thus, while optimizing land use, China must also strive to coordinate multiple objectives. The Ministry of Natural Resources will regularly evaluate the effectiveness of ecological protection redlines and promote collaborative efforts among various departments to strengthen supervision over ecological protection redlines. Ecological protection redlines cover most key ecosystems, including grasslands, important wetlands, coral reefs, mangroves, and sea grass beds.

2.4.2 Challenges Faced in Promoting Integrated National Spatial Governance in China

Existing literature on national spatial governance has focused more on urban construction and paid less attention to rural spatial governance, resulting in a lack of theoretical support to cope with the ever-changing urban–rural relationships. This, in turn, hinders the country's ability to meet the requirements of the current era for "multiple plans integration" national spatial planning [28]. The lack of effective implementation measures has presented significant challenges in scientifically controlling the decentralized, bottom-layer, and complex rural spatial aspects [22, 23].

An important goal of national spatial planning is to establish a unified spatial layout and comprehensive development and protection strategy nationwide [24]. However, the current irrational state of rural space development and utilization poses a significant obstacle to achieving the goal of integrated spatial planning. In the context of "multiple plans integration" spatial planning, efficient and equitable utilization and governance of urban and rural spaces remain crucial challenges that require attention and innovative solutions [25, 26].

Furthermore, while the institutional and policy framework for national spatial governance in China has been preliminarily planned, significant challenges remain in coordinating regulatory bodies. However, there is still ambiguity surrounding the boundaries of national spatial control by the Ministry of Natural Resources, including (1) the distinction and cooperation between law enforcement by the Ministry of Natural Resources and comprehensive ecological environmental law enforcement, (2) the division of responsibilities between the Ministry of Natural Resources and forestry departments in ecological protection and nature reserve management, and (3) the coordination of regulation between the Ministry of Natural Resources and the Ministry of Agriculture and Rural Affairs [27].

Despite some progress, optimizing the regulatory system is still an ongoing task. There is a lack of clear and unified technical standards and management systems for key policies such as spatial access and land use conversion [28]. Comprehensive regulations for various types of land conversion, particularly the conversion rules for different land types within agricultural and ecological spaces, are still lacking [29]. There are inconsistencies in the delineation and regulation of different natural reserves and ecological protection redlines [30]. In addition, there is a lack of a comprehensive monitoring and evaluation mechanism, and improvement is needed in feedback and in-process supervision to establish a more complete national spatial correction mechanism [31].

2.4.3 Studying the Progress of the National Spatial Planning and Identifying Opportunities for Synergies

To address complex issues such as land use, biodiversity loss, climate change, food security, water security, and environmental pollution, a more holistic perspective must be embraced, moving away from traditional industrial development thinking, and seeking systemic solutions tailored to China's specific situation. The following are some suggestions in China's context, aiming to achieve a resilient and balanced national spatial governance system with a vision of creating synergies between biodiversity protection, climate actions, ensuring food security and water security, etc.:

(a) Harmoniously Integrating Multiple Objectives to Establish a Coupled Pattern and Process for National Spatial Planning

Given China's vast and diverse regions, a comprehensive and systematic approach to national spatial governance is necessary. Decision-makers should recognize the interconnections between different objectives, such as biodiversity conservation, dual carbon goals, water and food security, and socio-economic development. By coupling "patterns (indicators and layouts)" with "processes (ecological processes)", a comprehensive optimization of the national spatial layout can be sought, leading to the formulation of reasonable spatial pattern allocation scenarios [32].

Another aspect of harmoniously integrating multiple objectives lies in reviewing the measures taken to address climate change. Policymakers need to analyze the effectiveness of measures taken in terms of farmland quantity, quality, and ecological aspects, including continuously improving farmland protection policies, implementing farmland renovation, and increasing land use efficiency.

(b) Coordinated Governance and Unified Planning for Synergizing National Spatial Planning Goals, Indicators, and Regulations

China's national spatial planning faces complex challenges, making it crucial to enhance coordination among regulatory departments. This can be achieved through clarifying responsibilities, simplifying decision-making processes, and strengthening policy implementation [32]. An important measure is to integrate various planning systems, such as the main functional zone planning, land use planning, and urban–rural planning, into a unified national spatial governance framework, thereby coordinating planning goals, indicators, and regulations [33]. Regarding the comprehensive regulation needs of the entire region and all elements of national spatial planning, policy-makers also need to further develop regulatory rules for ecological spaces and their different ecological functional zones [34].

Up to this point, the Ministry of Natural Resources has released several technical standards aimed at laying the foundation for territorial spatial planning. The Ministry has issued directives on the preparation of territorial spatial planning at the provincial and municipal levels. These guidelines play a crucial role in guiding and regulating the development of local plans.

(c) Strengthening Rural Spatial Governance for Balanced Urban–Rural Development

Amid rapid urbanization in China, rural spatial governance must be prioritized and given equal importance to urban development. A sound national spatial governance system must protect agricultural land and natural ecosystems while fully considering improvements to rural livelihoods.

2.4.4 Recommendations for Research

(a) Coordinating the Relationships Between Ecological Security, Food Security, and Water Security

i. The intense conversion between ecological land and arable land requires comprehensive consideration. The Third National Land Survey indicates that over the past decade, ecological land has increased overall, but frequent conversions between ecological land and arable land have been observed, with about 229 million acres of farmland flowing into regions with stronger ecological functions, while about 217 million acres of land in those regions have flowed back to farmland. This reflects the fact that the ecological construction pattern in some areas is not stable, with issues such as blind ecological construction and unreasonable ecological layout. The drastic land use conversion, to some extent, reflects policy conflicts between different periods, objectives, and value orientations. It is essential to balance ecological construction and farmland protection according to the principle of "cultivate suitable land, plant trees on suitable land, cultivate grass on suitable land, conserve wetlands on suitable land, leave uncultivated land uncultivated, and leave sandy land as sandy land."

ii. Water resource security is related to food security, ecological security, natural disaster mitigation and prevention, and the spatial matching needs improvement. China's water resources are unevenly distributed in time and space and do not match the distribution of population, economy, farmland, and energy. Therefore, seeking a balance point in quantity, quality, structure, and layout between ecological protection, farmland protection, and water resource security is of utmost importance.

(b) Fully Understanding Regional Resource Endowments, Environmental Backgrounds, and Socio-Economic Characteristics to Develop Differentiated Land Use Strategies

i. In Northwest China, focus on water resource security and ecological security. Further optimizing water use structure, improving water resource utilization efficiency, preventing the encroachment of ecological water, and enhancing wind and sand fixation ecological functions are essential.

Addressing grassland degradation and land desertification should also be prioritized, along with promoting the development of clean energy.

ii. In Northeast China, focus on ecological security and food security. Vigorously advancing sustainable use of black soil, consolidating the region's position as a nationally important commodity grain production base, protecting forests with significant water conservation functions in the Northeastern Forest Zone, and promoting the transformation of old industrial bases.
iii. In North China, focus on the matching of water resources and land resources. Emphasize the development of water-saving agriculture and address groundwater overexploitation.
iv. In East China, focus on water body pollution in specific areas. Promote regional integration development, facilitate industrial green transformation, and control eutrophication in lakes.
v. In Central China, focus on farmland protection. Given the concentration of high-quality farmland in the region, steps should be taken to prevent its loss and soil pollution.
vi. In South China, focus on ecological protection and environmental quality improvement. Utilize the ecological service functions of the southern hilly and mountainous areas, protect biodiversity, and address environmental pollution.
vii. In Southwest China, coordinate mineral resource development and ecological protection. Protect plateau lakes and plateau biodiversity, promote desertification control, commit to geological disaster prevention.

(c) Studying Policy Instruments that Strictly Adhere to Safety Bottom Lines while Balancing the Interests of Stakeholders

i. In national key ecological functional areas involving the relocation of farmers, herders, and enterprises, long-term livelihood considerations must be taken into account, encouraging them to become protectors, participants, promoters, and practitioners of ecological space, ecological construction, ecological protection, and the theory " Lucid waters and lush mountains are invaluable assets".
ii. In national main grain production areas, complementary policies should be improved to reflect national policies for strengthening agriculture and benefiting farmers, safeguarding the rights and interests of stakeholders such as large-scale grain growers.
iii. Land designated as permanent basic farmland should be solely used for planting food crops. Meanwhile, developing regenerative agriculture through reasonable planning of non-agricultural crops (such as trees and shrubs) around farmland and pastures friendly to pollinators can be a typical practice.

(d) Redefining Food Security and Aligning it with Appropriate Land Planning

Based on the latest dietary guidelines and future population trends, re-evaluate China's demand for crops and other kinds of foods and estimate the corresponding demand for agricultural land. With this information, re-examine current food production targets with the aim of guiding healthy, low-carbon, environmentally friendly, and zero-waste dietary habits and culture. Seek synergy among food security, ecological security, "dual-carbon" goals, water security, and other essential objectives.

2.5 *Weighing Land Use Through the Valuation of Natural Capital and Ecosystem Services*

The natural capital and ecosystem services provided by nature are the foundation of human society and economic development. Achieving multiple environmental and social goals, such as the "dual-carbon" goals, water and food security, also relies on natural capital and ecosystem services. Changes in land use can lead to variations in natural capital and ecosystem services, and assessing these can help measure the contributions of different land uses to economic and social objectives [1], reducing or avoiding adverse impacts resulting from inappropriate land use decisions.

2.5.1 Inadequate Consideration of Preserving and Enhancing Natural Capital in Current Land Planning and Decisions

Currently, the evaluation of land use primarily focuses on economic and social aspects, lacking comprehensive assessments of natural capital and ecosystem services. Current land use is often driven by singular demands, overlooking the impact on ecosystems and the maintenance of other services. In present national statistical and accounting practices natural capital and ecosystem services are still not valued at their true price; and are even treated as being available for free for companies [34]. Simultaneously, economic globalization has increased the influence of international capital flows on local land use decisions [35], weakening national policies aimed at preserving and increasing public goods. Therefore, by employing mature and applicable methods for natural capital and ecosystem services accounting, decision-makers can better understand the current and long-term impacts of their land use choices on the environment, society, and economy, and redefine the "value" in modern economic systems to include the value of nature.

2.5.2 The Current Status and Opportunities of Natural Capital Accounting and Ecosystem Services Evaluation

Because of the significance of natural capital and ecosystem services, biodiversity conservation is not only a global conservation goal but also a crucial foundation for synergizing the multiple crucial global objectives, including biodiversity targets, "dual-carbon" goals, food security, and water security.

(a) Progress in Natural Capital Accounting and Ecosystem Service Assessment

Research progress on natural capital accounting and ecosystem service assessment has been rapid both domestically and internationally. Natural capital accounting primarily focuses on the relationship between natural resource consumption and national debt, while ecological debts are not only national but affect all economic agents. However, due to excessive demand on ecosystem services and increased human activities, natural capital in many regions is depleting, and the capacity to supply ecosystem services is declining.

i. Progress in international research on natural capital accounting and ecosystem services evaluation.

Since 1993, the United Nations has successively released comprehensive environmental economic accounting frameworks, incorporating environmental assets into the national economic accounting system. In 2012, the "System of Environmental-Economic Accounting 2012—Central Framework" (SEEA-CF) was released, including environmental management costs and natural resource losses or gains in the national economic accounting, known as Green GDP accounting. Subsequently, the United Nations Statistical Commission published the System of Environmental-Economic Accounting—Ecosystem Accounting" (SEEA-EA) standard in 2021, describing the relationship between ecosystems and economic assets, integrating economic, environmental, and social data into a unified and coherent conceptual framework, providing a theoretical and methodological basis for conducting ecosystem asset accounting. This standard's release indicates the United Nations Statistical Commission's adoption of international standards for ecosystem services and ecosystem asset physical quantity accounting, recommending macroeconomic indicators such as Gross Ecosystem Product (GEP), and now further research on the valuation of ecosystem services, ecosystem asset, and marine ecosystem accounting is in progress.

ii. Progress in relevant research and practice in China.

Chinese scholars have localized the ecosystem services valuation principles and research methods proposed by Costanza, greatly promoting the development of such methods in China.

The assessment of ecological and environmental resources began with "Overall Plan for the Reform of Eco-civilization System" issued by the Central Committee of the Communist Party of China and the State Council in 2015. In 2022, the National Development and Reform Commission and the National Bureau of Statistics issued the "Guidelines for Gross Ecosystem Product Accounting". According to incomplete statistics, up to the present, various pilot projects for Gross Ecosystem Product (GEP) accounting have covered 18 provinces and 57 prefecture-level cities in China, with approximately 15 provinces implementing related policies to carry out the valuation of ecosystem products as a key initiative.

(b) Key Issues

The concepts of natural capital, ecosystem assets, and ecosystem services still need to be unified between international and domestic contexts and among different stakeholders to guide relevant practices and support the construction of a complete and unified evaluation system and indicators.

Natural capital accounting should ideally be an institutionalized information system, with documented data assurance and methods and regular production cycles. This means decision-makers can rely on information being available over the long term.

2.5.3 Research to Achieve Synergy, Exploring Systematic Land Use Through Models and Technologies

(a) Analyzing the Essence of Synergy and Strengthening Research on Multi-objective Synergy Models

Although land use and cover change (LUCC) often occur at the local level, their cumulative impact worldwide can severely impact the Earth system. The land use goals of different departments and stakeholders are not always compatible and often conflict. Choosing a specific land use function in a geographical region often requires balancing and negotiating between different temporal and spatial dimensions and stakeholders. The basis for such balancing and negotiation should be the optimal solution for natural capital and ecosystem services. However, converting vast areas of tropical rainforest into arable land for soybean production has also led to biodiversity loss, increased carbon emissions, and reduced carbon sequestration, causing overall damage to the forest's regulatory and supporting ecosystem services, with these negative impacts spilling over to the local, regional, and global levels.

However, defining the "optimal solution for natural capital and ecosystem services" is a significant challenge we face.

Trade is a critical dimension of sustainable land management. On the one hand, trade provides incentives for local stakeholders and investors to decide land use based on the natural resource endowment and comparative advantages of the land. This trade-facilitated division of labour and specialization may amplify ecological destruction through economies of scale. On the other hand, land forms the basis of many public goods, such as water quality, biodiversity, and stable climate, all of which can be traced back to land use. Therefore, ensuring the supply of public goods is a priority at the global level, but at the local level, local stakeholders will inevitably seek to increase production and improve livelihoods—these objectives may often conflict in many cases. Therefore, land use planning needs to not only balance different land functions but also consider and coordinate the interests of stakeholders at various scales.

Currently, short-term demands and economic gains are the primary motivations behind land use decisions, neglecting the long-term risks caused by unsustainable land use patterns. A scientific and unified framework with indicators is an essential tool for all stakeholders to promote dialogue and collaborate on transformational pathways.

(b) Enhancing Decision Quality Through Information Technology for Long-term and Cross-regional Impacts of Land Use

Natural capital accounting and ecosystem service assessments can serve as crucial support for comprehensive land use planning. By using integrated models and leveraging technologies such as artificial intelligence and the "metaverse", real-world multidimensional data can be simulated and presented in a visual manner to assist multi-stakeholders and decision-makers in visualizing the future scenarios and improving the quality of land use decisions.

Taking agriculture as an example, global climate warming has significantly affected the structure and distribution of agricultural crops in China. Rising temperatures have enriched heat resources in northern regions, extending the crop-growing season and shifting the accumulated temperature zone northward. The impact of climate warming on China's arable land utilization has become increasingly apparent. Since 1990, there has been a notable northward migration of rice cultivation in China, particularly in the northeastern region, where the core latitude of rice cultivation has shifted from 39–46 to 41–47° N. Additionally, Qinghai Province has experienced an increase in mild spring drought and a decrease in severe summer drought, significantly affecting agricultural production.[9] These changes were factors that past land use decisions could not adequately consider. By employing technology to display these trends and corresponding scenarios, decision-makers can formulate more scientifically and comprehensively planned policies. For instance, they can focus on climate and water resource constraints, dynamically adjust lands that are no longer suitable for cultivation due to climate change, and optimize the addition of new arable land within suitable climate zones.

(c) Natural Capital Preservation and Appreciation as a Crucial Basis for Achieving Synergistic Objectives

The research team believes that the essence of the "nature-positive" vision model lies in preserving and appreciating natural capital and ensuring the stable and balanced supply of ecosystem services. "Nature-positive" plays a vital role in multiple environmental sustainability agendas and economic development. It should be better recognized and integrated into fundamental considerations.

[9] Referred to content of the supporting reports on managing cropland and agro-ecosystem services in response to climate change.

i. About "Nature-positive".

Despite increased investments in nature conservation over the past decades, we have not successfully "reversed the curve of biodiversity decline" [36]. In response to biodiversity loss, several international institutions have jointly proposed the global "Nature-positive" conservation goal, which serves as a core concept for formulating biodiversity targets or industrial transformation processes [37]. "Nature-positive" seeks to slow down the rate of biodiversity decline compared to the state in 2020 through the efforts of various levels of governments, businesses, and the public. By 2030, it aims to surpass the state of 2020 and achieve full nature recovery by 2050, fostering harmonious coexistence between humans and nature [38].

ii. The Role of Nature-Positive Transitions in creating synergies.

The World Economic Forum's insight report "Seizing Business Opportunities in China's Transition Towards a Nature-positive Economy" released in early 2022 identified three economic systems most closely associated with nature loss: the food, land- and ocean-use system; infrastructure and built-environment system; and energy and extractives system [4]. By comparing the potential economic opportunities in these three major socio-economic systems under the "Nature-positive" scenario and the "business as usual" scenario, it was estimated that realizing all the "Nature-positive" transitions across the three systems could create approximately $1.9 trillion in business opportunities and 88 million sustainable jobs in China by 2030. This vision aligns closely with China's ambition for high-quality green development, emphasizing nature protection and restoration, as well as rational and sustainable use and management of natural resources, leading to substantial synergistic effects.

Moreover, because nature provides services such as food, water, energy, and climate regulation, nature-positive transformation is also a crucial pathway for achieving the crucial global objectives, such as food and water security and climate change mitigation. To realize this vision, it requires funding, technological and governance innovations, multi-stakeholder collaboration, and implementation in both macro- and micro-level land management and utilization.

(d) Strengthening Nature-Positive Transformation in Key Industrial Sectors' Green Development

According to the report "Seizing Business Opportunities in China's Transition Towards a Nature-positive Economy", 65% of China's GDP is at risk due to nature loss. The three major socio-economic systems identified in the report are closely related to about two-thirds of industrial sectors classified in China's national economic activities. This implies that about two-thirds of China's industries have the opportunity to support global nature conservation goals through implementing nature-positive transformation and seize the economic opportunities brought by this transformation by 2030. In December 2022, the "Kunming-Montreal Global Biodiversity Framework" was adopted by all parties, and its multiple objectives are closely related to businesses and financial institutions. For example, Target 15 requires businesses and financial institutions to assess, disclose, and manage the risks, dependencies, and impacts of their operations, supply chains, and portfolios on biodiversity. The framework will accelerate changes in policies, regulations, stakeholder expectations [39], and market conditions globally and signifies the transformation of various industries will become a common global trend.

The nature-positive transition of industrial sectors requires collaboration among policy-makers, industry associations, companies, and consumers, among other stakeholders. At the macro level, China has proposed the concept of "ecological civilization", including biodiversity protection to a national strategy and taking the lead in proposing ecological conservation redlines. Ambitious "dual-carbon" goals have also been set. China can break down barriers between climate and environmental actions and enhance synergy between "dual-carbon" goals and biodiversity goals, leading the way for a new type of high-quality development. In terms of practical transformation, more incentivizing policies, investments, and actionable tools and roadmaps are needed. Frameworks and tools being promoted internationally, such as Science-Based Targets for Nature, especially those related to land use, and the International Sustainability Standards Board (ISSB), can serve as crucial references for the transformation of relevant industries in China. In the future, industry-specific transformation paths that better align with China's socio-economic context will significantly contribute to advancing and implementing biodiversity goals and related objectives, such as climate goals, food security, and water security.

2.5.4 Recommendations for Research

(a) Investigate how to fully consider the preservation and appreciation of natural assets and the stable and sustainable supply of ecosystem services in land use planning and management decisions.

   i. Clarify the concepts of ecological assets and ecosystem services, refine the classification system for ecological assets and ecosystem services, and explore standardized indicators for valuing ecological assets and ecosystem services.
   ii. Study the current practices of using natural capital accounting and ecosystem service assessments for national spatial planning and land use decision making and improve the principles and methods for land use decisions that involve ecosystem service assessments and engage multiple stakeholders (at both macro and micro levels). Natural capital can contribute

to optimizing national spatial planning in several ways: (1) Provide a consistent, systematic, and transparent information system for development planners and policy-makers to effectively integrate nature into their decision-making processes for optimizing spatial planning; (2) Highlight the value of nature to decision-makers; (3) Provide direct information support to land use planning and zoning; (4) Evaluate the equity of access to ecosystem services and the benefits they provide; (5) Evaluate development, ecosystem restoration, and nature-based solutions investments on the basis of the value of expected ecosystem services return

iii. Conduct integrated research on ecosystem services, construct comprehensive models, quantitatively analyze the relationships between ecosystem services under different scenarios of natural and land use changes, and optimize combinations through exploration, optimization, and visualization using technologies such as artificial intelligence and the metaverse, aiming to find the best land use patterns.

(b) Research the use of natural capital accounting and ecosystem service assessments as the basis for land use decisions in industry's nature-positive transformation processes. For example:

i. Agriculture (or agriculture, forestry, animal husbandry, and fishery): Agricultural biodiversity is a critical component of biodiversity and provides various ecosystem services needed for human sustainable development. However, public awareness of its importance is far from sufficient for nature conservation. More specific topics could include: support regenerative agriculture, reduce harmful subsidies, promote the sustainable utilization of agricultural land resources through ecosystem service functions, innovate ecological compensation mechanisms for arable land, and improve mechanisms for dynamic balancing of arable land quantity, etc.

ii. Renewable energy: Evaluate land use decisions for renewable energy, etc. Research shows that at 25–80% penetration in the electricity mix by 2050, solar energy may occupy 0.5–5% of total land. The resulting land cover changes, including indirect effects, will likely cause a net release of carbon ranging from 0 to 50 gCO_2/kWh. Hence, a coordinated planning and regulation of new solar energy infrastructure should be enforced to avoid a significant increase in their life cycle emissions through terrestrial carbon losses [40].

(c) Develop specific measurable indicators for the nature-positive model that can be applied at multiple scales and advance the implementation of synergistic effects among multiple environmental objectives.

Given the importance of realizing nature-positive and related objectives, a key research and practical foundation is to assess and measure nature-positive objectives. However, comprehensive studies in this area are currently lacking. In the preliminary research, the research team plans to explore the construction of an assessment model for nature-positive objectives, followed by the development of a scientifically grounded comprehensive indicator evaluation method, baseline and tracking assessments, and methods that incorporate stakeholder knowledge. The goal is to implement actions to achieve multiple sustainable objectives through nature-positive transformation in industry sectors and business operations in the future.

References

1. IPBES. (2019, May 04). *The global assessment report on biodiversity and ecosystem service*, July 27, 2023. https://zenodo.org/record/6417333
2. World Bank. (2013, March). *Development research centre of the state council of the People's Republic of China*, July 27, 2023. https://elibrary.worldbank.org/doi/pdf/10.1596/9780821395455_CH03.
3. World Economic Forum. (2022, January). *Seizing business opportunities in China's transition towards a nature-positive economy*, July 27, 2023. https://www3.weforum.org/docs/WEF_New_Nature_Economy_Report_China_2022.pdf
4. Zhang, Y. (2023). Creating a new paradigm of modernization for human civilization. *Historical Review, 3*, 5–10.
5. Zhang, Y. (2020). Building harmonious coexistence between humans and nature in modernization. *Finance & Trade Economics, 12*, 18–21.
6. Zan, G., et al. (2023). Major results and analysis of the sixth national desertification and sandification survey. *Forest Resources Management, 0*(1), 1–7.
7. Adas, et al. (2011, April 04) *Meeting the challenge: Agriculture industry GHG action plan delivery of phase I: 2010–2012*, July 07, 2023. https://www.agindustries.org.uk/asset/3D9072B1-13F6-4EFF-8365ABD7BDBD7DB8/
8. Shafer, S., Walthall, C., & Franzluebbers, A., et al. (2014). Emergence of the global research alliance on agricultural greenhouse gases. *Carbon Manage, 2*(3), 209–214.
9. Jin, S., Zhang, Z., Hu, Y., Han, D., & Du, Z. (forthcoming). *Agricultural economic issues.*
10. Discussion Paper no. 6, Tokyo, Japan, September
11. Tang, H., & Li Z. (2012). Research on per capita grain demand based on China's balanced diet pattern for residents. *Scientia Agricultura Sinica, 45*(11), 2315–2327.
12. Willet, W., Rockström, J., Loken, B., et al. (2019). Food in the anthropocene: The EAT-lancet commission on healthy diets from sustainable food systems. *Lancet, 393*(10170), 447–492.
13. CPCNEW.CN. (2023, June 09). *Promoting modernization of a large population by building a high-quality livelihood*, July 27, 2023. http://theory.people.com.cn/n1/2023/0609/c40531-40009780.html
14. National Development and Reform Commission. (2023, February 07). *Comprehensive improvement of China's primary product supply security capability*, July 27, 2023. https://www.ndrc.gov.cn/xwdt/ztzl/srxxgcxjpjjsx/xjpjjsxjyqk/202302/t20230207_1348450.html
15. World Bank. (2008, June 19). *Rapid urbanization in China: Benefits, challenges, and strategie*, July 27, 2023. https://www.shihang.org/zh/news/feature/2008/06/19/chinas-rapid-urbanization-benefits-challenges-strategies
16. China Science Daily. (2020, May 07). *The next 15 years could be the most risky period for China's Farmland resource security*, July 27, 2023. https://news.sciencenet.cn/htmlnews/2020/5/439394.shtm
17. The State Council of the People's Republic of China. (2017, January 03). *State council notice on printing and distributing the national spatial planning outline (2016–2030)*, July 27, 2023. https://www.gov.cn/zhengce/content/2017-02/04/content_5165309.htm

18. Official Website of the Government of China. (2021, July 26). *Release of national land spatial planning in multiple places after "multiple plans integration" approach*, July 27, 2023. https://www.gov.cn/xinwen/2021-07/26/content_5627282.htm
19. National Development and Reform Commission. (2020, May). *Master plan of national important ecosystem protection and restoration (2021–2035)*, July 27, 2023. https://www.ndrc.gov.cn/xxgk/zcfb/tz/202006/P020200611354032680531.pdf
20. Ge, D., & Long, H. (2020). Rural spatial governance and urban-rural integration development. *Acta Geographica Sinica, 75*(6), 1272–1286.
21. Peng, J., Li, B., Dong, J., et al. (2020). Basic logic of territorial ecological restoration. *China Land Science, 34*(5), 18–26.
22. Liu, Y. (2020). The basic theory and methodology of rural revitalization planning in China. *Acta Geographica Sinica, 75*(6), 1120–1133.
23. Ge, D., & Lu, Y. (2021). Rural spatial governance for territorial spatial planning in China: Mechanisms and path. *Acta Geographica Sinica, 76*(6), 1422–1437.
24. Liu, Y., & Li, Y. (2017). Revitalize the world's countryside. *Nature, 548*(7667), 275–277.
25. Ye, C., & Liu, Z. (2020). Rural-urban co-governance: Multi-scale practice. *Science Bulletin, 65*(10), 778–780.
26. Zhang, Y., Long, H., Ma, L., et al. (2019). Research progress of urban- rural relations and its implications for rural revitalization. *Geographical Research, 38*(3), 578–594.
27. Yi, J., Guo, J., Ou, M., et al. (2023). Land use control: institutional changes, target orientation, and system construction. *Journal of Natural Resources, 38*(6), 1415–1429.
28. Lin, J., Wu, T., Zhang, Y., et al. (2019). Thoughts on unifying the regulation of territorial space use. *Journal of Natural Resources, 34*(10), 2200–2208.
29. Zhang, X., & Lv, X. (2020). Reform logic of territorial space use regulation and the response path of land spatial planning. *Journal of Natural Resources, 35*(6), 1261–1272.
30. Liu, J., Ma, S., Gao, J., et al. (2018). Delimiting the ecological conservation redline at regional scale: A case study of Beijing-Tianjin-Hebei Region. *China Environmental Science, 38*(7), 2652–2657.
31. Zhou, J. (2021). Thoughts on optimizing full-domain all-elements and whole-cycle land-use control mechanism of territorial spatial planning. *South Architecture, 2*, 18–25.
32. Qu, M., Ding, G., Guo, J., et al. (2020). Multi-objective collaborative governance mechanism of territorial space planning. *China Land Science, 34*(5), 8–17.
33. Li, M., Wang, J., Li, Z., et al. (2019). Preliminary understanding of national land spatial planning in the new era—Based on a national perspective. *Small Town Construction, 37*(11), 5–11.
34. Peng, J., Lv, D., Dong, J., et al. (2020). Processes coupling and spatial integration: Characterizing ecological restoration of territorial space in view of landscape ecology. *Journal of Natural Resources, 35*(1), 3–13.
35. Carlucci, E. (2023, January 04) *Addressing the nature financing gap: The role of natural capital accounting and natural asset companies*, July 23, 2023. https://www.iisd.org/articles/insight/addressing-nature-financing-gap
36. Lambin, E., & Meyfroidt, P. (2011, February 14). *Global land use change, economic globalization, and the looming land scarcity*, July 27, 2023. https://doi.org/10.1073/pnas.1100480108. https://www.pnas.org/doi/pdf/
37. Liao, W., Deng, H., Li, R., et al. (2018). Spatial characteristics and influencing factors of hydrological regulation services in the Yangtze River Basin ecosystem: A sub-basin scale analysis. *Acta Ecologica Sinica, 38*(2), 412–420.
38. Locke, H., Rockström, J., Bakker, P., et al. (2021). *A nature positive world: global goal for nature.*
39. World Economic Forum (2021, January). *The post-2020 global biodiversity framework and what it means for business*, July 27, 2023. https://www3.weforum.org/docs/WEF_Biodiversity_Targets_for_Business_Action_CN_2023.pdf. https://www3.weforum.org/docs/WEF_Biodiversity_Targets_for_Business_Action_CN_2023.pdf
40. Van de Ven, D. J., Capellan-Peréz, I., Arto, I., et al. (2022). The potential land requirements and related land use change emissions of solar energy. *Scientific reports, 11*(1), 2907.

Part III
Sustainable Production and Consumption

Chapter 7
Promoting Digitalization and Green Technologies for Sustainable Development

1 Theoretical Basis and Conceptual Framework

1.1 Introduction

The overarching topic of this study links two of the most relevant development trends of our time within a clear mission: digitalization as the dominating transformative force of the twenty-first century must be put at the service of sustainable development, as the most pressing transformation challenge of the twenty-first century.

Section 1 establishes a conceptual framework to research digital technology opportunities and related policy implications for leveraging digital technologies for sustainable development. Based on these initial considerations, the subsequent sections will investigate selected use cases in greater detail and elaborate in a four-step approach the theoretical basis and the conceptual framework.

Section 1 addresses the following questions. First, what is the nature of the sustainability challenges ahead of us, and what transformation capabilities are required for entering successful transition pathways in complex technological, economic, and social systems? Second, what is the key mechanism and the potential contribution of digital technologies and solutions to the sustainability transformation challenge, given their character as GPTs? Third, what are development patterns for digital technologies, related capabilities, and platform economies in China, and what is the state of the digital business ecosystem? Fourth, what are the strategic gaps to be tackled in order to fully benefit from the digital opportunity space, what are the characteristics of a sustainable development of digital economies and societies that require new policy thinking, and what are the generic categories of policy approaches to leverage the digital solution space for sustainability that should be investigated in the case studies?

Green Empowerment and High Quality Development,
https://doi.org/10.1007/978-981-96-4218-2_7

1.2 *Embracing the Sustainability Transformation Challenge*

1.2.1 Problem Statement: What Is the Nature of the Sustainable Development Challenge?

Entering the twenty-first century, humanity is confronted with severe and aggravating risks to the global Earth system and its ecological balance. Most prominently, climate change impacts are affecting societies around the globe, calling for a rigorous reduction in GHG emissions and—ultimately—aiming at climate neutrality around the middle of the century.

Hence, all countries worldwide are facing a joint challenge to setting the course for achieving the sustainable development of societies and economies while safeguarding the ecological foundations of our well-being and prosperity. From the broader perspective of the 17 sustainable development goals put forward in the United Nation's "Transforming Our World—2030 Agenda for Sustainable Development," the essence of the sustainability challenge is the need to coordinate and reconcile the development dynamics between the various dimensions of economy, society, and the ecological environment. While jointly experiencing numerous aggravating ecological crises, the world still faces divergent perspectives and trade-offs when entering zero-growth pathways of population and achieving equal economic prosperity on a global scale.

One holistic approach to this challenge is China's Ecological Civilization strategy. Facing the resource and environmental problems that occurred during the process of accelerating industrialization and urbanization, the Report of the 18th Communist Party of China (CPC) National Congress in 2012 proposed the strategy of Ecological Civilization Construction, believing that ecological civilization is the sum of material, spiritual, and institutional achievements made by human beings to protect and build a beautiful ecological environment. Also, it runs through the whole process of economic, political, cultural, and social construction. On March 24, 2015, the Political Bureau of the CPC Central Committee deliberated and adopted the Opinions on Accelerating the Construction of Ecological Civilization (hereinafter referred to as the Opinions). In October 2015, strengthening the construction of ecological civilization was included in the national "Five-Year Plan" for the first time. On March 11, 2018, the sixth item, "leading and managing economic work and urban and rural construction," in Article 89 of the Constitution "The State Council exercises the following functions and powers," was revised to "leading and managing economic work, urban and rural construction, and ecological civilization construction" through the amendment to the Constitution adopted at the first session of the 13th National People's Congress revised. The construction of ecological civilization is considered a fundamental change in China's development notion and mode, involving various aspects of economic, political, cultural, and social construction. Closely related to the layout of productive forces, spatial pattern, industrial structure, production mode, and living styles, as well as the value concept and institutional system, ecological civilization construction is a comprehensive and systematic project.

In this context, achieving sustainable development at a global scale is a historically unprecedented task, characterized by the following features:

- The increasing urgency to act as current trends and projections of ecological tipping points (e.g., in global climate systems) indicate that the room to manoeuvre is shrinking, and the time window for achieving sustainable system changes is closing.
- The complexity of reconciling and coordinating social and economic development processes in balance with ecological and environmental boundaries. These dimensions are intertwined, calling for holistic approaches to enhance human well-being and quality of life within a healthy natural environment. Importantly, this also requires global cooperation on climate change, biodiversity, artificial intelligence and other threats to human survival.
- The need for aspiring at bold ambitions, one of them being the decarbonizing of our societies within a few decades, calling for far-reaching system changes in the economy and society, such as the transition to a fully renewable energy supply and drastic reductions in global resource consumption patterns.
- The need for a system-thinking mindset at all levels. Individuals as well as institutions or societies must embrace, understand, and manage the multi-faceted interrelations, interdependencies, and dynamics of acting in complex systems. It is important to look for tangible action points and insights that are action relevant in this regard.

Therefore, there are some key implications for designing and implementing strategies for sustainable development.

1.2.2 How to Get It Solved: Cultivating the Core Transformation Abilities to Achieve Sustainable Development

Effective strategies to address the sustainability challenge aim at driving socio-techno-economic transformation processes. This requires the following key elements:

- A comprehensive system perspective covering the full range of sustainability transformation areas, such as energy, mobility, agriculture and food, manufacturing, consumption, cities, and urban transitions. Fighting climate change, for example, depends on GHG emission abatement in all sectors. Moreover, interdependencies and synergies need to be taken into account.
- Fundamental structural changes (system changes) that redefine the physical and institutional boundary conditions for individual and collective action, enhancement of public transport infrastructure, transforming the built environment of cities, (re)designing markets and economic rules, reform of institutional settings, mandates, and capabilities, etc.
- A process of understanding and acknowledging the dynamic nature of the problem as well as the growing evidence and knowledge on cause-impact relations and

solutions. There is no simple solution, and the whole world needs to advance step by step, taking the benefits from quick wins while investing in long-term solutions, continuously learning, and improving our capabilities.
- Collaboration and international cooperation as the system changes at the scale and impact needed cannot be achieved in isolation but require global effort, joined forces, and coordinated governance.

At the operational level, these requirements can be translated into core capabilities needed to initiate, drive, manage, and scale the multiple transition processes toward societal, economic, and environmental sustainability on all levels. Tackling the sustainability challenge requires the ability to realize multiple objectives related to a very high level of required changes.

1.3 Digital Capabilities and Sustainable Development: The Opportunity Space

Distinct from the previous three industrial revolutions, the ongoing digital transformation fundamentally changes the complex system interrelations between technology, economy, and society on an informational level. Whereas in former times those interrelations have been formed mainly by physical and monetary interactions, digital societies and digital economies are based on the emergence of cyberspace and the underlying data relations between objects, people, and activities—both in the physical space of manufacturing, infrastructures, and the built environment as well as in the economic and socio-cultural space. This offers new opportunities to create new technological economic and social systems oriented to sustainable development and, at the same time, requires new political and institutional regulatory capabilities (cf. Sect. 1.4).

This process is driven by the outstanding development dynamics of digital technologies, which are a new generation of information technologies, including the Internet, the Internet of Things (IoT), 5G/6G mobile communication, cloud or edge computing, big data analytics and artificial intelligence, etc. In addition, versatile digital platforms allow us to connect and integrate multiple digital services, thus accumulating digital capabilities, use cases, and user communities. This is of special relevance with regard to the interactions with startups. Whereas startup companies often succeed by solving specific "pain points" in economic, social, and ecological areas, the overarching platforms play a critical role in the selection and allocation of opportunities, technologies, human capital, and sources of financing for entrepreneurship.

Digital platforms, therefore, play a pivotal role in shaping the digital space by introducing new economic playgrounds and market coordination mechanisms. In this sense, digital transformation is a multi-dimensional phenomenon, combining physical hardware with virtual software and services.

This offers great opportunities for sustainability. Through sensing, data collection, connection, and sharing, data analysis driven by growing computing power and performant algorithms and the data-based interference into the physical world by controls, data-driven operation regimes, and ultimately even automated devices such as robots, human beings can continuously monitor, evaluate, reorganize, and optimize economic and social activities. Digital transformation offers a powerful solution space of digital capabilities that now need to be matched with the urgently required transformation capabilities introduced in the previous Sect. 1.1.

The key question is now: how to leverage the digital opportunity space for the sustainability challenge? Digital transformation and its technologies are not limited to specific use cases or sectors. In line with historical precursors such as the introduction of electricity at the end of the nineteenth century, digital transformation represents the next level of GPTs that lay the foundation for a broad boost of technological innovation and socio-economic development across civilizations. Digital technologies are versatile and ubiquitous, their impact diffuses into almost every aspect of human life. The emergence of such cyberspace and its interaction with the physical and social spaces turns the core elements of "data" and "computing" into strategic assets for creating and managing complex technological, economic, and social systems.

Moreover, digital transformation is characterized by unique patterns of economic cost degression. Technical progress is continuously squeezing the cost of hardware, illustrated by the famous Moore's Law for computation performance. Once the physical basis of infrastructure and devices is established and the initial costs of software development are covered, any further expansion of digital functionality and outreach can be realized at very low or even close to zero marginal costs. This is a fundamental difference from traditional manufacturing paradigms, and it opens up so-far unseen opportunities for scaling digital services within user groups and application areas as well as for the transfer between application areas and sectors.

With these aspects in mind, the implications for sustainability policy-making are straightforward. The availability and accessibility of green digital technologies need to be expanded to strengthen the digital capabilities and put them into service for growing our transformation capabilities through spreading and scaling digital solution use cases for sustainability.

Hence, the task is twofold:

- Technology push. Ongoing technical progress and innovation are boosting the performance of digital technologies and, hence, are expanding the versatility and functionality of the digital toolbox. This potential needs to be leveraged and directed toward the pressing sustainability challenges. Part of this task is to reduce the environmental footprint of digital technologies themselves regarding energy demand, GHG emissions, and e-waste (Sect. 2).
- Transformation pull. From the transition perspective, the single-use cases need to be scaled, disseminated, and transferred, aiming at growing critical momentum

and self-sustaining demand for system change. Hence, the notion of the sustainability transformation capabilities helps to systematically search, scan and investigate the opportunities to benefit from the digital capabilities—and hence spurring the sustainable deployment and commercialization of technologies and solutions.

This section provides a generic framework for describing the functional relations between digital technologies and the sustainability transformation capabilities introduced above in Sect. 1.1. It sheds light on general aspects of how to shape the complex socio-technical innovation system required for sustainability in technologies, economy, and society.

1.4 Sustainable Development Driven by Digital Technology: China's Practices and Prospects

After 2015, it has become a strategic choice of China to achieve sustainable development through the development of the digital economy. This section illustrates innovative applications of digital technology driving the achievement of economic, ecological, and social goals of sustainable development through the building of core competencies. The section refers to quantitative data analysis of technology partnerships among 2200 Chinese digital backbone enterprises.

1.4.1 Characterization of China's Digital Economy

By classifying the core business of 2200 digital key enterprises, these enterprises are distributed in 20 business areas. Among these, two business areas, enterprise technology integration and solutions together with smart business and retail, account for the highest proportion of the total number of enterprises, which are 17.20% and 10.31% respectively.

These enterprises are centred around a few dominant platform nodes. In 2021, in the value network of China's digital economy, the 10% nodes with the highest degree of centrality accounted for about 70% of the links in the network. Innovative platform enterprises, including Huawei, Baidu, Alibaba, Tencent, JD, iFLYTEK, and SenseTime, and industrial innovative ecology guided by them are playing leading roles in the development of China's digital economy. The platform, together with research universities, research institutions, technology-based SMEs and startups, governments, and other organizations, form an innovation ecosystem and are carriers of digital technologies and capabilities.

1.4.2 Technology System and Application Areas

Digital technology is a complex technology system, including 17 key technologies, which are big data and cloud computing, the Internet of Things, intelligent robots, intelligent recommendation, 5G, blockchain, speech recognition, virtual/augmented reality, AI chips, computer vision, natural language processing, biological recognition, space technology, photoelectric technology, automatic driving, human–computer interaction, and knowledge map.

Digital technologies are widely applied to 19 applied fields in China, which are intelligent management of enterprises, intelligent marketing and new retail, scientific finance, smart city, smart medical treatment, new media and digital content, intelligent manufacturing, smart education, intelligent transportation, network security, intelligent logistics, intelligent cultural tourism, smart governance, smart energy, intelligent hardware, intelligent connected vehicle, smart home, smart agriculture, and intelligent security.

Viewing from the results of the analysis of application complexity of China's digital technology in 2021, knowledge map, blockchain, human–computer interaction, natural language processing, and intelligent recommendation systems are the top five complex key digital technology for technology application. The results of the calculation of the complexity of a technical system of the application field showed that intelligent manufacturing, smart home, intelligent marketing, new retail, intelligent hardware, and intelligent management of enterprises are the top six technically complicated applied fields.

1.4.3 Innovative Applications of Digital Technology in the Economy

The application of digital technology in the economy first became large-scale in the Consumer Internet and then extended to the Industrial Internet. Since 2016, the upgrading of the Consumer Internet and the diffusion of industrial ecosystem in cyberspace to small- and medium-sized cities and rural areas have not only driven the development of tertiary industry but also radiated the transformation and upgrading of secondary industry.

In the field of the Consumer Internet, the innovative application of the new generation of information technology has given rise to the dominance of trading platforms. The development of trading platforms has produced vast amounts of data that are online, shared, and traded in real time. China's Consumer Internet sector is mature in metropolitan areas, including Beijing, Shanghai, Hangzhou, Chongqing, and Shenzhen. Since 2016, the trend has moved down to small- and medium-sized cities and rural areas. It not only allows consumers to enjoy more convenient services but also creates a lot of employment and entrepreneurial opportunities for low- and middle-income people, especially rural residents.

The Industrial Internet is a new type of infrastructure deeply integrated with the new generation of information technology and the industrial economy. Through the interconnection of equipment, material flows, and parts in the value chain, new value systems, including flexible manufacturing, service, and services, emerge. For the development of the Industrial Internet, AI plays a pivotal role. Production intelligence includes the development and production of intelligent factories, intelligent production lines, and intelligent equipment. From the current situation of integrated development of AI and the manufacturing industry, in 2021, among 32 AI technology cooperation between core industry sectors and the manufacturing industry, the top five are all related to the equipment manufacturing industry. Intelligent equipment is at the forefront of deep integration of artificial intelligence and manufacturing. From the practice of enterprise digitalization and intelligent transformation, the deeply integrated development of artificial intelligence and the manufacturing industry can continuously improve the total factor productivity, making China's economy enter the development stage of high economic growth.

1.4.4 Application of Digital Technology in Social Fields

The application of digital technology in the social sector is mainly to solve the "social pain points" that emerged during the industrialization and urbanization of China. Among these applications, deploying digital technologies to solve urban problems and achieve poverty eradication are successful examples.

Smart Cities

China's urbanization entered a rapid development stage in the twenty-first century. The rapid development of cities has also brought about social pain points such as traffic, safety concerns, pollution, and access to education and medical care. Since 2009, the combination of edge computing and cloud computing systems has given birth to the city brain concept, which laid the digital capability in the construction and development of smart cities. Section 3 will elaborate on the topic of cities.

Digital Poverty Alleviation

In 2016, the Chinese government released the Online Poverty Alleviation Action Plan, which calls for the implementation of five major online poverty alleviation projects: network coverage, rural e-commerce, online intellectual support, information services, and online public welfare, giving full play to the role of digital technology in boosting poverty alleviation and achieving targeted poverty alleviation and eradication objectives. The plan aimed to lift more than 70 million people out of poverty, where the innovative application of digital technologies was expected to play a significant role.

1.5 Leveraging the Digital Transformation Opportunity Space—Closing the Strategic Gap Between Digital Technology and the Sustainability Transformation

The previous section revealed the great opportunities to foster sustainable development by deploying digital technologies and leveraging digital capabilities for manyfold transformation purposes in our economies and societies. To fully benefit from this potential, mechanisms of how to implement digital solutions and explore the diversity of channels for dissemination must be better understood.

The starting point is promising in that a lot of digital capabilities are present. Digital technologies and solutions are already omnipresent in everyone's lives, offering versatile applications, and increasingly shaping the world. Many players are already engaged in moving technology borderlines through R&D, driving business innovation, and creating new markets. At the same time, however, digital transformation is still far from being a self-fulfilling sustainability promise.

Technology innovations, business models, and use cases too often still serve short-sighted patterns of profit maximization at the price of resource consumption and environmental damages or even perpetuate prevailing unsustainable fossil path dependencies. Whereas digitalization can improve the lives of women, e.g., through e-commerce in rural areas, there is an increasing risk of distorted and gender-biased data foundations of AI modelling resulting in flawed results, discrimination, inequalities, etc. of such solutions. To ensure that data is used appropriately, the governments need to gain knowledge of data biases, study them systematically, and adhere to international guidelines [1].

Therefore, digital transformation needs to be politically framed and guided in order to provide directions and purposes to technological innovation and deployment, which includes both the trigger for exploiting currently untapped benefits and limiting adverse effects of today.

Hence, a holistic policy approach is needed to tackle the strategic gaps across the various stages of the technology implementation and deployment cycle. Such an approach needs to consider stakeholders' divergent roles and responsibilities within the complex socio-technical systems. Aiming at innovation and business-friendly frameworks to leverage private enterprises and market dynamics, clear guidance on overarching ecological and social values and practical incentives for directing economic activities toward decarbonization and resource efficiency are in demand, which sheds light on the role and effectiveness of (public) institutions and highlights the responsibility of politics to enforce adequate coordination mechanisms.

The following three core building blocks will be essential for policy initiatives striving for a digital-driven sustainability transformation.

Fostering a Green Technology Push

Given the increasing availability of digital technologies and, thus, growing technological innovation momentum (cf. Sect. 1.3), the main task is to diminish the environmental footprint of digital technologies. Therefore, "greening" and decarbonizing the digital sector must become a policy priority in the early stages of technology provision, including less resource use and enhanced circularity of digital devices.

Accordingly, academic, public, and private R&D funding will be beneficial but not sufficient without enforcing ecological quality requirements. In this context, early-stage technology development will benefit from proper incentives for sustainability during the commercialization phase, such as regulation on the clean energy supply for data centres and telecommunication infrastructure, as well as the proper reuse and recycling of devices (Sect. 2).

Guiding and Governing Sustainable Digital Value Propositions

Enterprises, private actors, and platform players play a vital role in integrating digital technologies and accumulating digital capabilities, e.g., within digital ecosystems. They set the rules and boundaries for access to solutions and define the conditions for enjoying the digital value propositions offered.

Usually, the commercial interests of those players and prevailing market characteristics determine the outcome, again, too often not yet oriented toward sustainability goals. Hence, in this stage, policy and effective institutions for market surveillance can play an important role in guiding and governing the scope and rules for offerings, e.g. This requires a holistic approach to institutional design as well as policy coordination among various government agencies.

Growing Market Momentum and Strengthening the Sustainability Transformation Pull

At the same time, private engagement and market forces represent powerful drivers for sustainability transformation. Hence, all kinds of private and public stakeholders need to articulate their increasing demand for realizing digital use cases with sustainability impact, which triggers technology innovation, solution development, and the commercial engagement of players aiming at capturing the growing green solution markets.

Moreover, a very important role for policy-making is to provide complementing market frameworks and market incentives for establishing and growing a self-sustaining commercial dynamic in deploying sustainable digital solutions. It is equally important to break the supply pull for digital value propositions that are associated with unsustainable outcomes, such as the fostering of GHG-intensive consumption or fossil fuel exploration.

The following sections will discuss the strategic gaps and relevant policy requirements in greening the digital sector, sustainable cities, and climate change adaptation in detail. Therein, a specific focus will be given to the following aspects: (1) the opportunity for governance innovation and related growth of policy capacities through digital solutions, i.e., digitalization as enabler and driver for policy reforms; (2) the pivotal role of politics in creating and shaping markets and, hence, preparing the seed ground for private business model innovation and entrepreneurship; (3) the importance of overarching environmental policy frameworks of climate change mitigation, resource management, nature preservation etc., such as enforcing the growth of renewable energy sources (RES) as key to achieve decarbonized energy supply to data centres or implementing adequate CO_2 and resource pricing for better integrating harmful environmental externalities into private commercial rationalities; and (4) the contributions of government-led digital projects aiming at environmental protection that serve as drivers for digital capability accumulation and, thus, can ignite multiplication and growth effects in adjacent digital solution arenas.

The Taihu Lake Economic Zone is one of the most economically developed regions in China. Since 2005, it became obvious that industrialization and a continuous increase in population density are putting pressure on the Taihu Lake ecosystem. On May 29, 2007, a cyanobacteria contamination incident caused by water pollution in Taihu Lake resulted in tap water pollution in Wuxi and a severe shortage of domestic and drinking water. The key to solving pollution problems in Taihu Lake was adopting the Internet of Things (IoT) technology to develop a multi-source data-based cyanobacteria monitoring and early warning system, which had shown an important opportunity for developing the IoT-based digital economy in Wuxi.

With the successful implementation of the pollution control, in 2012, the number of employees of the IoT in Wuxi exceeded 100,000; in 2018, Wuxi officially dominated the Chinese IoT standard; in 2021, Wuxi Internet of Things cluster was selected and approved as a national advanced manufacturing cluster; in December 2022, the total number of IoT enterprises in Wuxi exceeded 4000.

1.6 Conclusion

Section 1 provides the conceptual framework to map the strategic gaps, related policy approaches, and selected major policy options for leveraging the transformative power of digitalization for the sustainability challenge. The digitalization process offers new capabilities for understanding, monitoring, and managing our socio-economic systems and, thus, enlarges the capabilities for the urgently required sustainability transformation in society and the economy. Based on data-related and increasingly virtual interactions between individuals, companies, and institutions, a new kind of so-called socio-economic cyberspace is emerging that defines specific conditions for both commercial and political actions. As one prominent feature, cyberspace serves as a dynamic economic innovation ecosystem centred around platforms with growing technology performance, which is boosting business model innovation and entrepreneurial momentum.

The following sections will elaborate on this interplay between the digital opportunity space for sustainability and the related strategic gaps and policy requirements to achieve such a transition. This report will focus on fighting climate change and the urgent need to reduce GHG emissions. As the world's second-largest economy, the largest GHG emitter, and the nearly largest population, China has both a global responsibility and a national need to redirect its economy toward carbon-free renewable resources and, ultimately, decarbonize the entire society and economy.

In this context, the report will highlight two prominent use cases of Sustainable Cities (Sect. 3) and Climate Change Adaptation (Sect. 4), representing the complex and challenging transformation arenas of global relevance that will extensively benefit from the contribution of innovative digital solutions. As a fundamental prerequisite, however, the environmental footprints of any digital solution and infrastructure must be minimized to prevent detrimental rebound effects and to safeguard the ecological benefits of deploying digital technologies. As a baseline for the subsequent discussions, Sect. 2 will provide insights and recommendations for ensuring a sustainable and climate-friendly digitalization.

2 Green the Digital Sector and Accelerate Digitalization for Green Transformation

2.1 Introduction

As the international communities initiated the sustainable development concept, low-carbon-oriented economic and social transformations surged. Guided by the *United Nations 2030 Sustainable Development Agenda*, more than 130 countries and regions have committed to carbon peaking or carbon neutrality, sending a strong signal of green and low-carbon transformation.

The initiation of the digital era results in extensive and profound revolutions in both economy and society, triggering new modes of thinking, production, and life. The digital sector plays an active role in nearly all fields and sectors by promoting green technology innovation, improving economic efficiency, enhancing energy conservation, emission reduction, and carbon reduction, and is one of the drivers for the green transition.

The digital sector links with the green transformation of the economy through three channels: (1) the digital industry directly reduces the carbon emission throughout the entire value chain; (2) the digital industry indirectly benefits the greening process (green development) by driving technical and social innovation; and (3) the greening of the digital sector has a rebound effect, which means that the energy savings achieved by improving energy efficiency are partially or completely offset by expansionary energy consumption. However, the hope that digitalization will reduce overall energy consumption and resource use may not be fulfilled, as digitalization not only improves efficiency but also increases the options and activities of consumption that cancel out any efficiency gains. The international community has detected new opportunities for digitalization to accelerate the greening process. For instance, the European Commission launched the "European Green Deal," where the dual-track transformation of "greening" and "digitalization" is highlighted in the EU's work in the next 5 years, and some practice cases have emerged in European countries such as Germany (Appendix 2).

The lack of standardized definitions for the terms "digital economy," "digital sector," and "digital infrastructure" continues to pose significant challenges to statisticians worldwide, hindering accurate measurement and analysis. According to the

U.S. Bureau of Economic Analysis, digital economy activities are organized by digital infrastructure (hardware and software), e-commerce (B to B and B to C), and priced digital services, including telecom services, Internet and data services, cloud services, etc. The SDIA (Sustainable Digital Infrastructure Alliance) defines digital infrastructures as follows: "The total physical and software-based infrastructure necessary to deliver digital goods, products and services. This includes data centres, fibre infrastructure, server hardware, personnel, IT virtualization and infrastructure software, operating systems, etc." However, it is crucial to recognize that digital infrastructures encompass not only the data centres where data is processed but also the networks through which it flows. These networks play a vital role in the overall framework of digital infrastructure. The digital sector encompasses various elements, including intelligent and efficient computing centres that serve as digital facilities. Digital companies prioritize green development as a fundamental aspect of their operations, while digital services contribute to energy consumption reduction through algorithm optimization. This section focuses on the greening strategies within the ICT fields, the ways in which the digital sector promotes sustainability, and provides relevant policy recommendations.

2.2 *Present Development and Problems*

2.2.1 Present Green Development of the Digital Sector

China has made breakthroughs in the development of the digital economy. According to the prediction by the China Academy of Information and Communications Technology (CAICT), in 2021, the scale of the digital economy reached CNY45.5 trillion, with a nominal growth rate of 16.2% year-on-year, 3.4 percentage points higher than that of GDP in the same period. The digital economy contributed to 39.8% of GDP. The added value of the digitization industry reached CNY 8.4 trillion, with an increase of 11.9% year-on-year, contributing to 7.3% of GDP.

As the rapid development of the digital economy leads to the rise of additional demands and business, the carbon emissions of the digital sector also increase rapidly. According to official statistics, from 2012 to 2017, the total carbon emissions of the digital sector, including the above-mentioned subsectors, increased by 61% in China, ranking first among all economic subsectors. The significant development of digital services, including the Internet, software, and telecommunications, contributed mainly to the growth of carbon emissions in the digital sector. Overall, as the digital sector boosts support to social and economic development, its carbon emission tends to increase in the short term. Nevertheless, carbon emissions can potentially decline in the medium to long term due to the following factors: (1) advancements in energy conservation and carbon reduction technologies within the digital sector; (2) adjustments and optimizations in China's energy mix; and (3) a growing proportion of non-fossil energy sources.

2.2.2 Integration of the Digital Sector and Green Development

Digital Infrastructure: High Pressure on Green Upgrading

According to the *Energy and Environmental Efficiency Framework Initiative* issued by the ICT sector of the EU, the information and communication industry uses almost 10% of energy and releases 4% of carbon dioxide. China's digital industry, especially digital infrastructure construction, is at the initial stage of green transformation. The high cost associated with carbon emissions reduction accounting poses a challenge in objectively and effectively measuring the energy conservation and carbon emission reduction benefits resulting from the construction and transformation of digital infrastructure. Consequently, it becomes difficult to directly apply the accounting results to assess and incentivize such construction and transformation projects.

A significant issue in the realm of digital infrastructure is the lack of transparency in both the public and private sectors. This lack of transparency permeates all aspects, including data centres and software, making it difficult to ascertain their energy consumption levels. To efficiently implement targeted and meaningful sustainable measures in digitalization, it becomes essential to establish transparency across these domains. One potential approach is to systematically categorize different types of digital infrastructures, such as data centres, networks, and software applications, and define specific action areas that require transparency and key performance indicators to access their sustainability.

Digital Companies: Facing Higher Costs and Stronger Regulation

Digital companies currently face a challenging operating environment characterized by two major factors. Firstly, they encounter significant costs associated with green upgrading, research and development of green ICT products and services, implementing low-carbon operations, as well as marketing and promotion efforts; Secondly, they must navigate international carbon tax policies, green trade barriers, and other regulatory measures. Furthermore, the absence of standardized green product certification and labeling weakens regulation within the market for green products. Consequently, digital companies face difficulties in managing their green products, and even the environmental impact figures published by major Internet giants lack verifiability. In such a chaotic landscape, consumers are unable to discern which digital products or services are environmentally friendly.

Digital Services: Strengthening Social Consensus to Empower Green Transformation

The public sector plays a crucial role in regulating and supporting environmental pollution control and other aspects. However, the difficulty lies in accurately measuring the green GDP contribution enabled by digital services, which hinders effective investment in research and development and social support promoting green ICT products and services. This, in turn, presents challenges in terms of R&D and financing. Consequently, green ICT products and services tend to have relatively higher pricing compared to conventional products. However, due to the limited promotion of the green consumption concept in society, public awareness regarding green ICT products and services remains low. As a result, consumers find it challenging to make informed decisions and actively choose green ICT products and services, considering factors such as price, quality, and other reasons.

Furthermore, software plays a significant role in determining the environmental friendliness of the deployed information and communication technology. It directly impacts energy consumption and can contribute to the premature replacement of hardware. Therefore, it is crucial to acknowledge the issue of software-related obsolescence and its environmental implications.

At the global level, the rapid growth of digital infrastructure and data traffic poses challenges in regulating their environmental impacts. In order to understand and mitigate the environmental impacts associated with the digital economy, it is essential to have knowledge of the specific electricity consumption, greenhouse gas emissions, and other environmental effects that stem from it.

2.3 Green Development Path in the Digital Fields

2.3.1 Digital Infrastructure and Facilities

Present Development

In 2022, China's telecommunications sector witnessed significant growth, with its business revenue reaching CNY1.58 trillion, marking an impressive 8% increase compared to the previous year. Notably, emerging sectors like data centres, cloud computing, big data, and the Internet of Things (IoT) experienced rapid development, contributing to a total business revenue of CNY307.2 billion in 2022. This figure represents a substantial growth rate of 32.4% compared to the previous year. Furthermore, China's mobile communication infrastructure also witnessed substantial expansion. By the end of 2022, the country had a total of 10.83 million mobile communication base stations, of which 2.312 million were 5G base stations. These 5G base stations accounted for 21.3% of the total, indicating a notable increase of 7 percentage points from the end of the previous year.

With the rapid expansion of digital infrastructure and facilities, there is a corresponding increase in the pressure to reduce carbon emission reduction. Improper disposal of materials from decommissioned digital infrastructure and facilities can have detrimental effects on soil and the surrounding environment. Moreover, it is projected that by 2035, the combined electricity consumption of China's data centres and 5G base stations will amount to 5–7% of the country's total electricity consumption, while their total carbon emissions will contribute to 2–4% of China's overall carbon emissions. Considering the average service life of data centres and 5G base stations, which is approximately 10 years, and taking into account other factors, the "lag effect" in carbon reduction within digital infrastructure poses a significant challenge for China in achieving carbon-peaking targets.

- 5G base station. As of September 2022, global 5G subscribers had reached 853 million, marking a remarkable year-on-year increase of 113.5%. This achievement corresponds to a 10.5% penetration rate among mobile subscribers. In China alone, by the end of November 2022, over 2.287 million 5G base stations had been deployed, accounting for more than 60% of the global total. Consequently, the peak power consumption of 5G base stations is approximately 3 to 4 times that of 4G base stations. Currently, the energy consumption of the 5G network remains high. In 2021, the total energy consumption of three major communication operators in China amounted to 13.69 million tons of standard coal, including 105.3 billion kWh of power consumption. Furthermore, as the deployment of 5G base stations continues to expand rapidly, and with the initiation of 6G construction anticipated by 2030, energy consumption is expected to rise for an extended period.
- Data centres. In 2018, data centres accounted for approximately 1% of global energy consumption. However, by 2021, the total computing power of computing devices worldwide had surged to 615EFlops, experiencing a substantial growth rate of 44%. It is projected that global computing power will reach 56ZFlops by 2030, with an average annual growth rate of 65%. In 2021, China's computing devices contributed significantly to the total computing power, reaching 202EFlops, which accounted for approximately 33% of the global share. It is expected that the total energy consumption of data centres in China will reach 380 billion kWh by 2030, accompanied by a carbon emission growth rate exceeding 300%. In addition to energy consumption and carbon emissions, it is crucial to address other issues, such as the use of halogenated cooling liquids, e-waste management, and other related environmental concerns.
- Chip manufacturing. Based on the statistics provided by the China Semiconductor Industry Association, the sales of China's integrated circuit (IC) industry reached CNY1,054.83 billion in 2021, showing a year-on-year growth of 18.2%. Specifically, the manufacturing sales accounted for CNY317.63 billion, experiencing a significant year-on-year growth rate of 24.1%. In the future, the chip market size in China is expected to continue expanding, driven by the ongoing wave of intelligent upgrading within the manufacturing industry. It is important to address the concerns and adopt appropriate measures to mitigate the negative environmental effects associated with IC production.
- Photovoltaic device manufacturing. The newly installed solar power capacity is estimated to be 87.41 GW in 2022, an increase of 60.3% year-on-year. Based on the standard 25-year service life in the market, the number of discarded PV modules will be huge in a few years. Production technology, regulatory efforts, economic benefits, and other relevant factors determine the degree of the negative impacts.

Future Path

Increase the proportion of green energy applications. In regions with favourable climatic conditions, it is encouraged to implement locally generated renewable energy sources, such as small-scale wind power and rooftop photovoltaic (PV) systems. Additionally, exploring low-cost green power options through power transactions, such as direct purchases by large-scale users in areas with competitive green power prices, is also encouraged. For smaller data centres, optimizing the energy supply pattern can be achieved through the utilization of modular hydrogen fuel cells, solar panel houses, and other relevant technologies. In an effort to reduce energy consumption in 5G base station construction, China has actively promoted the co-construction and sharing of 5G base stations. Furthermore, the country has facilitated green electricity transactions related to 5G infrastructure, ensuring a more sustainable approach to powering these base stations and associated facilities.

Strengthen the role of industry standards and the supervision of green production. Industry standards, technical support, industrial incentives, etc., can be leveraged to promote the integration and innovation of information and energy, implement the manageability and controllability of energy terminals, and enhance product carbon footprint management. Other actions include, such as formulating relevant management measures for green production and operation, improving the rules and regulations for monitoring, statistical work, supervision, review, reporting, and disclosure of energy consumption and carbon emissions, forming a green and low-carbon development management system throughout the whole value chain, strengthening the efforts on the supervision of green production of ICT products, and punishing the companies that randomly discharge pollutants.

Promote the recycling of discarded electronic equipment products. In addition to the establishment of a compulsory recycling list and associated management measures for waste electronic products and packages, there is a need to further focus on research and application of recycling technologies. Specifically, advancements are required in the disassembly, transportation, recycling, and reuse of retired electronic equipment, which can be effectively carried out by third-party professional organizations. The formulation of standards and technical specifications, process environment management, intelligent and refined disassembly, etc., shall be implemented more strictly. It shall be ensured that the standard requirements for component recycling are met.

Carry out carbon footprint accounting for electronic equipment products. In managing carbon assets, the electronic equipment industry must consider several factors, such as resource consumption, efficiency, life cycle, repairability, and recyclability of the electronic equipment through integrating data from the entire production process of the electronic equipment and the evaluation of carbon emissions generated by electronic equipment products throughout their life cycle. Promoting product carbon labels requires quantitative evaluation of carbon emissions generated throughout the material consumption, manufacturing, transportation, usage, and waste disposal involved in products and services. This evaluation shall be displayed on product carbon labels along with information about the carbon-reduction measures taken by companies and their carbon neutrality actions.

2.3.2 Green Digital Services

Present Development

In the context of the digital economy's development, the demand for algorithm model training to support the development and application of AI technologies has grown significantly. However, this growth has also resulted in increased energy consumption and carbon emissions within the digital sector. In recent years, the emergence of the super-large-scale AI model industry has opened up a new development path of AI with "high computing power + big data + large parameter quantity" and further promoted the rapid growth of the model training scale and computing volume required. Therefore, algorithm researchers should carefully select algorithm models and take the energy efficiency of these models as an index to evaluate compute-intensive models [2].

At the same time, China's ultra-large-scale AI model industry is experiencing rapid growth. In July 2021, Baidu launched ERNIE 3.0, a large-scale knowledge-enhanced model with tens of billions of parameters and 4 TB data trained.

Future Path

The carbon emission problems resulting from the growth of the ultra-large-scale AI model industry and the expansion of data training scale pose a significant challenge for the digital sector within the context of green development.

Explore and optimize the operational energy efficiency of large models. In June 2021, the Alibaba DAMO Academy released M6, a Chinese multimodal pre-training model with trillions of parameters. During the R&D process, the model was trained with 480 V100 GPUs. Compared with Google, NVIDIA, and other institutions, it saved computing power resources by over 80% and improved training efficiency by nearly 11 times.

Use "Tiny AI" to reduce power consumption. Through tiny technologies, tiny AI platforms can train and run AI algorithms with lower energy consumption, thus maximizing hardware capacities. Tiny AI improves the efficiency of chips, platforms, and algorithms, enables low-power AI training and application deployment, and allows intelligent operation without interacting with the cloud.

However, it is important to note that solely focusing on the efficiency and optimization of AI algorithms is not the sole solution to address the environmental challenges arising from the digital sector. While improving efficiency is crucial, it is equally essential to aim for a general reduction in overall consumption and promote sufficiency to prevent potential rebound effects.

2.3.3 Green Development of Digital Companies

Present Development

In addition to fostering green development in the electronic equipment manufacturing industry, communication industry, and software and information service industry, it is crucial to prioritize the green development of digital companies as well. Digital companies play a pivotal role as the comprehensive carriers of the digital sector. However, as digital businesses continue to rapidly grow and industrial ecological complexity increases, there has been a concerning upward trend in carbon emissions resulting from these companies' operations and supply chains. The primary sources of carbon emissions generated from digital companies encompass both direct emissions and fugitive emissions from their greenhouse gas emission sources (Scope 1), as well as indirect emissions arising from purchased energy such as electricity and heat (Scope 2). Additionally, emissions from office spaces, employee commuting, business travel, and other related factors within the supply chain contribute to the overall carbon footprint (Scope 3). Recent data indicate that Scopes 2 and 3 account for the majority of greenhouse gas emissions produced by leading digital companies, as elaborated below.

In the fiscal year 2022, Alibaba's total greenhouse gas emissions amounted to 13.25 million tons. These emissions can be categorized into different scopes based on their sources and impacts. The direct greenhouse gas emissions by Alibaba entities (Scope 1) accounted for 927,000 tons. This includes emissions from the combustion of stationary sources (such as natural gas), fugitive emissions (such as refrigerant escape), and emissions from mobile sources (such as company-owned vehicles in retail business). The greenhouse gas emissions from electricity and heat purchased by Alibaba (Scope 2) were approximately 4.445 million tons. This electricity and heat primarily served the operational needs of cloud computing data centres, retail stores, warehouses, and offices. Indirectly, the greenhouse gas emissions generated from the upstream and downstream within the value chain (Scope 3) that can be accurately measured were around 7.877 million tons, which were mainly generated from the fuel consumption in the transportation and distribution services purchased by e-commerce operators, purchased electricity for company-leased data centres, packaging materials and consumables, and staff travel.

In 2021, the direct carbon emissions from the greenhouse gas emission sources owned or controlled by Tencent (Scope 1) amounted to 19,000 tons, representing approximately 0.4% of the total emissions. These emissions primarily originated from company-owned vehicles, diesel generators, and refrigerants. The indirect emissions of greenhouse gases resulting from the generation of power and other energy purchased for Tencent's operations (Scope 2) reached 2.349 million tons, constituting about 45.9% of the total emissions. These include emissions arising from electricity consumption in company-owned or jointly constructed data centres and office buildings. Furthermore, all other indirect carbon emissions generated in Tencent's supply chain (Scope 3) accounted for 2.743 million tons, representing around 53.7% of the total emissions. These emissions primarily stem from capital goods, company-leased assets, and staff travel.

Baidu's carbon dioxide emission equivalent amounted to 490,841.4 tons in 2020 and 1,791,607.8 tons in 2021. The direct emissions (Scope 1) accounted for 16,407 tons, including combustion emissions from stationary sources and fugitive emissions involving boiler equipment, restaurant equipment, refrigerators, and other similar sources. The indirect emissions (Scope 2) were 601,740.2 tons and primarily resulted from the consumption of purchased electricity, steam, and heat. Other indirect emissions (Scope 3) totalled 1,173,460.6 tons, including emissions from electricity consumption of employees commuting and emissions associated with company-leased data centres [3].

Future Path

Emphasize the social responsibility of green development and encourage sustainability reporting. Since 2020, prominent Internet platforms companies such as Tencent, Ant Group, Alibaba, Baidu, Huawei, and others have initiated carbon neutrality action plans. These companies have demonstrated a commitment to transparency by disclosing their green development initiatives, pathways, and targets. They have utilized various means, such as releasing roadmaps, environmental information disclosure (ESG) reports, and corporate social responsibility (CSR) reports to communicate their progress. To align their strategies with sustainability goals, it is essential for companies to enhance the integration of digital and green development strategies. Large-scale digital companies are taking the lead in moving rapidly toward sustainable development, setting an inspiring example for others to emulate.

In April 2021, Ant Group announced its *Carbon Neutrality Roadmap,* outlining three primary objectives. Firstly, the company committed to enhancing energy efficiency and reducing emissions in its office buildings and transportation by leveraging renewable energy sources. Secondly, Ant Group pledged to collaborate with data service providers that prioritize low energy consumption or renewable energy usage, to achieve a 30% renewable energy power consumption rate by 2025. Furthermore, the company aimed to encourage its suppliers to establish and implement their carbon neutrality targets. Lastly, in cases where emissions reduction measures are insufficient, Ant Group committed to offsetting the remaining emissions by either establishing carbon sink forests or directly purchasing carbon credit products.

In December 2021, Alibaba released the *Alibaba Carbon Neutrality Action Report* and put forward three major targets: to achieve carbon neutrality in its operations no later than 2030; to halve the carbon emission intensity of upstream and downstream value chains no later than 2030 and take the lead in achieving carbon neutrality in cloud computing and promoting the green development of this sector; to achieve 1.5 billion tons of carbon reduction through the platform economy ecosystem in 15 years. Energy transformation, scientific and technological innovation, and fostering a stakeholder economy will become the core factors in achieving carbon neutrality in the future.

In February 2022, Tencent released the *Tencent Carbon Neutrality Target and Roadmap Report*, unveiling its carbon neutrality action plan (Appendix 3). Tencent pledged to achieve comprehensive carbon neutrality not only within its operations but also across its supply chain. By 2030, the company aims to ensure that 100% of its electricity is sourced from renewable and green power. The realization roadmap encompasses three dimensions: energy-conservation and efficiency improvement, increased utilization of renewable energy, and carbon offsetting initiatives.

Shift to renewable energy to promote the private sector's action to reduce carbon emissions. Carbon emissions reductions at the internal operation level are crucial for the green development of digital companies. Companies can drive innovation and implement energy conservation and emissions reductions measures across management and technology domains to reduce their carbon footprints. On the one hand, companies can achieve low-carbon emissions at the source by introducing green energy, such as photovoltaic power generation, wind-solar hybrid systems, and solar heating, within their self-built office parks. On the other hand, companies can promote green office practices by leveraging digital technologies, such as intelligent sensors and the Internet of Things, to realize the intellectualization of office parks and buildings. By building a visual and intelligent energy-management and control centre, companies can refine energy consumption management and achieve green office practices.

Promote the green transformation of the supply chain and industrial ecosystem. Digital companies play a significant role in shaping the industrial supply chain and have the potential to collaborate with upstream and downstream partners to drive low-carbon transformations of relevant products and supply chains. Through these efforts, digital companies can contribute to creating a more environmentally friendly and sustainable business ecosystem. To achieve this, digital companies can make efforts in two aspects: On the one hand, they can leverage their digital technology expertise to build and optimize a comprehensive green supply chain, including solutions such as green product packaging, environmentally conscious logistics and transportation, energy-efficient warehousing, and paperless e-procurement to achieve energy conservation and emission reductions. On the other hand, digital companies can promote the circular economy by optimizing the production, sale, use, recovery, and reuse of ICT equipment and terminal facilities, and maximizing the reuse of components (such as obsolete server racks) to reduce the carbon emissions of ICT equipment waste.

Guide the formation of green habits on the consumer side. Digital companies pose a unique opportunity to utilize their influence on users and promote environmental awareness through their widely accessible digital services. By guiding consumers toward adopting a green lifestyle and fostering a positive social environment that supports carbon neutrality, these companies can make a significant contribution to advancing environmental protection. The ongoing efforts of digital companies have already led to significant carbon-reduction effects and have successfully cultivated green consumption habits in various domains, such as travel, home, office, study, catering, shopping, and electronic product recycling. Their contributions toward embracing green value have been noteworthy.

Support consumers to reduce their data and hardware consumption. Additional measures can be taken to promote environmentally friendly behaviour, such as minimizing the use of digital resources. Optimizing video content to match the display size of end devices is one way to ensure that the default resolution aligns with the device's capabilities. Additionally, disabling the automatic playback of video content on web pages by default can help reduce unnecessary data consumption. To prevent false incentives that may encourage excessive data usage, flat rates for large data volumes should be avoided.

2.4 Mode of Digital Technology-Based Greening

Use digital services to promote green development. Companies operating in the software and information technology services industry can provide a wide array of solutions, including software development, data analysis, and intelligent transformation, to assist businesses in different sectors in managing and optimizing their environmental impacts on production and business activities. Large-scale digital service companies and leading firms across industries can also provide information technology consulting services or integrated solutions to small and medium-sized companies in traditional industries, such as manufacturing, to facilitate the coordinated digital and green transformation and bring technology into the green development of industrial companies.

Apply key clean technologies to promote green development. Digital technology holds significant potential to enhance efficiency and optimize resource allocation within traditional industries. It drives innovation in processes and services while facilitating intelligent and green development, particularly in areas such as energy optimization and decision control. The emergence of new network technologies, such as 5G, enables every production unit to be perceivable, communicable, connectable, and computable. AI-powered analysis technologies transform decision-making processes and empower intelligent decision-making capabilities. Cloud computing and big data technologies have enabled new applications across various fields. Leveraging the massive data collected by sensors, these technologies enable the effective utilization of data resources and unlock their full value. According to IDC research, the continued adoption of cloud computing between 2021 and 2024 has the potential to reduce carbon dioxide emissions by over 1 billion tons. To achieve carbon neutrality by 2060, the contribution of AI-related technologies to carbon reduction will increase year by year, with a minimum target of 70% by 2060.

Apply software and information technology services in industries. Software and information technology services play a crucial role in promoting green economic and social development by providing networked, digitized, and intelligent technology solutions. These solutions empower industrial transformation and upgrading, optimize structures, modernize government supervision and social services, accelerate the adoption of green production methods and lifestyles, and contribute to the overall reduction of social energy consumption. The next-generation information technology can effectively be applied in energy-intensive sectors to drive the clean transformation of energy structures. It enhances energy efficiency, reduces environmental impact, and promotes resource recycling, thereby directly contributing to the reduction of carbon emissions and the achievement of carbon-peaking and carbon neutrality goals. This is particularly important in critical carbon-emitting sectors such as energy and electricity, industry, construction, and transportation. By strengthening the deep integration of digital technology applications in these sectors, energy and resource consumption can be reduced, leading to simultaneous improvements in production and carbon efficiency. Clarity is needed in both the key components and the path ahead. For example, the World Economic Forum has developed a framework with the key transformational stages of the software-defined vehicle, including the main strategic decisions required and the related impact that can be achieved in each stage.

3 Digital Technology and Sustainable Development of Cities

3.1 Introduction

Digitalization contributes to several urban "pain points" and offers sustainable solutions. Ongoing urbanization drives planetary change and shapes economic and social development. Capturing their dynamics is demanding, as cities are complex systems. Digital technology offers the opportunity to manage complexity, adding a new layer of governance. While unchecked digitalization is an acceleration force of unsustainable outcomes, digitalization can also become an enabler to accelerate the systemic transformation in cities for carbon neutrality and inclusiveness through circularity and sufficiency.

Cities are places characterized by exchange and communication among a vast multitude of individuals. A central theory of the emergence of the first cities suggests that cities were firstly trade hubs, in which further functions were increasingly built. This penultimate social role is realized by a physical infrastructure offering shelter for many in limited space and mobility, enabling people to meet and deliver goods. Cities are attractive precisely for the opportunities of exchange that often translate into economic, social, and cultural opportunities and diverse ways of living. However, this glorious role of cities is compromised by its success, at least when implemented the way it currently is. Large cities suffer from congestion, and the physical infrastructure and transport systems support unsustainable outcomes, from air pollution to climate change, traffic congestion, housing shortages, inadequate water supply, and energy shortages. Generalized congestion results in a "pain point" of cities, where the environment degrades and livelihoods are compromised (increasing the suffering of many struggling to afford high living costs), limiting the potential for economic development.

The role of cities matters from a global sustainability perspective [4]. Direct and indirect energy use and consumption are the main drivers of anthropogenic climate change [5]. Urbanization contributes to land loss and, thus, marginally also to food insecurity and biodiversity loss, notably in the Pearl River Delta [6]. Inside cities, air pollution, noise, and unsafe traffic conditions compromise well-being and place quality.

Digitalization can support sustainability transitions, and it can do so in three urban sectors: transport, buildings, and urban planning. First, digitalization can replace resource-demanding activities with services, e.g., when trips replace video meetings. Second, digitalization can support the optimization of activities, e.g., through smart homes with optimized HVAC or pooling trips in shared taxis. Third, it can intensify the use of resources, thus creating a lower resource demand per service, e.g., supporting the high use of flexible spaces and avoiding resource use for both construction and operations. Fourth, digitalization can communicate sustainability implications, initiating more sustainable decisions.

With all that potential, a note of caution: digitalization and associated efficiency gains lead to increased consumption levels, so-called rebound effects, that may even overcompensate for the efficiency effects. Information and communications technologies also have a significant and rapidly growing environmental footprint. This observation indicates the importance of proper evaluation of systemic effects, a focus on options that promise high sustainability wins, and the avoidance of digitalization associated with a high environmental burden.

The population gathers with the economic and social development of cities, which also gives birth to innovation and thus to new ways of service provisioning that ameliorate undesired outcomes and offer pathways to sustainability. Digitalization and artificial intelligence are general-purpose technologies that can find applications in all domains of life. However, digital and green technologies have also created conditions for solving urban development problems. Central to this is the potential to focus digital applications on increasing quality but avoiding quantity—additional material turnover. All those examples require that the positive effects are not outnumbered by indirect effects, such as more traffic when there is more space on the roads, more lighting when light is cheaper, or more leisure travel when saving time on commuting.

3.2 State of Smart Cities in China

China's 12th Five-Year Plan from 2011 covers digital/smart cities. In 2011, China's 12th Five-Year Plan promoted infrastructure construction, urban informatization, and refined management to improve urban digital capabilities. China has been advancing smart city construction since 2012. The Ministry of Housing and Urban–Rural Development of China selected 290 cities, districts, and towns as smart city construction pilots between 2013 and 2015 and has established an indicator system to guide the smooth implementation of the pilot areas. To ensure the healthy and sustainable development of smart cities, the General Office of the China State Council has proposed guidelines and objectives, which include promoting the development and upgrading of the information industry and infrastructure (2013), increasing the development and sharing of information resources and scientifically formulating the top-level design of the smart city construction (2014), and building a geospatial framework for digital cities (2015).

China's 13th Five-Year Plan from 2016 further promotes digital/smart cities. During China's 13th Five-Year Plan, China focused on providing intelligent technologies related to smart cities and industries to enhance productivity and resource efficiency. The main goal was to integrate intelligent technologies with urban planning and industry and to achieve highly-visible results in Digital China by 2020, such as forming distinctive smart cities. Core technologies include cloud computing, big data, 5G, and Internet of Things. The innovative projects were developed in the areas of smart agriculture, smart energy systems, smart oceans, smart transport, and geospatial information systems.

China's 14th Five-Year Plan from 2021 continues and amplifies a focus on digital/smart cities. The 14th Five-Year Plan continues building digital China and smart cities, realizing the dual path of technology development and online applications. Technology development includes cloud computing, big data, the Internet of Things, Industrial Internet, blockchain, artificial intelligence, virtual reality, and augmented reality. To promote the digitization of management services, China has proposed the construction of a digital village to digitize China's rural areas, break down data barriers, and improve collaborative processing capabilities and operational efficiency. At the same time, China continues to build smart cities and realize the intelligent transformation of lifestyle services, including property, elderly care, childcare, tourism, health care, logistics, and transportation, to build a convenient environment for digital living and innovate service models and products.

"Guiding opinions on promoting the healthy development of smart cities"—the basic Chinese smart city construction principles. The basic principles are fourfold. First, with human needs at the core, smart cities innovate urban management models and public services to provide a wide-coverage, multi-level, personalized, and high-quality public service system to urban residents. Second, smart cities apply advanced technologies based on regional characteristics, such as geographic location, history and culture, resources and industries, and socio-economic settings. Smart cities can be tested first in regions or districts with better comprehensive conditions and gradually roll out nationally while avoiding large-scale and redundant construction. Third, smart cities focus on stimulating market vitality and encouraging social capital to participate in investment, construction, and operation to leverage the sustainable development path of smart cities. Fourth, smart cities should follow secured information management principles to avoid data leaks and protect privacy.

China's New Urbanization Policy is achieving remarkable progress. The urbanization rate of permanent residency is 65% and continues to grow, requiring further reform of the household registration system. Urban clusters serve as centres of spatial patterns of urbanization, linked by the world-leading high-speed railway network that is rapidly expanding. Cities are increasingly designed around high-tech manufacturing and information technologies and improved public services and infrastructures while renovating old communities and shanty towns. Development is similarly advanced in rural areas, with a focus on water, electricity, roads, and the Internet, and the income gap between rural and urban settings is getting smaller.

These developments encounter several challenges. Environment, transportation, urban planning and construction, and land use are areas of increasing pain points, exacerbated by unbalanced development and utilization of resources [7]. The biggest challenge of new urbanization is promoting balanced development and reducing the pressure on big cities. The measures currently proposed by China are rural revitalization and green development of urban agglomerations through technological efficiency [8].

Shenzhen and Hangzhou are two examples of digital cities in China. Shenzhen's smart city development model is called "1 + 4," supported by new infrastructure construction in four areas: public services, urban governance, digital economy, and security prevention and control. As a national pilot for high-quality infrastructure development, Shenzhen has built 70,000 5G base stations and achieved city-wide coverage of 5G networks to support smart city development. Shenzhen Tong, an APP, provides the basis for residents' daily life and government affairs. The so-called One City Map envisages the integrated supervision of the city. Shenzhen has established a 5G industry cluster to promote the development of the digital economy and implemented Shenzhen Special Economic Zone Data Ordinance to protect data security.

The structure of Hangzhou smart city is core system + sub-platform (district, county, and city) + data system + application scenario. The core system connects all sub-platforms and databases to realize multi-platform data and business interoperability. Application scenarios are built around the core system and sub-platforms to achieve intelligence in health care, transportation, housing management, industrial development, and other areas. In large cities in the central and western regions, such as Chengdu, innovative measures have also been taken to promote digitalization and green development (Appendix 4).

Chinese smart cities offer opportunities for efficient system management of GHG emission reduction. Carbon neutrality goals also require the replacement of fossil fuel hardware. Digital and AI-based applications can support residents, city managers, and industry in reducing their GHG emissions. Digital monitoring of water quality, air pollution, traffic flows, and online services for citizens forms the backbone for digital applications, many of which are AI-based. For residents, AI-based improvements in logistics, if supported by real incentives, such as CO_2 pricing in transport, can reduce packaging and return rates and improve consumer satisfaction. AI-based services, especially in shared mobility, offer substantial improvements in mobility, especially if accompanied by higher occupancy per vehicle [9, 10]. For city managers, AI enables real-time adjustment of operational plans and fast implementation of city-level climate goals and emergency measures, thus reducing waste of resources and CO_2 emissions caused by disasters, operation errors, and poor scheduling methods. For industries, AI enables real-time supervision of all production processes, including the selection of raw materials for products, production, transportation, and sales, to maximize the use of production resources and reduce production costs and CO_2 emissions. AI calculates and predicts market demand in real time, adjusts production volumes with production and sales plans in real time, and prevents CO_2 emissions from overproduction [11]. However, efficiency gains are often overcompensated by accelerated consumption and GHG emissions, requiring systematic measures in addition to AI-based efficiency to achieve climate goals [12].

The following session focuses on three sectors: transport, buildings, and urban planning. Section 3 focuses on options for how digital solutions offer the potential for sustainability and a healthy economy, with three policy recommendations.

3.3 Mobility: Avoid-Shift-Improve Approaches

Transitions in urban mobility toward sustainability can be categorized into "avoid, shift, and improve" [13]. Digitalization can support the transition in all three areas and can, in addition, provide opportunities for new business models. The three categories are characterized as follows:

- Avoid. This action area focuses on avoiding travel altogether by providing high-level services with reduced need for travel and transport. Key options include home office policies, digital work environments, and urban planning that provide high accessibility but low levels of inefficient motorized transport.
- Shift. It focuses on switching from polluting to fewer polluting modes of transport, which often translates into a change from private motorized modes to environmentally friendly modes, such as public transport or active modes (cycling and walking). Digital is central in new modes that can provide the flexibility and accessibility of individual transport while keeping some of the efficiency of public transit—shared pooled mobility and multimodal routing.
- Improve. It focuses on technological efficiency, both in terms of the efficiency of vehicles and traffic flows. Especially for the latter, digital options can provide crucial support.

Three options were detailed where digitalization can leverage substantial potential for sustainability transitions. The first is shared or pooled mobility. There are many smart mobility options, all of which are present in Chinese cities. However, it is beneficial to understand their differential environmental footprints. Some smart mobility options belong to those with the worst environmental footprint, especially ride-hailing. Ride-hailing includes rides without passengers (so-called deadheading), which makes ride-hailing highly emission-intensive and adds to congestion. In contrast, bike and e-scooter sharing demonstrates environmental benefits.[19] Also, ride-pooling services, such as Didi's carpooling option, offer substantial benefits. The International Transport Forum demonstrated in a modeling study of Helsinki that replacing commuter trips with shared pooled mobility (prohibiting city entry by private vehicles) enables a 37% reduction in congestion and a 33% reduction in GHG emissions—even before electrifying car fleets. Shared pooled mobility not only relies on digital smartphone services but also optimal routing algorithms and thus necessitates advanced digitalization business models.

The second is online work and education. The home office prevents travel trips and saves time for employees. Accordingly, a study of Chinese employees during the pandemic lockdown finds that the quality of jobs is experienced higher in working from home [14]. Online education, forced upon China and most other countries globally by the pandemic, leads to a drastically reduced carbon footprint from students [15]. However, to reach the full educational potential of students and meet their social needs, online education must be complemented with in-person seminars.

The third is efficient management of traffic flow. Hangzhou's Urban Brain, a collection and combination of available urban data processes with artificial intelligence, had urban traffic flow as its first showcase. Even though Hangzhou's population grew by about 1.5 million between 2010 and 2020, about 30%, average travel speeds increased by 15.3%, and the peak hour congestion rate was reduced by 9.2% [16]. As a result, Hangzhou's traffic congestion ranking dropped from 2nd to 57th place in China [17]. Public transport can also be operated more efficiently. With a combination of Radio Frequency Identification (RFID), GPS, and video analytic technologies, light-rail speed and position monitoring systems can improve railway operational safety as demonstrated in Hong Kong [18].

In general, all three measures—shared transport, teleworking, and traffic flow management—allow for better traffic flows. Unless combined with complementary measures to limit private car traffic, such as an inner-city toll, there is a high risk the benefits are eaten by the latent demand that exists in cities with congestion.

3.4 Buildings: Avoid-Shift-Improve Approaches

Sustainability in the building sector involves consideration of indoor air quality, avoiding the high material and GHG emission footprints of construction, and energy-efficient use, as part of overall urban infrastructure. Relying inter alia on a study on the more efficient use of buildings [19], this includes:

- Avoid. Reduce the need for space and use space more efficiently (shared desks and space). Use modular design to adapt space for new purposes. Renovate old buildings instead of replacing them with new construction. Apply digital sufficiency to ICT use in households.
- Shift. Provide heating service via private heat pumps, rooftop solar heating, geothermal energy, and large-scale heat pumps for district heating. Decentralize energy supply with solar PV on rooftops, and install more double-glazed windows, better insulation, and LED lighting, etc.
- Improve. Make use of more efficient energy. Create virtual load management with smart meters in coal and gas power plants to reduce air pollution and GHG emissions. Optimize energy use through intelligent buildings. Retrofit existing buildings, optimize material-saving construction for passive housing, e.g., based on 3-D printing; make use of the digital economy and home office to optimize existing building use.

Three digital case studies were selected, including the big data-based management of building stock, digital sufficiency, and smart metering and feedback. First, big data-based approaches can optimize the use and modification of existing building stock for sustainability. A key application is data sourcing of thermal properties of existing building stock to allocate optimal strategies for retrofitting, saving precious energy and gas. A data-based strategy could include scanning building stock with thermal cameras in winter. Merging with climate data and climate projection could then enable the computation of prioritization in building stock retrofitting. Another strategy involves reallocating floor space, inter alia as demanded by the digital economy. A reduction of bank offices and retail is accompanied by an increase in demand for logistical warehousing. Big data tools could help optimally redistribute and reduce the need for floor space.

Second, digital sufficiency can support the sustainable lifestyles of households and building users [20]. Digital sufficiency includes hardware, software, user, and economic sufficiency. A central motivation is that increasing energy use and GHG emissions are more and more attributed to digitalization and the abundant purchase and use of digital devices.

Third, smart metering and feedback can create virtual power plants, creating a sustainable and economically advantageous instead of new constructions of gas and coal power plants. Smart metering allows for dynamic pricing and load reduction in times of high electricity prices. It can save household budgets and release utilities from the burden of constructing economically non-viable power plants, releasing the rarely used capacity. For high-consuming households, smart metering, and information feedback can reduce energy bills by 16%, even if current implementations usually do not reach that high numbers. Still, smart metering and information feedback, combined with price signals, can reduce at least a gigaton of GHG emissions if persistently applied over several years globally [21].

3.5 *Spatial Planning: Use AI for Sustainable Urban Design*

Spatial planning refers to the advanced long-term management of cities. Spatial layout predetermines accessibility, housing affordability, job market opportunities, air pollution, noise nuisance, and GHG emissions from transport and building use. Spatial planning is a forward-looking strategy to realize vast sustainability gains. The main dimensions for sustainability evaluation include the energy use and GHG emissions committed by urban form, local air quality and ambience, water absorption capacity, and quality of place, including the enjoyment of street life for seniors and children.

Digitalization and machine learning can support urban planning for sustainability [10]. Computing energy use and GHG emissions as a function of spatial layout and street networks enables planners to estimate the relevance of urban form in driving air pollution and GHG emissions [22]. High-resolution maps can also visualize the contribution of the built environment in amplifying heat waves and associated heat stress, thus enabling targeted urban resilience plans and urban greening strategies.

Spatial and transport planning are also areas where technical solutions intertwine with gender equity and inclusion. Spatial planning and smart mobility solutions must be designed in an inclusive manner to avoid repeating past mistakes that sideline women, children, and the elderly.

4 Digital Technology and Climate Change Adaptation

The effects of climate change are increasingly transforming human living environments. While mitigation action remains a priority, government efforts must also focus on helping people adapt to today's climate impacts. Emerging digital technologies, which provide more efficient, rapid, and reliable risk monitoring and forecasting, enable better decision making based on quantitative, actionable indicators [23] and can play an essential role.

This section aims to provide policy recommendations for deploying digital tools to promote climate change adaptation in China. It first identifies the major climate change adaptation challenges in China and then discusses the high-performing digital solutions developed and deployed in countries outside of China. Afterward, the section analyzes these solutions in the Chinese context and formulates policy recommendations for advancing similar approaches. In keeping with the overall theme of the report, the section focuses its discussion on adaptation in urban areas, with a broader discussion of digital adaptation solutions and policies to be developed in future research.

4.1 Major Climate Change Adaptation Challenges in China

An analysis of climate change risk and adaptation challenges must begin with a typology against the situation in China that can be assessed. Following the risk framing in IPCC AR6 [24], the analysis in Sect. 4 defines risks as the potential for adverse consequences for ecological or human systems and impacts as the consequences of realized risks on natural and human systems. Climate change risks and impacts result from dynamic interactions between climate-related hazards and the level of exposure and vulnerability of the ecological or human system.

The main climate-related hazards in China are increase in climate-related extreme weather events on the one hand and increased vulnerability because of more exposure, etc. on the other. Significant changes in China's annual mean temperature and precipitation characteristics over the past century have been observed: The average surface air temperature increased by 0.98 °C between 1901 and 2010, with a particular acceleration in warming since 1980; Mean temperature changes are projected to be higher than the global average, at an estimated of 5 °C increase by the end of the century. Although the annual average precipitation did not exhibit any significant changes, notable shifts were observed in the spatial and temporal distribution of precipitation. Changes in climate conditions also alter the severity of weather and climate extremes in China.

The main exposures include two categories, namely natural ecosystems and socio-economic systems. In China, the vulnerability of natural ecosystems lies mainly in the fact that socio-economic development creates strains on water resources, terrestrial ecosystems, and coastal ecosystems.

Although China has taken some adaptation actions to reduce the above-mentioned hazards, exposure, or vulnerability, there are still many residual risks that have not been coped with and adapted. According to the Climate Risk Country Profile of China, the major impacts (realized risks) include water scarcity; soil degradation, dryland expansion, and desertification; coastal erosion rates and saltwater intrusion; productivity loss of the agriculture sector and food security; disruption of the operation of urban infrastructures and broader economies; increase in heat-related mortality; increase in water- and vector-borne diseases; and malnutrition. These are the main risks that China needs to adapt to.

China has always attached equal importance to mitigation and adaptation and implemented a national strategy to promote adaptation to climate change. The Chinese National Climate Change Adaptation Strategy 2035 defines three adaptation priorities to mitigate climate change risks in China: (1) strengthening climate change monitoring, early warning, and risk management; (2) improving the climate change adaptation ability of natural ecosystems, including water resources, terrestrial ecosystems, and coastal zones; and (3) improving the climate change adaptation ability of socio-economic systems, including agriculture, urban systems, and human health.

4.2 *Potential Digital Solutions for Climate Change Adaptation*

Given the major climate change adaptation challenges and priorities in China discussed above, this section explores international cases and best practices, keeping the Chinese needs in mind. Extensive literature reviews focus on the potential role that digital technologies and tools can play in promoting climate change adaptation. Some of these tools can assist in long-term planning for adapting to climate change, while others can aid in forecasting and reacting to immediate climate change hazards. As shown in Sect. 3, spatial planning can help cities build resilience to climate change challenges. This sub-section explores two other main areas: climate change monitoring and early warning; urban water management in response to flooding—the most frequent immediate climate hazards China is facing.

4.2.1 AI Techniques for High-Accuracy Precipitation Forecasting

Climate change monitoring and hazard early warning rely on precise climate and weather extremes forecasting. China has made substantial efforts and progress in developing and implementing extreme weather forecasting and risk management systems. Current weather forecasting approaches rely on physics-based techniques. Such methods are sensitive to approximations of the physical laws on which they are based and are constrained by their high computational requirements [25]. Therefore, the weather forecasting systems could further improve efficiency and accuracy.

A promising new approach for precise precipitation forecasting is currently in development, utilizing AI techniques [26]. Unlike traditional methods that rely on explicit physical laws, AI weather forecast models learn to predict weather patterns directly from observed data, resulting in faster and more accurate predictions [25].

These approaches, which include Google's MetNet2, Pangu-weather, and others, could potentially improve the scope and resolution of the predicted forecasts [27]. Although it is still in the laboratory stage, MetNet-2 has shown its potential for accurate and efficient precipitation forecasting [26].

These models have only recently been published in journals such as *Nature Communications* and have not yet been implemented. Several obstacles hinder the implementation of such approaches. The first is that the technologies need to be further developed to expand the scope of weather phenomena considered and extend the forecasting horizon to days and weeks, while there are no mechanisms and incentives for multiple participation of universities and corporate research institutes in China. The second is that they are more data-intensive, which creates issues in data collection, integration, and governance [25].

4.2.2 Operational Digital Twins in the Urban Water Sector

Changes in the spatio-temporal distribution of precipitation driven by climate change create challenges for water management in Chinese urban areas. The substantial uncertainty associated with precipitation events results in an increased frequency of high flows within sewer systems, causing a significant number of sewer overflows. These overflows discharge untreated wastewater into the environment, posing serious threats [3] and exacerbating water pollution, water scarcity risk, and water-borne disease. The Swedish cities of Gothenburg and Helsingborg have implemented a digital twin approach to water resource management, which has great potential to reduce the vulnerability of the Chinese urban water sector to climate change.

The digital twin approach implements a decision support system with online flow prediction and suggestions for control strategies. Utilizing the digital twin approach provides greater confidence in decision making by enabling quick visualization of the effects of changes in control strategy [3]. Sewer overflows are common in Chinese cities after heavy rainfall [28]. The untreated wastewater, possibly carrying pollutants from the ground, enters the urban environment, causing serious water pollution, freshwater shortages, and an increase in water-borne diseases. Therefore, scaling up digital twin solutions will generate great impacts on urban water management in China.

While drawing on experiences from the Gothenburg and Helsingborg cases, implementing the digital twin presents several challenges [3]. One major challenge is integrating data reflecting water flow fluctuations from multiple sources, including weather forecasting, water-intensive entities, sewage treatment sector. The insufficiency of high-quality data collection and management, lack of inter-institutional data transfer coordination mechanisms, or absence of privacy protection laws could prevent a smooth integration of multiple source data. Another challenge is establishing trust and confidence in the digital twin system among control room operators.

4.3 Governance Innovation for Leveraging Digital Adaptation

As demonstrated in the case studies earlier, digitalization plays a significant role in promoting climate change adaptation. Meanwhile, there are many challenges in developing and deploying such digital solutions, which can be grouped into four categories: (1) the development of digital technologies, especially research that promotes techniques from labs to practical application; (2) data collection, management, and integration; (3) coordination issues between multiple institutions, and (4) legal issues on the permissibility of sharing data or developing novel approaches. AI shall play an important role in this regard by helping identify these long-term risks and support decisions on how to build resilience.

China has initiated a series of research and development programs to foster the creation of digital solutions for climate change adaptation. For instance, the Ministry of Science and Technology is backing studies on extreme weather and forecasting technologies and research on digital adaptation decision support systems. However, funding in this area is relatively limited, resulting in such research typically being conducted on a small scale at universities and other public research institutions, and the findings often remain in the laboratory stage. The "last-mile problem" of practical applications remains unresolved. Moreover, there is a deficiency in comprehensive incentive policies that encourage participation from various sectors to develop digital solutions for climate adaptation. Private sector research institutions also possess robust capabilities in this aspect. Given that adaptation efforts are long-term in nature and often have positive externalities, appropriate incentive policies are required to address market failures and promote private sector involvement.

Deploying digital technologies for climate change adaptation requires support from multiple data sources, including data from individuals, businesses, governments, and non-profit organizations. These multi-sources of data are not easily put together, from either a technical, legal, or management perspective. First, the different statistical standards and the varying quality of data from different departments result in data islands and poor information sharing in China. Second, privacy and data security cannot be guaranteed, which undermines the willingness of individuals and businesses to share data. Third, there is a lack of standardized sharing methods and policies. The Chinese government has also made many attempts in this regard.

Deploying digital technology for climate change adaptation requires coordinated efforts from multiple institutions across sectors. For example, agriculture, urban planning, emergency management, and meteorological departments need to exchange information on digital solutions for climate change adaptation. China is actively establishing cross-sectoral coordination mechanisms for climate change adaptation, promoting the coordinated advancement of policies and technologies.

Legal issues related to liability and accountability may arise when deploying novel digital approaches for climate change adaptation. Determining who is responsible for potential uncertainties and harm caused by digital adaptation technology is essential to address these concerns. Since deploying digital solutions are still in the initial stages, relevant government policy support is needed.

5 The Gender Perspective

The last 50 years of record-breaking economic growth that benefited large parts of Chinese society have not yet been able to sufficiently improve women's status relative to their male counterparts. In most countries that move from an industrial to a service economy, the equality of women follows a U-shaped path, from higher-level of equality in pre-industrial society to lower equality in industrial society to, again, higher equality in the service economy. In China, where services already account for more than half of GDP, this pattern does not hold, and the equality of women follows an L-shaped path [29]. Women's labour force participation rate declined from 73% in 1990 to 61% in 2021, compared to that of men which declined from 86% in 1990 to 78% in 2021. Thus, even though China has made huge strides in economic growth and general welfare to its population, some tendencies point to growing inequalities between men and women.

Artificial intelligence (AI) and digital infrastructures more generally are not only technologies. "At a fundamental level, AI is technical and social practices, institutions and infrastructures, politics, and culture. Computational reason and embodied work are deeply interlinked: AI systems reflect and produce social relations and understandings of the world" [30]. Today, women make up only 22% of AI workers globally [31]. This picture is also evident in science, technology, engineering, and mathematics (STEM) fields in general, where women are in the vast minority. Women are underrepresented in the STEM field throughout the world, be it in higher education, tertiary education, or the workforce [32]. They make up roughly 29% of scientific research positions globally, and 23% in East Asia and the Pacific [33]. Tech companies in Silicon Valley estimate that only 1% of their applicants for technical jobs are women [34]. UNESCO reported that women make up 31% of the research and development positions globally and that women are 25% less likely than men to operate basic digital technology.

In China, less than 30% of the students in STEM fields are female, less than 20% of acclaimed tech positions are held by women, and the Chinese Academy of Science counts only 6% of women in STEM faculties [35]. The UNDP Resident Representative estimates that 80% of high-paying STEM jobs in the green economy in China will be held by men [36].

Although disruptive technologies bring new jobs and potential if handled correctly, they also harbour many risks for social inequalities: The people who benefit most from the effects of Industry 4.0 are those who are able to take advantage of digital solutions and services. Those who may be left behind are those who are digitally disconnected, those who cannot afford digital solutions or who do not have or are not granted access, those who lack basic knowledge and educational options, and those who don't trust their digital skills. The largest marginalized group affected by these barriers and the resulting digital divide is women [37]. Disadvantages can increase if discrimination is compounded by other factors, which are referred to as intersectionality. Consequently, women living in remote and rural areas, Indigenous women, women with disabilities, or young or elderly women face additional and probably different forms of barriers and marginalization [38].

To exclude women from design, policies, and social interventions based on digital technologies has harmful consequences for effective climate change mitigation and prevention. First, women are more vulnerable to climate change. Climate change is expected to increase gender-based violence, child marriage, and school dropout rates of girls, and force women to travel longer distances to fulfill daily tasks [39]. If digital solutions do not take women's needs into account properly, they risk exacerbating existing inequalities and limiting the effectiveness of those solutions.

Second, women are effective agents, arguably more effective than men, of climate change adaptation and prevention measures. However, women are underrepresented in "green jobs" and the STEM fields, which are crucial to successfully applying digital tools for sustainable development goals. Furthermore, women in decision-making positions shift the companies they lead decisively in the direction of sustainable growth [40]. Women have also been found to be more efficient drivers of ESG programs and to improve sustainable business practices. If women remain rare in the STEM fields, which are crucial to developing digital solutions, chances are that these industries will not tap into the full pool of talents to find efficient solutions for a green transition of the digital sector that works for all [41].

6 Policy Recommendations

The Chinese economy has undergone a rapid process of digitalization over the past decades, significantly contributing to China's economic development and extending its influence beyond the national border. Against this background, this study aims to explore the link between digitalization and sustainability with a clear mission: digitalization as a dominating transformative force of the twenty-first century must be put at the service of sustainable development. The study assesses the relationship between digitalization and sustainability in the Chinese context, reviews global best practices, and provides policy recommendations concerning four major areas of action: greening the digital sector, building smart sustainable cities, leveraging digitalization for climate change adaptation, and mainstreaming gender in digital transformation.

6.1 Greening the Digital Sector

- Promote the green transformation of digital infrastructure and facilities to reduce carbon emissions.

First, promote energy conservation, energy efficiency, and low carbonization of digital infrastructures. Specifically, policy actions should start from two aspects: to keep the construction of new digital infrastructures under strict, sustainability-oriented supervision and promote the energy conservation and low-carbon transformation of existing new infrastructure. However, there is a concern regarding the potential for greenwashing, where terms such as “clean energy” or “green infrastructure” can become subjective and open to interpretation, which is sometimes observed in the practices of digital companies. To ensure genuine progress toward sustainability in digital infrastructures, transparency is an essential prerequisite.

Second, optimize the energy mix of existing data centres through green power trading, improve the efficiency of wind, solar, and other new energy, and reduce the carbon emissions of these data centres. Building on the self-commitments of Alibaba, Ant Group and Baidu, a binding 100% renewable energy target for all data centres and digital companies above a certain size is suggested. Information regarding energy consumption, as well as indicators like PUE and other metrics discussed in earlier sections, should be made publicly accessible. Digital companies should demonstrate transparency by openly showcasing how they have achieved these metrics.

Third, manage the carbon emission data of electronic equipment in their full life cycle. Manufacturing companies should be encouraged to establish a “carbon asset management system” to monitor the carbon emission data of different links and products, standardize carbon data management and accurately identify the actual situation by accounting. Upstream companies engaged in raw materials, components, and parts should be encouraged to transmit carbon emission data to the downstream, through the green supply chain management, to provide high-quality databases for downstream companies to track their product carbon footprint and carbon emission, using indicators such as Carbon Productivity and Carbon Production to evaluate green production.

- Actively promote the low-carbon development of digital companies and encourage green social responsibility.

First, advocate digital companies to shoulder social responsibility, establish corresponding incentive mechanisms, encourage large digital companies to track, measure, and publish their energy consumption and carbon emissions, and pay attention to the positive and negative impacts of digital technology on the environment.

Second, build an ESG evaluation system with Chinese characteristics, promote the establishment of a scientific and standardized index system with Chinese characteristics on the basis of realizing the unification of international and domestic standards, and effectively guide digital companies to pay more attention to environmental and social responsibilities during investment. Company ESG reporting standards should be set up to strengthen the evaluation of government regulators on the company's ESG compliance.

Third, explore the application of digital tools such as "green passport for digital products" to record the carbon emission footprint based on the supply chain, and design incentive measures and price mechanisms under the guidance of government procurement to promote consumers to purchase more eco-friendly products and services.

Fourth, use digital platform services to leverage green consumption, and cultivate public green living habits, for example, supporting consumers to reduce their data and hardware consumption. Algorithms may prefer consumption of high-quality and long-livability products but demote the consumption of high-turnover, high material-intensity products. Improve the policy environment conducive to shared economic services modes, such as shared travel, online office, and transactions of idle items.

- Improve the support service capabilities of digital technologies to empower industries to manage and reduce carbon emissions.

Policy-makers should adapt to the development needs of various industries for accurate carbon emission management and green production with lower carbon emission levels, accelerate the R&D and promotion of basic, cutting-edge, and applicable software and information technology services, and enhance the support of digital carbon management and reduction innovation capability.

In terms of carbon reduction via digital technology deployment, it is essential to keep optimizing algorithms because more efficient machine learning algorithms (such as sparse models) can improve the efficiency and reduce the energy consumption of model operation, as well as use special chips for machine learning for large model training. Compared with general-purpose processors, special chips and systems optimized for machine learning training, if used, can significantly improve the performance and energy efficiency of large models. In addition, cloud computing should be used instead of internal computing resources to reduce energy consumption.

To empower other industries through the application of digital technology to reduce carbon, it is important to enhance the research on sensing technologies such as real-time data exchange and information processing and integration, improve the comprehensive performance of carbon sensors, drive breakthroughs in big data technologies such as data mining, machine learning, and modeling analysis, and promote carbon-efficient data management, analysis, and prediction. Furthermore, specific policy actions are needed to support the construction of promotion centres targeting systematically integrated digital and green transformations.

In areas such as smart energy and intelligent manufacturing, digital companies and relevant agencies can carry out demonstration applications of digital carbon management and reduction, provide digital services or solutions for green upgrading and transformation for industrial companies, and advance the digitalization, intelligence, networking, and low-carbon development of traditional industries to better promote digital carbon reduction in various fields.

6.2 Building Smart Sustainable Cities

Smart sustainable solutions can create win–win-win opportunities at the city level: they can promote a modern economic sector; they make cities more livable and advance the quality of life; and they help reduce the local and global environmental burden. However, realizing these goals requires substantial policy and technological efforts and expertise in both digital infrastructure provision and sustainability goals. A central risk is misunderstanding "smart" as being equal to "sustainable." Digital solutions are general-purpose technologies making applications, technologies, and business activities more efficient, but environmental consequences also depend on rebound effects and area of application. It is hence paramount to double down on efforts in sectors and applications that promise the highest sustainability gains, particularly in reducing GHG emissions, improving air quality, encouraging sustainable lifestyle, and achieving a high quality of life. The following policies can help achieve win–win-win situations in Chinese cities.

- Develop a "Smart Sustainable Cities Audit" System

The role of audits is to evaluate the performance outcomes, to ensure, in this case, that opportunities created by digitalization are fully utilized to support sustainable development within mobility, buildings, and spatial planning. In cases when weaknesses are found, suggestions in terms of such things as incentives, regulations, and further research should be presented. The "Smart Sustainable Cities Audit" could be generalized to a "Sustainable Digitalization Audit," covering all parts of society.

- Advance Urban Digital Governance in the Context of Climate Solutions

Cities like Hangzhou and Shenzhen are world-leading in digital infrastructure and offer the potential for highly adaptive urban governance. It is suggested that the cities' urban climate management makes the best use of these infrastructures to promote climate neutrality. This requires coordinated plans, evaluation of the most effective options in reducing GHG emissions, and the development of digital applications in mobility, building, and urban planning areas consistent with climate goals. Digital innovations have the potential to improve sustainable infrastructures as well as urban governance practices.

- Promote Special Economic Sustainability Zones Experimenting With Highly Efficient Shared Pooled Mobility

Provide carpooling and similar services that transport several people in one vehicle (minibus) with tax credits and preferred access to urban street networks. At the same time, limit and regulate on-street parking and introduce inner-city tolls for private vehicles. The expected result is a much more efficient door-to-door mobility system without congestion, better air quality, and smoother economic functioning of cities. With the support of digital platforms, cities such as Hangzhou would be in an excellent position to advance this policy and become a global pioneer in this regard.

- Advance Sustainable Urban Planning With Artificial Intelligence

Map building stock and its energy-relevant features and street networks and compute sustainability metrics of urban form. Apply results to sustainable urban planning to make urban use more accessible, less energy- and resource-intensive, healthier, and resilient to high precipitation and extreme heat and cold events. Build a scalable service for all Chinese municipalities.

6.3 Leveraging Digitalization for Climate Change Adaptation

- Strengthen Support for Scientific and Technological Innovation and Formulate More Resilient and Targeted Digital Adaptation Solutions

Comprehensively evaluate the scientific, technological, economic, and social research results related to China's adaptation to climate change, starting with digital adaptation in agriculture, cities, and ecosystem management, and systematically strengthen the application of digital technologies in agricultural production, urban disaster prevention and mitigation, and ecological protection.

- Improve Multi-Source Data Integration and Management and Multi-Institutional Coordination and Cooperation Mechanism

Developing data standards, data-sharing mechanisms, and coordination policies in the field of climate change adaptation, including especially (1) creating data-sharing centres (data marketplace) that bridge the data gap between different digital projects; (2) developing appropriate data regulation policies to remove disincentives for data sharing; and (3) establishing information exchange and cooperation mechanisms between agriculture, urban planning, emergency management, and meteorological departments.

- Accelerate Legislation to Encourage the Digital Climate Change Adaptation Solutions

Accelerate the construction of supporting laws for developing and deploying digital climate change adaptation solutions, including especially (a) establishing laws to protect trial-and-error activities, which is essential for the deployment of new technologies; and (b) improving laws on data sharing and privacy protection to remove disincentives for data sharing and multistakeholder participation.

6.4 Mainstreaming Gender in Digitalization

As a typical general-purpose technology, AI both reflects and produces social relations and understandings of the world. Therefore, certain questions must be asked when promoting AI and digital infrastructure for general use in society, such as which resources are necessary, how they are extracted, and who stands to benefit and who does not. To address these questions, the report makes the following recommendations: first, make gender analyses part of policy planning and monitoring; second, empower women and diversity experts in planning, implementation, and monitoring processes; third, make gender-disaggregated data part of every data collection activity and promote data-based research on gender; fourth, include bias detection in risk assessments for new digital technologies and solutions and implement measures to mitigate biases. To deal with gender-related biases related to AI, the government can consider the following policy actions:

- Raise Awareness. Gain knowledge of biases, study them systematically, and adhere to international guidelines. The implementation of policies to address gender bias in data and AI systems, e.g., regulations that encourage companies to conduct gender impact assessments or the promotion of algorithmic transparency, can accelerate the process.
- Detect biases in datasets. Risk analyses help to identify weaknesses and threats in datasets. Biases can be made visible with the help of software, by using explainable AI techniques contrary to black-box models and to help understand and identify any gender bias, through evaluating bias with disaggregated analyses and by drawing on scientific studies.
- Reduce existing biases. Existing weaknesses in datasets can be fixed by improving the process of data collection or combining data sources together with debiasing, e.g., balancing data and removing stereotypical features.
- Communicate. It is important to be transparent about methods and data models. The disclosure of results and the project design helps to enable others to understand and improve the model.

References

1. Berkley Haas Center for Equity, Gender and Leadership. (2020). *Mitigating bias in artificial intelligence. An equity fluent leadership playbook*. UCB_Playbook_R10_V2_spreads2.pdf. berkeley.edu. Access March 13, 2023.

2. Patterson, D. et al. (2021). *Carbon emissions and large neural network training*. arXiv preprint arXiv:2104.10350
3. Digital Water. (2023). *Operational digital twins in the urban water sector*. International Water Association, March 16, 2023. https://iwa-network.org/publications/operational-digital-twins-in-the-urban-water-sector-case-studies/.
4. Skea, J., et al. (2022). *Climate change 2022. Mitigation of climate change. Summary for policymaker*. IPCC.
5. Lwasa, S., et al. (2022). Urban systems and other settlements. In *IPCC, 2022: Climate Change 2022: Mitigation of Climate Change. Contribution of Working Group III to the Sixth Assessment Report of the Intergovernmental Panel on Climate Change* (p. 158). Cambridge University Press.
6. Bren d'Amour, C., et al. (2016). Future urban land expansion and implications for global croplands. *Proceedings of the National Academy of Sciences*, 201606036.
7. Jiang, H., et al. (2021). An assessment of urbanization sustainability in China between 1990 and 2015 using land use efficiency indicators. *Npj Urban Sustain, 1*, 1–13.
8. Liu, T., & Li, Y. (2021). Green development of China's Pan-Pearl River Delta mega-urban agglomeration. *Science and Reports, 11*, 15717.
9. Guo, Y., Xin, F., & Li, X. (2020). The market impacts of sharing economy entrants: Evidence from USA and China. *Electronic Commerce Research, 20*, 629–649.
10. Milojevic-Dupont, N., & Creutzig, F. (2021). Machine learning for geographically differentiated climate change mitigation in urban areas. *Sustainable Cities and Society, 64*, 102526.
11. Johnson, M., et al. (2021). Impact of big data and artificial intelligence on industry: Developing a workforce roadmap for a data driven economy. *Global Journal of Flexible Systems Management, 22*, 197–217.
12. Creutzig, F., et al. (2022). Digitalization and the anthropocene. *Annual Review of Environment and Resources, 47*, 479–509. (Volume publication date October 2022). First published as a Review in Advance on September 2, 2022. https://doi.org/10.1146/annurev-environ-120920-100056.
13. Bongardt, D., et al. (2013). *Low-carbon land transport: Policy handbook*. Routledge.
14. Qu, J., & Yan, J. (2023). Working from home vs working from office in terms of job performance during the COVID-19 pandemic crisis: Evidence from China. *Asia Pacific Journal of Human Resources, 61*, 196–231.
15. Yin, Z., et al. (2022). The impact of online education on carbon emissions in the context of the COVID-19 pandemic—Taking Chinese universities as examples. *Applied Energy, 314*, 118875.
16. Caprotti, F., & Liu, D. (2022). Platform urbanism and the Chinese smart city: The co-production and territorialisation of Hangzhou City Brain. *GeoJournal, 87*, 1559–1573.
17. Yan, X., & Li, T. (2022). Construction and application of urban digital infrastructure—Practice of "Urban Brain" in facing COVID-19 in Hangzhou, China. *Engineering, Construction and Architectural Management*.
18. MTR Lab. (2023). *Innovative solutions*. https://www.mtrlab.com.hk/tech-solution?lang=en
19. Höjer, M., & Mjörnell, K. (2018). Measures and steps for more efficient use of buildings. *Sustainability, 10*, 1949.
20. Santarius, T., et al. (2022). Digital sufficiency: Conceptual considerations for ICTs on a finite planet. *Annals of Telecommunications*. https://doi.org/10.1007/s12243-022-00914-x
21. Khanna, T. M., et al. (2021). A multi-country meta-analysis on the role of behavioural change in reducing energy consumption and CO_2 emissions in residential buildings. *Nature Energy, 6*, 925–932.
22. Silva, M. C., et al. (2017). A spatially-explicit methodological framework based on neural networks to assess the effect of urban form on energy demand. *Applied Energy, 202*, 386–398.
23. Argyroudis, S. A., Mitoulis, S. A., & Chatzi, E. (2022). Digital technologies can enhance climate resilience of critical infrastructure. *Climate Risk Management, 35*, 100387. https://doi.org/10.1016/j.crm.2021.100387

24. Masson-Delmotte, V., Zhai, P., & Pirani, A. (2021). Climate change 2021: The physical science basis. In *Contribution of Working Group I to the Sixth Assessment Report of the Intergovernmental Panel on Climate Change* (p. 2).
25. Kalchbrenner, N., & Espeholt, L. (2021). *MetNet-2: Deep learning for 12-hour precipitation forecasting*, March 15, 2023. https://ai.googleblog.com/2021/11/metnet-2-deep-learning-for-12-hour.html
26. Espeholt, L., Agrawal, S., & Sønderby, C. (2022). Deep learning for twelve hour precipitation forecasts. *Nature Communications, 13*(1), 5145. https://doi.org/10.1038/s41467-022-32483-x
27. Bi, K., Xie, L., & Zhang, H. (2022). *Pangu-weather: A 3D high-resolution model for fast and accurate global weather forecast*. arXiv, 2022, March 15, 2023. http://arxiv.org/abs/2211.02556. https://doi.org/10.48550/arXiv.2211.02556
28. Talamini, G., Shao, D., & Su, X. (2016). Combined sewer overflow in Shenzhen, China: The case study of Dasha River. In *Sustainable development and planning 2016* (pp. 785–796). Penang, Malaysia, March 16, 2023. http://library.witpress.com/viewpaper.asp?pcode=SDP16-066-1. https://doi.org/10.2495/SDP160661
29. Brussevich, M., et al. (2021). *China's rebalancing and gender inequality*. IMF. imf.org
30. Crawford, K. (2021). *Atlas of AI. Power, politics, and the planetary costs of artificial intelligence* (p. 8). Yale University Press.
31. UN Women. (2023). *In focus: International women's day*. UN Women—Headquarters. Access March 13, 2023.
32. Ortiz-Martinez, G., et al. (2023). Analysis of the retention of women in higher education STEM programs. In *Humanities and social sciences communications* (p. 10). Catalyst. (2022). *Women in science, technology, engineering, and mathematics (STEM) (Quick Take)*. Catalyst. Access March 17, 2023. SWE. (2022). *Global STEM workforce*. Society of Women Engineers. swe.org. Access March 8, 2023.
33. UNESCO. (2020a). *STEM education for girls and women: Breaking barriers and exploring gender inequality in Asia*. UNESCO Digital Library. Access March 8, 2023.
34. UNESCO. (2020b). *Gender biases in AI and emerging technologies*. UNESCO. Access March 7, 2023.
35. UNDP. (2021). Designing a fairer future: why women in tech are key to a more equal world. United Nations Development Programme. undp.org. Access March 17, 2023.
36. UNDP. (2022). *Women in science can change the world*. United Nations Development Programme. undp.org. Access March 17, 2023.
37. GSMA. (2022). *The mobile gender gap report 2022*. Access March 6, 2023.
38. UNESCO. (2020c). *Artificial intelligence and gender equality: Key findings of UNESCO's global dialogue*. UNESCO Digital Library. Access March 6, 2023.
39. UNFCCC. (2022). Dimensions and examples of the gender-differentiated impacts of climate change, the role of women as agents of change and opportunities for women. In *Synthesis report by the secretariat*. UNFCCC. Access July 16, 2023.
40. Deininger, F., & Gren, A. (2022). *Green jobs for women can combat the climate crisis and boost equality*. worldbank.org. Access March 14, 2023.
41. Giner-Reichl, I. (2023). *This is how women can power the green transition*. World Economic Forum. weforum.org. Access March 14, 2023.
42. Black, J. S., & van Esch, P. (2020). AI-enabled recruiting: What is it and how should a manager use it? *Business Horizons, 63*, 215–226.
43. Bradley, A. J. (2020). *Brace yourself for an explosion of virtual assistants*. https://blogs.gartner.com/anthony_bradley/2020/08/10/brace-yourself-for-an-explosion-of-virtual-assistants/. Access March 7, 2023.
44. Collett, C., Neff, G., & Gouvea Gomes L. (2022). *The effects of AI on the working lives of women* (p. 46). UNESCO.
45. Encarnacion, J., Emandi, R., & Seck, P. (2022). *It will take 22 years to close SDG gender data gaps*. https://data.unwomen.org/features/it-will-take-22-years-close-sdg-gender-data-gaps. Access January 30, 2023.

46. Falchetta, G., & Noussan, M. (2021). Electric vehicle charging network in Europe: An accessibility and deployment trends analysis. *Transportation Research Part D: Transport and Environment, 94*, 102813.
47. Höjer, M., & Wangel, J. (2015). Smart sustainable cities: definition and challenges. In *ICT innovations for sustainability* (pp. 333–349). Springer.
48. Lambrecht, A., & Tucker, C. (2019). Algorithmic bias? An empirical study of apparent gender based discrimination in the display of stem career ads. *Management Science, 65*(7).
49. Madgavkar, A., et al. (2019). *The future of women at work: Transitions in the age of automation.* McKinsey. Access September 2, 2023.
50. McKinsey. (2021). *A conversation on artificial intelligence and gender bias*. McKinsey. Access March 17, 2023.
51. Meteorological Technology Development Leading Program 2020–2035. (2019). http://www.gov.cn/zhengce/zhengceku/2019-11/04/5456909/files/c7c2e1cfb36d4817ba6f6d8fd293f7f7.pdf
52. OECD. (2022). *The effects of AI on the working lives of women* (p. 21). OECD.AI. Access March 7, 2023.
53. Patenall, H. (2022). *Who are knowledge workers and how does AI technology speed up their work?* Aiimi. Access March 14, 2023.
54. Ramboll. (2021). *Gender and (smart) mobility*. Green Paper.
55. Rockström, J., et al. (2009). A safe operating space for humanity. *Nature, 461*(7263), 472–475. https://doi.org/10.1038/461472a
56. Rockström, J., & Figueres, C. (2018). Exponential climate action roadmap. In *Global climate action summit*. February 02, 2023. https://exponentialroadmap.org/exponential-roadmap/
57. Santarremigia, F., Molero, G., & Malviya, A. (2022). *A methodological approach to reveal fair and actionable knowledge from data to support women's inclusion in transport systems: The Diamond approach.*
58. Silva, M. C., et al. (2018). A scenario-based approach for assessing the energy performance of urban development pathways. *Sustainable Cities and Society*. https://doi.org/10.1016/j.scs.2018.01.028
59. Strubell, E., Ganesh, A., & McCallum, A. (2019). *Energy and policy considerations for deep learning in NLP*. arXiv preprint arXiv:1906.02243
60. UN Women. (2018). *Making women and girls visible: Gender data gaps and why they matter*. Issue-brief-Making-women-and-girls-visible-en.pdf. unwomen.org. Access March 14, 2023.
61. UN Women. (2022). *Creating safe and empowering public spaces with women and girls*.
62. UNESCO, OECD, IDB. (2022). *The effects of AI on the working lives of women*. iadb.org. Access March 13, 2023.
63. Vargas-Solar, G. (2022). Intersectional study of the gender gap in STEM through the identification of missing datasets about women: A multisided problem. *Applied Sciences 12*(12). mdpi.com
64. Wagner, F., et al. (2022). Using explainable machine learning to understand how urban form shapes sustainable mobility. *Transportation Research Part D: Transport and Environment, 111*, 103442.
65. Wall, S., & Schellmann, H. (2021). *LinkedIn's job-matching AI was biased. The company's solution? More AI*. MIT Technology Review. Access March 7, 2023.
66. World Economic Forum. (2022). *Open source data science: How to reduce bias in AI*. Access March 13, 2023.
67. World Economic Forum. (2023). Global new mobility coalition. *Global New Mobility Coalition*. https://initiatives.weforum.org/global-new-mobility-coalition/home
68. Zhang, D., et al. (2021). *The AI index 2021 annual report* (p. 138). 2021-AI-Index-Report_Master.pdf. stanford.edu. Access March 7, 2023.

Chapter 8
Trade and Sustainable Supply Chains

1 Context

A suite of Chinese government aspirations and external developments will impact the degree to which China can achieve trade and supply chain security in a manner that is also long-term sustainable.

(1) Chinese Government Aspirations

Three Chinese government aspirations or ambitions are particularly relevant for the nexus of trade, supply chains, and sustainability.

Aspiration 1: Achieve carbon peaking before 2030 and carbon neutrality before 2060.

In 2020, the Chinese government announced an aspiration to peak national carbon emissions before 2030 and achieve carbon neutrality before 2060 [1]. In this context, "carbon" means carbon dioxide equivalent (CO_2e), which encompasses all greenhouse gases (GHGs). These climate aspirations cover all major sectors of the economy, including energy generation, transportation, food systems, land use, and more. The Ministry of Ecology and Environment also is studying the development of green trade policies [2].

Aspiration 2: Achieve food and energy security and resiliency.

Food security and resiliency is critical to China's national security. In 2020, President Xi Jinping stated that "food security is an important foundation for national security" [3] and that every actor should take responsibility for securing food supplies [4]. Moreover, China's new "dual circulation strategy" encourages China to reduce its international supply chain uncertainties [5]. Together, these are calls for an appropriate combination of self-sufficiency and open trade.

Likewise, energy security is critical to China's national security. China's 14th Five-Year Plan for Modern Energy System prioritized the establishment of a modern energy system that addresses both sustainability and supply security issues. It requires

Green Empowerment and High Quality Development,
https://doi.org/10.1007/978-981-96-4218-2_8

the promotion of green and low-carbon energy transformation through strengthening clean energy industry, implementing renewable energy substitution actions, promoting the construction of a new power system, and gradually increasing the proportion of new sources of energy [6].

Aspiration 3: Transform from manufacturing country to a manufacturing power.

China's industrial green upgrading refers to the transformation and upgrading of traditional industries into green and sustainable industries. This development blueprint for high-tech industries is a key part of China's industrial green upgrading. The Chinese government has implemented various policies and initiatives as well as increasing investment for R&D and tax incentives for these sectors.

(2) External Developments

A number of recent external developments will impact the nexus of trade, supply chains, and sustainability, as well. For instance, new international trade policies are putting a price on carbon and tackling deforestation. New international agreements are requiring signatories (including China) to improve the sustainability (including avoided deforestation) of its economic (including trade) activities. Accelerating corporate trends are signaling that companies and financial institutions will increasingly focus on reducing greenhouse gas emissions from their global supply chains. Finally, consumer trends indicate that domestic Chinese consumers are increasingly desiring products (whether domestically sourced or imported) to be sustainably produced.

International Trade Policies

- EU CBAM

The Carbon Border Adjustment Mechanism (CBAM) is a policy proposed by the European Union (EU) that would impose a carbon border tax on certain imports from countries that do not have equivalent carbon pricing policies. The aim of the tax is to level the playing field between domestic producers and foreign producers who are not subject to the same carbon costs. The CBAM would cover a range of goods, including steel, cement, aluminum, fertilizers, and electricity.

CBAM means challenges as well as opportunities for China. As the world's largest exporter, the CBAM could result in higher costs for Chinese exporters and consequently lower competitiveness in the European market, if China does not take significant action to reduce its carbon emissions. Meanwhile, the policy also presents opportunities for China to accelerate its transition to a low-carbon economy and promote its green industries. The policy could encourage China to invest more in renewable energy and other green technologies, which would create new opportunities for the green transformation of supply chains.

- EU Deforestation Regulation

The EU Regulation on Deforestation-free products (EUDR) prohibits the placing on the EU market covered commodities (soy, cattle, palm oil, coffee, cocoa, rubber and wood) and certain derivatives (such as chocolate and beef) that were produced on land deforested or degraded after December 31, 2020. Covered commodities and products also must be produced in accordance with local laws. The European Parliament voted to pass the regulation in April 2023, and a final vote in the European Council is expected shortly. Companies are required to comply with the regulation starting 18 months after the regulation enters into force. Small and medium-sized enterprises have 24 months after entry into force before they need to comply.

While the regulation applies to companies not countries, those countries that produce or process covered commodities that are placed on the EU market will likely receive increasing requests for clarification of local laws governing production and processing, as well as requests for information about where products processed in China were sourced from in order to comply with the geolocation requirement.

- Regional Trade Deals

The regional trade deals, such as the Regional Comprehensive Economic Partnership (RCEP) and the Comprehensive and Progressive Agreement for Trans-Pacific Partnership (CPTPP) have significant impact on China.

The RCEP agreement involves 15 countries in the Asia–Pacific region, accounting for around 30% of the world's population and GDP. As a founding member, China will benefit from reduced tariffs, increased market access, and improved trade and investment flows within the region. However, China also will face intensified competition from other member countries, particularly in areas where it has traditionally been strong, such as manufacturing.

The CPTPP agreement, which includes 11 countries across the Asia–Pacific, is a high-standard free trade agreement that covers a wide range of sectors, including goods, services, and intellectual property. China is not a member of the CPTPP but has expressed interest in joining. Joining the agreement would require China to meet the high standards of the agreement.

Furthermore, the RCEP and CPTPP agreements are seen as a response to China's growing economic influence and its Belt and Road Initiative. These agreements provide other countries in the region with an alternative option for economic integration, which could reduce China's regional dominance.

While the RCEP and CPTPP agreements present opportunities for China to expand its economic influence and deepen its economic ties with other countries in the region, they also pose challenges in terms of the need for economic reforms and the need for conforming with related environment standards and requirements.

- The Glasgow Leaders' Declaration on Forests and Land Use

The Glasgow Leaders Declaration on Forests and Land Use represents a commitment by country leaders to collectively halt and reverse forest loss and land degradation by 2030 while promoting sustainable development and an inclusive rural transformation. The Declaration was launched at the November 2021 United Nations Climate Change Conference (COP 26) and was signed by 141 countries, including China. As part of this commitment, countries agreed to facilitate trade and development policies—internationally and domestically—that promote sustainable development, sustainable commodity production, and sustainable commodity consumption that work to the mutual benefit of signatories and that do not drive deforestation and land degradation.

- Kunming–Montreal Global Biodiversity Framework (GBF)

The GBF is a new global agreement designed to safeguard the world's biodiversity [7]. It was signed by 196 countries [8], including China, in December of 2022 at the Conference of Parties to the Convention on Biological Diversity. GBF goals are to halt biodiversity loss, sustainably use biodiversity, equitably share biodiversity's benefits, and adequately implement financial resources and technology. Underpinning all of this, GBF calls for sustainable production systems and legal trade practices that are aligned with biodiversity conservation goals in order to prevent further degradation and biodiversity loss in exporter countries. Given the amount of imports into China from countries with high levels of biodiversity, the GBF will have implications for Chinese international trade.

Corporate Trends

- SBTi

The Science Based Targets Initiative (SBTi) is a partnership between CDP, the United Nations Global Compact, WRI, and WWF. This initiative provides a framework for companies to set science-based targets to reduce their greenhouse gas emissions, and also provides support and technical assistance to those companies. As of May 2023, nearly 2500 companies representing one-third of the world's market capitalization have set science-based targets through SBTi [9].

Science-based targets validated by the SBTi must include all Scope 1 emissions (from assets owned by the company) and Scope 2 emissions (from purchased electricity) as defined by the Greenhouse Gas Protocol. If a company's Scope 3 emissions (those from its supply chains) comprise 40% or more of its total emissions, then those emissions also must be included in targets [10]. This attention to scope 3 emissions means that these companies will increasingly focus on reducing greenhouse gas emissions from their global supply chains, which will include the agricultural products, manufactured goods, and other raw material they purchase.

- ESG Investing

ESG investing, sometimes referred to as "sustainable investing" or "responsible investing", describes investing that incorporates environmental, social, and governance-related issues [11]. ESG investing balances traditional investing with ESG considerations to form a longer-term perspective, considering both a company's financial performance as well as its societal impacts.

The PRI (Principles for Responsible Investment) is a UN-supported organization that represents the largest coalition of organizations committed to ESG investing and also supports signatories in their efforts to engage in responsible investing. As of year-end 2022, the PRI had 5319 global signatories, representing $121 trillion in assets under management [12]. As of November 2022, more than 30 financial institutions with combined assets under management of more than US $8.7 trillion have already signed up to the commitment to use best efforts to eliminate agricultural commodity-driven deforestation (for palm oil, soy, beef, pulp and paper) from their investment and lending portfolios by 2025 and publish credible progress—a critical step toward reversing deforestation globally and aligning the sector with a Paris Agreement-compliant 1.5 °C pathway [13].

- Consumer Trends

"As outlined in a recent SPS on greening soft commodity value chains [14], tomorrow's markets" are increasingly demanding more sustainable food consumption and production. This trend is not relegated solely to European and North American consumers; domestic Chinese consumers are moving in this direction, too. According to the survey, 69% of consumers expressed that they accept green products at higher prices than regular products, 79% of consumers will incorporate their moral values into their daily shopping, and 82% of consumers express willingness to purchase sustainable branded products [15]. Particularly with regard to food, according to a survey conducted in 2021, more than 90% of consumers are willing to pay a premium for low-carbon food, and more than half of consumers are willing to pay a premium of more than 10% [16].

2 Possible Implications for Several Soft Commodities Important to Chinese Trade

The confluence of these Chinese government aspirations and recent external developments will likely have a number of implications for several key soft commodities that are important for Chinese trade. "Soft commodities" are raw materials and their derivatives that are grown or produced by the agriculture and forestry industries. These include plant- and animal-derived material for use as food, fiber, feed, medicines, cosmetics, detergents and fuels [14]. In this study, we focus on three such commodities for which China is a major global importer: soybeans, beef, and palm oil.

(1) Soybeans

Soybeans are vitally important to the Chinese economy. As the largest processor of soybeans in the world, China processes more than 80% of the soybeans it produces and imports into oil and meal for animal feed. Around 15% of China's soy consumption is for direct human food and derivative human food products [17].

China is a major player in the global soybean trade. China is the world's largest soybeans importer [18], accounting for 60% of global soybean trade. These imports met 86% of Chinese consumption needs in 2021 [19]. China's imports have steadily grown since 1996, mainly to meet the need of its domestic livestock industries and are expected to continue growing through 2030.

China imports soybeans mainly from Brazil, the United States, and Argentina, which combined consist of around 95% of China's total soybean imports. In 2020, China brought in 64 million tons (accounting for 62% of China's total soy import) from Brazil, 26 million tons (accounting for 25% of China's total soy import) from the United States and 8 million tons (accounting for 7% of China's total soy import) from Argentina [20]. Soy expansion is a large driver of conversion of forests and grasslands [21]. There are two types of impacts: "direct impacts" when forests and savannas are immediately converted to soy production and "delayed impacts" when forests are first cleared for other lower-economic-value land uses (mostly cattle grazing) and then later those pastures are converted into soybean fields [22]. From 2001 to 2015, soy directly converted 4 million ha of forest and had a delayed impact on another 4 million ha, mainly in the South American countries Brazil and Argentina [23]. In 2019, one third of South American soybean planted area was located in the Cerrado [24], the most biodiverse savanna ecosystem in the world. In 2020, 264,000 ha of soy was harvested from land deforested within the past five years In the Cerrado [18]. Soy-driven deforestation has large greenhouse gas consequences. In 2020, Brazilian soybean-driven deforestation and conversion resulted in 28 million tonnes of CO_2e from native vegetation [17]. Therefore, reducing the deforestation and savanna conversion associated with soy production will be an important component of China's efforts to align its soy sourcing and trade with its goals of carbon neutrality and meeting global agreements on biodiversity conservation and climate.

(2) Beef

China is the world's largest beef importer. From 2010 to 2020, beef imports to China grew 110% to 3.4 million tons per year, accounting for 33% of the world's total exported beef [25]. China's import of beef is projected to continue to grow for the rest of this decade.

From 2016 to 2021, the annual growth rate of beef consumption in China was 7.5% [26]. In 2022, China's beef consumption reached 1.1 million tons, ranking second among nations in global beef consumption [26]. From 2011 to 2021, per capita beef consumption of China grew from 4.53 to 6.95 kg/capita/year, a growth of about 50% [26]. This growth could be attributed to the dietary transition in China driven by growth in GDP per capita.

The largest source of imported beef into China is Brazil, accounting for more than 40% of imports by weight. Next in line are Argentina (15%) and Uruguay (10%). Brazil's exports to China were valued at 7.5 billion in 2022 [27]. Cattle ranching for beef, however, is by far the largest direct driver of deforestation in the Brazilian Amazon [28], and the Brazilian Amazon is the area of the planet experiencing the largest levels of deforestation per year [29]. Tropical deforestation in the Amazon releases significant amounts of greenhouse gas emissions and severely threatens biodiversity.

(3) Palm Oil

Palm oil is a versatile and important commodity to China, with 80% of its domestic consumption used for food and 20% used for industrial purposes [30]. Palm oil is particularly popular in the food industry—accounting for 17% of China's vegetable oil consumption—due to its high saturated fat content which makes it resistant to high cooking temperatures and makes it stable. It is essential for a wide range of food products such as instant noodles, traditional snacks, fast food, ready-made products, industrial bakery, candy, chocolate, and edible oils. Additionally, palm oil is utilized for industrial oleochemicals such as soap, candles, make-up, and lubricants.

Since China does not produce palm oil, imports make up 98% of the country's total consumption [31]. China has become the third largest consumer and the second largest importer of palm oil in the world [32], with its imports accounting for 14% of global palm oil imports in 2020 [31]. China's imports grew rapidly through 2009, followed by a decline by 2016 and a subsequent resurgence in growth. Indonesia and Malaysia supplied 71% and 27%, respectively, of China's imported palm oil in 2019 [33]. And, in fact, 17% of Indonesia's exports in 2021 were to meet Chinese consumption demand [34].

Oil palm cultivation in Southeast Asia is the dominant factor in deforestation and peat conversion in the region, with the associated greenhouse gas emissions and loss of biodiversity habitat in one of the most biodiverse regions of the planet. The carbon and biodiversity consequences associated with clearing tropical forests are significant. For Indonesia, one third (3 million ha) of its primary forest loss in the past 20 years is due to oil palm expansion [35]. Fortunately, Indonesia has reduced deforestation associated with palm oil between 2018 and 2020 to only 18% of the level in 2008–2012, despite continued increase in palm oil production. This proves the possibility to balance the demand for palm oil products with conservation of tropical ecosystems [35].

(4) Impetus for "Deforestation- and Conversion-Free" Soy, Beef, and Palm Oil

EU Regulation on deforestation-free products: The new regulation requires due diligence from Chinese companies to ensure that the products (processed in China) including soy, beef, or palm oil placed on the EU market are not linked to post-2020 deforestation, as well as not violating local laws on production and processing. Failure to meet these standards could result market access restrictions, limiting the ability to sell products to markets within the EU and in other regions.

- Kunming–Montreal Global Biodiversity Framework: The GBF calls for sustainable production and trade aligned with biodiversity conservation and the prevention of degradation and biodiversity loss in exporter countries. Since China was a co-host country of the GBF (which was adopted by nearly all countries in the world), global expectations are that China will fulfill GBF goals in its domestic activities and international trade.
- The Glasgow Leaders' Declaration on Forests and Land Use: The Declaration includes a commitment among signatories to "facilitate trade and development policies, internationally and domestically, that promote sustainable development, and sustainable commodity production and consumption, that work to countries' mutual benefit, and that do not drive deforestation and land degradation". China is a signatory to this Declaration, and the country has already publicly committed

to avoiding trade deals for soy (and other products) that drive deforestation or conversion of other natural ecosystems [36].

- SBTi: The Science Based Targets for Forest, Land, and Agriculture (FLAG) will help Chinese companies set reduction targets for Scope 3 GHG emissions in line with Paris Agreement [37, 38]. Yum China has pledged to decrease its Scope 3 GHG emissions from purchased goods by 66.3% per ton of goods purchased by 2035 relative to 2020 [39, 40].

3 Possible Implications for Chinese Industry Supply Chains

We have already discussed the overall impact of the CBAM in Sect. 1. By imposing additional carbon tariffs on foreign products with higher emission but lower levels of environmental regulations, CBAM resolved the environmental externalities of international trade to some extent, ensured fair competition between domestic and foreign suppliers and facilitated global low-carbon development.

However, given the different levels of development and situations of different countries, there are quite large differences in carbon market prices and measurement standards among countries. The implementation of the CBAM remains a big challenge and needs to overcome several obstacles. Based on this, this section introduces another way of reducing carbon emissions from exports, which is to emphasize the role of the market and raise the price of unclean energy, to reduce China's carbon emissions.

Therefore, we define and pursue the carbon emissions of the value added of export commodities in China, and further study the correlation between China's export carbon emissions and energy price to make further policy implications.

(1) Definition

Firstly, it is important to establish a clear definition of "carbon emissions of the value added of export commodities". This measure calculates the amount of carbon emissions generated during the production of the total value added of an exported commodity. Total value added refers to the difference between the market value of a product or service and the sum value of its inputs.

To illustrate this concept, let us consider the example of an iPhone X. The lens may be made in Japan, the screen in South Korea, the audio processors in the US, the chips in Chinese Taipei, the buttons in Chinese Mainland, and the assembly process in Chinese Mainland. The Chinese value added in the export of an iPhone X is the value of the parts made and assembled in China. If the total value of the iPhone X exported from China is $409, and only $104 is attributed to Chinese value added, then the carbon emissions of value added for this iPhone X would only include the carbon emissions generated during the production of the $104 worth of Chinese value added.

(2) Stylized Facts of China's Export Carbon Emissions and Carbon Intensity

Once this definition is established, we can calculate the carbon emissions of value added for different export industries using the world input–output table and energy consumption formula. For instance, we find that China's carbon emissions of added value in the electric industry's exports, such as the iPhone X, were approximately 4.05 million tons in 2014.

(3) Quantitative Analysis of Factors Influencing Export Carbon Emission

In this section, we apply the econometrics model and study the impact of fuel prices on export emissions.

According to China's energy structure, coal is one of the most important fuel in China. It is the main source of energy, and also the main source of carbon emission. Therefore, we use the coal prices as the main indicator in this regression. Our regression model is shown as follows.

$$\ln(Export\ Carbon\ Emission)_{ijt} = \beta_0 + \beta_1 \times \ln p_{China,t} + \gamma \times X_{ijt} + t + \alpha_i + \varepsilon_{ijt}$$

$$\ln(Export\ Carbon\ Emission)_{ijt} = \beta_0 + \beta_1 \times \ln \mathrm{wind}p_{China,t} + \gamma \times X_{ijt} + t + \alpha_i + \varepsilon_{ijt}$$

In this regression, i means country, j means industry and t means year. $\mathrm{Ln}(Export\ Carbon\ Emission)_{ijt}$ measures the log value of export carbon emission in year t country i and industry j. The independent variable is $\ln p_{China,t}$, the log value of coal price in China in year t. X_{ijt} relates to a series of control variables, including the log value of per capital income, the GDP growth rate, and the different ratio of secondary industries. We put all these control variables into the regression to alleviate the confounding problem which may be caused by missing some key variables. Finally, we put t and α_i into the regression to control the time trend as well as all the invariant country characteristics.

Besides, we also replace the independence variable to $\ln \mathrm{wind}p_{China,t}$, the wind price of year t, to test the clean energy's impact on export carbon emissions.

Our main interest is β_1 among all the estimated coefficients. It represents the price elasticity of export carbon emission and shows that one percent change in coal price will result in β_1 percent change for export carbon emission in total. Regression shows that the β_1 equals to − 0.129, which means that when the coal price goes up 10%, China's export carbon emissions go down 1.29%, while for clean energy, the according coefficient is positive and insignificant. The results show that the price restrictions on unclean energy are more efficient compared with subsidy on clean energy.

This sub-regression gives us strong and intuitive interpretations. Firstly, the carbon emissions of exports in manufacturing industries (basic metal, machinery et al.) have negative correlation with the coal price. It is very intuitive. All the industries are industry which highly rely on coal. When the coal price goes up, the production costs of these industries go up simultaneously, which drives the entrepreneurs in these industries to use cleaner technology and reduce carbon emission. Secondly, the carbon emissions of exports in mining or substitutes (wood or petrol) have positive correlation with the coal price. When the coal price goes up, more entrepreneurs will try to find other alternatives, driving larger demand in these industries. Therefore, the additional demand will drive more production and increase the total carbon emission in this case.

4 Policy Recommendations

In light of Chinese government aspirations, emerging external developments, and implications for commodities critical to Chinese trade, we propose several policy recommendations: (A) integrate sustainability (or "green") criteria within global supply chains, (B) secure a sustainable soy and beef trade agreement with Brazil, (C) secure a sustainable palm oil agreement with Indonesia and Malaysia, (D) leverage the power of both the market and public policies to drive the low-carbon transformation of trade patterns of industries, and (E) develop incentives for green products in the regional trade agreements. The first applies to Chinese trade overall. The second and third focus on soft commodity trade. The fourth and fifth focus on trade in industrial commodities.

(1) Integrate Sustainability Criteria Within Global Supply Chains

China could integrate sustainability or "green" criteria into all its global supply chain arrangements. The 14th Five-Year Plan (FYP) for High Quality Trade Development provides a foundation for this. For instance, the 14th FYP calls for:

- Establishing green and low-carbon trade standards and certification systems;
- Improving green standards, certification, and labeling systems and promoting international cooperation and mutual recognition;
- Promoting the integration of domestic and international green and low-carbon trade rules and mechanisms;
- Exploring the establishment of a carbon footprint tracking system for the whole life cycle of foreign trade products; and
- Conducting green and low-carbon trade cooperation, among other things.

One concrete step for doing this is for China to incorporate collaboration on sustainable trade and supply chains into existing frameworks for regional economic, trade, and environmental collaboration. A good instance of this is the signing of a green value chain partnership among China and ASEAN countries (scheduled for September/October 2023 at the China-ASEAN Environment Collaboration Forum).

(2) Secure a Sustainable Soy and Beef Trade Agreement with Brazil

China could negotiate and sign a trade agreement with Brazil to secure long-term supplies of legal and sustainable soy and beef. To give such a landmark trade agreement the profile it deserves, China and Brazil could jointly announce the agreement at either the G20 Ministerial Meeting on Agriculture to be held in mid-2024 in Brazil (where sustainable agriculture will be a focus topic) or at the 30th Conference of the Parties to the UNFCCC to be held in late-2025 in Belem, Brazil. The trade agreement would be a natural evolution of the historic meeting in Beijing between Chinese President Xi Jinping and Brazilian President Lula in mid-April 2023, which resulted in the Brazil-China Joint Statement on Combatting Climate Change that included:

> We commit to broadening, deepening and diversifying our bilateral cooperation on climate issues, in areas such as, transition to a sustainable and low carbon global economy … We intend to engage collaboratively in support of eliminating global illegal logging and deforestation through effectively enforcing their respective laws on banning illegal imports and exports.

Such a trade agreement is in the national interests of China and of Brazil. It would ensure long-term, stable supplies of soybean and beef (and thus improve Chinese food security) in a manner that is aligned with emerging international trade policies, meets international agreements signed by China (e.g., Glasgow Declaration on Forests and Land, Kunming–Montreal Global Biodiversity Framework), satisfies corporate trends (e.g., SBTi), and meets growing consumer trends (see Sect. 2.2). Such an agreement would be aligned with the ambitions of Chinese agricultural companies, too. For instance, COFCO's Sustainable Soy Sourcing Policy states, "We expect suppliers to collaborate in increasing our soy supply chain traceability, eliminating deforestation throughout our supply chain and transitioning towards soy production free from native vegetation conversion, so as to protect critical ecosystems such as the Amazon, Cerrado and Gran Chaco".

Such a trade agreement is in the national interests of Brazil, as well. It would help the country eliminate illegal conversion of forests and other natural ecosystems (e.g., President Lula has publicly stated that ending illegal deforestation is one of his top priorities) and bring in much needed finance and know-how for boosting supply of sustainably grown soy and beef. As a result, such a trade agreement aligns with Brazil's national sovereignty, national laws, and national ambitions. Moreover, Brazil is already working to meet similar trade arrangements now being put forth by the European Union (i.e., the EU Deforestation Regulation) and the United Kingdom (i.e., the revised Environment Act), so a China–Brazil trade agreement would not place any additional burdens on Brazil.

Building blocks of a China–Brazil sustainable soy and beef trade agreement could include:

- *Standards and certification*—The agreement would define what qualifies as "legally" produced and traded soy and beef and, ultimately, what qualifies as natural ecosystem "conversion-free" soy and beef. The agreement could build on the learnings and infrastructure developed for voluntary efforts to create pragmatic regulatory standards or a public sector certification system. Fortuitously, voluntary definitions, standards, and associated certification systems have already been developed (or are in the process of being developed) with industry input. For example, the Consumer Good Forum's Forest Positive Coalition—led by 21 companies with a market value of US $2 trillion—has developed Soy and

Beef Roadmaps that lay out commitments and actions the group will implement to remove soy-driven and cattle-driven deforestation and ecosystem conversion from their supply chains. The Soft Commodities Forum, led by the World Business Council for Sustainable Development, is a collaboration of six leading agribusinesses that identifies solutions to eliminate soy-driven deforestation and conversion of native vegetation in the Brazilian Cerrado. Members also have established procurement commitments, including COFCO's requirement that suppliers collaborate to eliminate deforestation and transition towards soy production free from native vegetation conversion. Furthermore, the Roundtable on Responsible Soy offers certification for responsibly produced soy.

- *Due diligence and traceability*—The agreement would articulate the means of traceability and due diligence. In this context, due diligence is the process of assessing and reducing the risk that soy or beef imports are linked to illegal or unsustainable practices. Traceability is the ability to follow a product from production/harvest all the way to the distribution stage of the supply chain. An array of tools already are available to support due diligence and traceability. When used in combination, due diligence and traceability can verify a commodity's origin, the chain of custody, and compliance with the trade agreement (e.g., legality, sustainability). Voluntary traceability systems are already in place and being used by numerous companies with operations in Brazil. Brazilian government-led efforts at the state and national level can also provide stepping-stones. For instance, the state of Pará successfully implemented the public–private partnership "Green Protocol of Grains" to eliminate illegal deforestation associated with soy, rice, and maize—which covers 96% of production [41]. Brazil also has successfully implemented a national control system for the origin of forest products (SINAFLOR) to provide a federal oversight system over the forest sector across all states [42].
- *"Restore, produce and protect packages"*—If Brazil is to increase its supply of soy and beef to China over time without converting forests or other natural ecosystems into agricultural land, then Brazil will need to increase yields on existing croplands and grazing lands. In other words, Brazilian farmers and ranchers will need to simultaneously restore productivity to degraded areas, boost production yields, and protect the nature that remains. Numerous scientific studies demonstrate that this is possible in Brazil [43–49], and the country has a track record of productivity improvements. The trade agreement could include provisions for exchange of agricultural know-how, inputs, and financing to support sustainable intensification of Brazilian soy and beef production. Programs that demonstrate the feasibility of doing this already exist.

(3) Secure a Sustainable Palm Oil Agreement with Indonesia and Malaysia

China could negotiate and sign a trade agreement with Indonesia and Malaysia to secure long-term supplies of legal and sustainable palm oil. The trade agreement would build upon recent progress by China with both nations. For instance, in November 2022, Chinese Vice Minister and China International Trade Representative from the Ministry of Commerce (MOFCOM) called for green trade of palm oil at the China-Indonesia Agricultural Trade Promotion Event. In April 2023, the China Chamber of Commerce of Import and Export of Foodstuffs, Native Produce and Animal By-Products (CFNA) signed a Memorandum of Understanding (MOU) with the Malaysian Palm Oil Board regarding increasing the stability and sustainability of palm oil supply chains. The MOU includes a call to jointly explore and implement palm oil traceability systems. The CFNA also signed the MOU with RSPO to co-work on sustainable palm oil in China. A natural evolution of these developments would be a sustainable palm oil trade deal between China and each of these countries. As with the aforesaid China–Brazil trade agreement, optimal timing of a palm oil trade deal announcement would be within the next two years given the focus of G20 2024 and the UNFCCC COP in Brazil.

Such a trade agreement would be in the national interests of Indonesia and Malaysia, too. Both nations seek to halt illegal clearing of forests and draining of peatlands for oil palm plantations. Indonesia even has a moratorium on all conversion of primary forest and peatlands to oil palm. Furthermore, both nations are demonstrating that they can remain palm oil export superpowers while dramatically driving down rates of deforestation. As evidence, in recent years Indonesia has had the greatest reduction in primary forest clearing among nations, while in Malaysia primary forest loss has leveled off [29]. This recent performance indicates that adherence to a sustainable palm oil agreement with China would be feasible. Moreover, as with Brazil, Indonesia and Malaysia will already need to meet similar trade arrangements now being put forth by the European Union and have set up a joint task force with the EU to work towards implementing the requirements.

Building blocks of sustainable palm oil trade agreements with Indonesia and with Malaysia would be similar to those described for soy and beef for Brazil, namely:

- *Standards and certification*—The trade agreement would define what qualifies as "legally" produced and traded palm oil and, ultimately, what qualifies as forest and peat "conversion-free". Fortunately, definitions, standards, and associated certification systems have already been developed. Both Indonesia, via the Indonesian Sustainable Palm Oil (ISPO) system, and Malaysia, via the Malaysian Sustainable Palm Oil (MSPO) system, have mandatory palm oil certification standards. In addition, the Roundtable on Sustainable Palm Oil (with more than 5400 members globally) has a voluntary standard for ensuring no deforestation, no peatland conversion, and fair treatment of farmers that has achieved strong industry uptake. A trade agreement could include measures to support smallholders to achieve certification and thus ensure access to the Chinese market. As an example of such a measure, RSPO is providing support to the Indonesian province of Jambi to enable smallholders to gain ISPO certification as a stepping stone towards the more stringent RSPO certification. The Palm Oil Collaboration Group is generating industry alignment around an independent reporting framework and independent verification of sustainable palm oil. Trade agreements between China and Indonesia and Malaysia could specify standards for legality and sustainability that build on existing systems such as ISPO, MSPO and RSPO.

- *Due diligence and traceability*—The agreement would articulate the means of traceability and due diligence. Systems are already in place, including numerous voluntary systems applied by companies, and government-led traceability systems that are built into regulatory systems. The system is recognized by the EU as compliant with the EU Timber Regulation. The EU waived due diligence requirements for SVLK-licensed timber to facilitate market access. Similarly, a trade agreement on palm oil could recognize operational traceability systems in Indonesia or Malaysia as sufficient for the conduct of due diligence by companies importing palm oil into China.
- *"Produce and protect packages"*—If Indonesia and Malaysia are to increase their supply of palm oil to China over time without converting forests or peatlands into oil palm plantations, then the two countries will need to increase yields on existing plantation area, including on smallholder plots (which tend to have lower yields than industrial-scale plantations).
- *Lower import tariffs*—China could have lower import tariffs on palm oil shipments that demonstrate they are legal and conversion-free (via the certification and traceability provisions described above).

(4) Leverage the Power of Both the Market and Policies to Drive the Low-Carbon Trade Patterns and Industry Supply Chains

- Invisible Hand—Market

Based on our studies in last section, the first policy implication is that the market mechanism is our tool to achieve environmental goals. We find that when the coal price goes up 10%, China's export carbon emissions go down 1.29%. Precisely, when the price of coal goes up, the China's export of carbon emission goes down. This is because as the price of coal goes up, Chinese exporters find coal is relatively more expensive than other kinds of energy, and therefore they use less coal but more other relatively cheaper energy. Since coal is a nonrenewable energy, the more we use, the less we have and price of coal is in an ascending channel. With the help of market mechanism, the booming coal price will lead to a lower export carbon emission in China. To sum up, the raising price of energy from fossil fuels will eliminate industries with heavy pollution and lower the carbon emission in export by market mechanism.

- Visible Hand—Government

Market is an important force to lower carbon emission, while government should not only rely on market but also take actions actively.

First and foremost, government should lower carbon emission in a way that "destruction" comes after "construction". In other words, government should promote the "construction" of green industry in the first step. After the green industry is established and developed, government then started to limit the traditional "dirty" industry and reshape the economy. More precisely, we need a mature green industry as a preparation for the abolish of fossil energy. If we do not fully prepare the green industry but close the dirty industry in first step, there will cause severe economic imbalance, such as forced power rationing to achieve green targets.

Second, we need infrastructure to back up the foundation of the green industry. The development of green industry is important, but the development of supporting infrastructure for the green industry is more important.

Take the survey in Gansu Province as an example. Gansu as an inland province with deserts has abundant sunshine and wind power resources. Local government has only developed a few solar power plants and wind power plants. Limited development of green industry is caused by the limited consumers to buy their green energy. Local energy consumers are limited due to the deserts and culminate. Furthermore, the transmission line is not enough to send those green energy to the east part of China, where power demand exceeds supply. To sum up, infrastructure backing the green industry is vital for the development of green industry. It is important to build a single energy market from Xinjiang to Shanghai, which be an incentive to develop green energy.

Third, it is inevitable that an economic shift from dirty industries to green industries will lead to unemployment. A cut on heavy industries and mining could result in job loss in those industries. Government needs to prepare for the unemployment during the economic transformation from dirty to green.

(5) Develop Incentives for Green Products in the Regional Trade Agreements

We recommend that China cut the import tariff on green products, and further advocates the green tariff cut in the world trade organization and other regional trade agreements, such as RCEP and CPTPP. In addition, we recommend impose high tariffs on products with high energy and environment footprints.

References

1. United Nations. (2020). *Enhance solidarity to fight COVID-19, Chinese president urges, also pledges carbon neutrality by 2060*. https://news.un.org/en/story/2020/09/1073052
2. Yicai. (2023). *MEE: Promote the study and introduction of green finance, trade and industrial development policies*. https://baijiahao.baidu.com/s?id=1758131486461662264&wfr=spider&for=pc
3. Xinhuanet. (2020). Why did this matter become an "eternal topic" in the mind of the General Secretary. https://news.china.com/zw/news/13000776/20200723/38531294.html
4. People's Daily. (2021). Xi: Ensuring grain security for Chinese people's "rice bowl". http://politics.people.com.cn/n1/2021/0923/c1001-32234793.html
5. China Council for International Cooperation on Environment and Development (CCICED). (2021). *Global green value chains: China's opportunities, challenges and paths in the current economic context*. http://en.cciced.net/POLICY/rr/prr/2021/202109/P020210917469069544512.pdf
6. National Energy Administration (NEA). (2022). *14th five-year plan on modern energy system planning*. http://www.nea.gov.cn/1310524241_16479412513081n.pdf
7. Convention on Biological Diversity (CBD). (2022). *Kunming-Montreal global biodiversity framework*. https://www.cbd.int/doc/decisions/cop-15/cop-15-dec-04-en.pdf

8. Vandvik, V. (2023). *Cheat sheet to the Kunming-Montréal global biodiversity framework.* https://www.uib.no/en/cesam/159846/cheat-sheet-kunming-montr%C3%A9al-global-biodiversity-framework
9. Science Based Targets Initiative (SBTi). (2023, May 31). *Companies taking action.* https://sciencebasedtargets.org/companies-taking-action
10. Science Based Targets Initiative (SBTi). (2023). *SBTi criteria and recommendations for near-term targets.* https://sciencebasedtargets.org/resources/files/SBTi-criteria.pdf
11. Organisation for Economic Cooperation and Development (OECD). (2020). *ESG investing: Practices, progress and challenges.* www.oecd.org/finance/ESG-Investing-Practices-Progress-and-Challenges.pdf
12. Principles for Responsible Investment (PRI). (2023). *Signatory update: October to December 2022.* https://www.unpri.org/download?ac=18057
13. United Nations Framework Convention on Climate Change (UNFCCC). (2022). *Frequently asked questions: Financial sector commitment letter on eliminating agricultural commodity-driven deforestation.* https://climatechampions.unfccc.int/wp-content/uploads/2022/11/FAQ_FI-commitment-letter_COP27.pdf
14. China Council for International Cooperation on Environment and Development (CCICED). (2020). *Global green value chains: Greening China's "soft commodity" value chains.* https://cciced.eco/wp-content/uploads/2020/09/SPS-4-2-Global-Green-Value-Chains-1.pdf
15. SynTao. (2022). *In-depth interpretation of China's low-carbon consumption status and development path.* https://mp.weixin.qq.com/s?__biz=MzI4NTc0NDc3NA==&mid=2247493834&idx=3&sn=47782c545fedd2e78ffd5a2a6cd1122e&chksm=ebe52323dc92aa35b77afaf12e89c850a9adbc0852cefb29e8e71e357af554770f7d2a3ca523&scene=27
16. Xinhuanet, et al. (2022). *Insight report on plant-based meat in China.* http://www.news.cn/tech/download/2022zgzwrjtdcbg.pdf
17. United States Department of Agriculture (USDA). (2021). *USDA agricultural projections to 2030.* https://www.ers.usda.gov/webdocs/outlooks/100526/oce-2021-1.pdf?v=587
18. FAOSTAT and USDA. (2023). *Record U.S. FY 2022 agricultural exports to China.* https://www.fas.usda.gov/data/record-us-fy-2022-agricultural-exports-china
19. FAOSTAT. (2022). *The FAOSTAT domain emissions totals.* http://www.fao.org/faostat/en/#data/GT
20. Feng, F., Zhang, Z. N., Gu, Y. Z., et al. (2022). Discussion on approaches to improving soybean supply capacity in China. *Bulletin of Chinese Academy of Sciences, 37*(9), 1281–1289. https://bulletinofcas.researchcommons.org/cgi/viewcontent.cgi?article=2078&context=journal
21. Song, X. P., Hansen, M. C., Potapov, P., et al. (2021). Massive soybean expansion in South America since 2000 and implications for conservation. *Nature Sustainability, 4*, 784–792. www.nature.com/articles/s41893-021-00729-z
22. Schneider, M., Goldman, L., Weisse, M., et al. (2021). *The commodity report: Soy production's impact on forests in South America.* https://www.globalforestwatch.org/blog/commodities/soy-production-forests-south-america/
23. Weisse, M., & Goldman, E. (2021). *Just 7 commodities replaced an area of forest twice the size of Germany between 2001 and 2015.* https://www.wri.org/insights/just-7-commodities-replaced-area-forest-twice-size-germany-between-2001-and-2015
24. Stockholm Environment Institute (SEI). (2022). *Connecting exports of Brazilian soy to deforestation.* https://www.sei.org/featured/connecting-exports-of-brazilian-soy-to-deforestation/
25. FAOSTAT. (2023). *Food balances.* https://www.fao.org/faostat/en/#data/FBS
26. China Animal Agriculture Association (CAAA). (2023). *Review of China's beef industry development in 2022 and outlook for 2023.* http://www.chinafeedm.com/h-nd-20058.html
27. China Customs. (2023). http://gdfs.customs.gov.cn/customs/syx/index.html
28. Searchinger, T., Waite, R., Hanson, C., et al. (2019). *Creating a sustainable food future—A menu of solutions to feed nearly 10 billion people by 2050.* https://research.wri.org/sites/default/files/2019-07/WRR_Food_Full_Report_0.pdf
29. Global Forest Watch (GFW). (2023). *Forest pulse: The latest on the world's forests.* https://research.wri.org/gfr/latest-analysis-deforestation-trends

30. Oilcn. (2019). *Instant noodles are popular again, palm oil imports surge: Understanding the status quo and prospect of palm oil consumption in China*. https://www.oilcn.com/article/2019/09/09_69647.html
31. FAOSTAT. (2021). *Production and trade balance*. https://www.fao.org/faostat/en/#data/QI
32. Jiang, Y. F. (2020). *Sustainable palm oil seeks breakthrough in China*. https://chinadialogue.net/en/food/sustainable-palm-oil-seeks-breakthrough-in-china/
33. Chain Reaction Research (CRR). (2021). *China, the second-largest palm oil importer, lags in NDPE commitments, transparency*. https://chainreactionresearch.com/report/china-the-second-largest-palm-oil-importer-lags-in-ndpe-commitments-transparency/
34. Statista. (2023). *Export volume of palm oil from Indonesia to China from 2012 to 2021*. https://www.statista.com/statistics/1037682/indonesia-palm-oil-export-volume-to-china/
35. Stockholm Environment Institute (SEI). (2022). *Indonesia makes progress towards zero palm oil deforestation: But gains in forest protection are fragile*. https://www.sei.org/featured/zero-palm-oil-deforestation/
36. World Economic Forum (WEF). (2022). *China's role in promoting global forest governance and combating deforestation*. https://www.weforum.org/reports/china-s-role-in-promoting-global-forest-governance-and-combating-deforestation
37. Science Based Targets Initiative (SBTi). (2022). *Forest, land and agriculture science based target setting guidance*. https://sciencebasedtargets.org/resources/files/SBTiFLAGGuidance.pdf
38. Yum China. (2021). *Yum China commits to the science based targets initiative to reinforce its climate action efforts*. https://ir.yumchina.com/news-releases/news-release-details/yum-china-commits-science-based-targets-initiative-reinforce-its
39. Yum China. (2022). *Yum China's science-based targets approved, aiming for over 60% GHG emissions reduction by 2035*. https://ir.yumchina.com/news-releases/news-release-details/yum-chinas-science-based-targets-approved-aiming-over-60-ghg
40. Food and Agriculture Organization of the United Nations (FAO). (2023). *Environmental and social standards, certification and labelling for cash crops*. https://www.fao.org/3/y5136e/y5136e00.htm#Contents
41. Planeta Campo. (2022). *Pará produces 96% of its soy compliant with green protocol for grains*. https://planetacampo.com.br/para-produces-96-of-its-soy-compliant-green-protocol-for-grains/
42. Food and Agriculture Organization of the United Nations (FAO) and World Resources Institute (WRI). (2022). *Timber traceability—A management tool for governments. Case studies from Latin America*. Rome. https://doi.org/10.4060/cb8909en
43. Cohn, A. S., Mosnier, A., Havlík, P., et al. (2014). Cattle ranching intensification in Brazil can reduce global greenhouse gas emissions by sparing land from deforestation. *Proceedings of the National Academy of Sciences, 111*(20), 7236–7241. https://doi.org/10.1073/pnas.1307163111
44. Cardoso, A. S., Berndt, A., Leytem, A., et al. (2016). Impact of the intensification of beef production in Brazil on greenhouse gas emissions and land use. *Agricultural Systems, 143*, 86–96. https://doi.org/10.1016/j.agsy.2015.12.007
45. Ermgassen, Z., Erasmus, K. H. J., Alcântara, M. P., et al. (2018). Results from on-the-ground efforts to promote sustainable cattle ranching in the Brazilian Amazon. *Sustainability, 10*(4), 1301. https://doi.org/10.3390/su10041301
46. Yap, P., Rosdin, R., Abdul-Rahman, A. A. A., et al. (2021). Malaysian sustainable palm oil (MSPO) certification progress for independent smallholders in Malaysia. *IOP Conference Series: Earth and Environmental Science, 736*(1), 012071. https://doi.org/10.1088/1755-1315/736/1/012071

47. Nurfatriani, F., Ramawati, R., Sari, G. K., et al. (2022). Oil palm economic benefit distribution to regions for environmental sustainability: Indonesia's revenue-sharing scheme. *Land, 11*(9), 1452. https://doi.org/10.3390/land11091452
48. Musim Mas. (2021). *Future ready: Sustainability report 2021*. Musim Mas. https://www.musimmas.com/wp-content/uploads/2022/10/Musim-Mas-SR2021.pdf
49. Sime Darby. (2022). *Sime Darby oils global supply chain Q1 2022 NDPE IRF profile*. https://www.simedarbyoils.com/documents/SDO-Global-NDPE-Q1-2022.pdf

Part IV
Low-Carbon and Inclusive Transition

Chapter 9
Innovative Mechanism of Sustainable Investment in Environment and Climate

1 Introduction

Globally, an increasing number of governments have announced net-zero carbon emission targets. As of the end of 2022, 21 countries have enacted net-zero carbon emission legislation, 81 countries have issued policy documents on net-zero carbon emission objectives, 22 countries have made net-zero carbon emission declarations and commitments, and 53 countries have proposed and discussed carbon emission targets.[1]

Achieving the goal of carbon neutrality needs huge amount of funding. According to the *Global Landscape of Climate Finance 2021* released by the Climate Policy Initiative (CPI), an estimated US $4.35 trn per year is needed by 2030 to meet the internationally agreed climate goals and transition to a sustainable, net-zero, and resilient future. However, in 2021, climate finance only reached the scale of US $0.85 trn [1], leaving a substantial gap. The latest estimate by the International Energy Agency (IEA) also suggests that to achieve net-zero emissions by 2050, global investments in clean energy technologies need to double by 2030 from the recent average of US $1.2 trn per year to US $4.2 trn [2]. According to CICC's *Guidebook to Carbon Neutrality in China*, achieving carbon neutrality in China requires approximately RMB139 trn in total green investments. A survey by the International Renewable Energy Agency (IRENA) indicates that over the past two decades, only about 20% of institutional investors have invested in renewables through funds, and less than 1% of institutional investors have directly invested in renewable projects [3].

As upstream capital players, state-owned investors are characterized by their long-term and large-scale investments, and can play a greater role in channeling more funds into green industries. As the majority of their funds comes from public assets, regulators can also guide more funds to cleaner asset categories through rule-making and restrictions (Fig. 1).

[1] Net Zero Tracker, https://zerotracker.net/.

Green Empowerment and High Quality Development,
https://doi.org/10.1007/978-981-96-4218-2_9

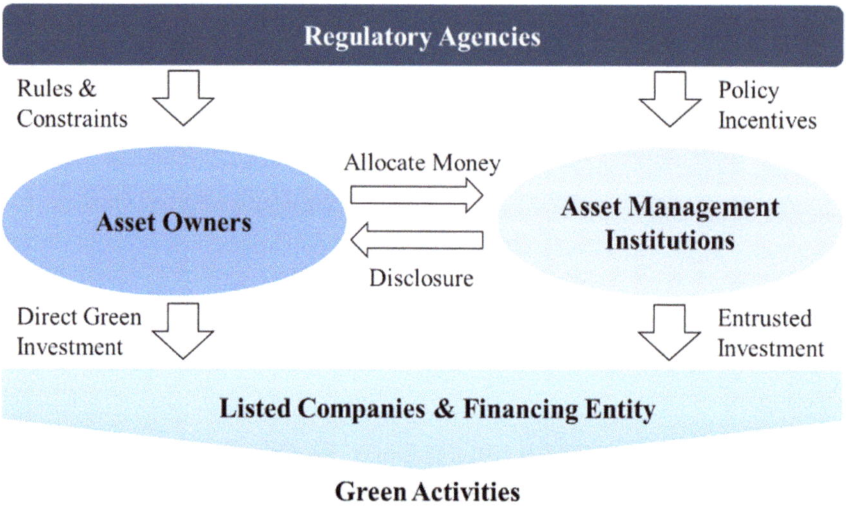

Fig. 1 Transmission mechanism of the oversight on ESG investing by asset owners. *Source* CICC Research

Globally, state-owned investors have made progress in sustainable development. In comparison, China's state-owned investors remain in an early stage of sustainable investing. By the end of 2022, the NCSSF opened its ESG portfolio to public funds for bidding. In April 2022, the China Securities Regulatory Commission (CSRC) issued the *Opinions on Accelerating the Promotion of High-quality Development of the Public Fund Management Industry*.

2 Trends and Motivations of Sustainable Investing by State-Owned Investors

According to statistics from Global SWF, the total assets under management (AuM) of global state-owned investors surpassed US $30 trn in 2020, equivalent to one third of the global GDP. If just 1% of the US $30 trn is invested in sectors related to climate change, this would be approximately 3.7 times the current climate investment commitments of multilateral development banks. In terms of national statistics, the United States leads the world by a significant margin with US $10.9 trn in total assets, followed by China and the United Arab Emirates as the second and third largest asset owners in the world, with approximately US $3.5 trn and US $1.9 trn in total assets, respectively. With respect to the scale of individual sovereign wealth funds and public pension funds (Fig. 2), the top three state-owned investors in the world are China Investment Corporation (CIC, US $1.4 trn), Japanese Government Pension Investment Fund (GPIF, US $1.3 trn), and Norges Bank Investment Management

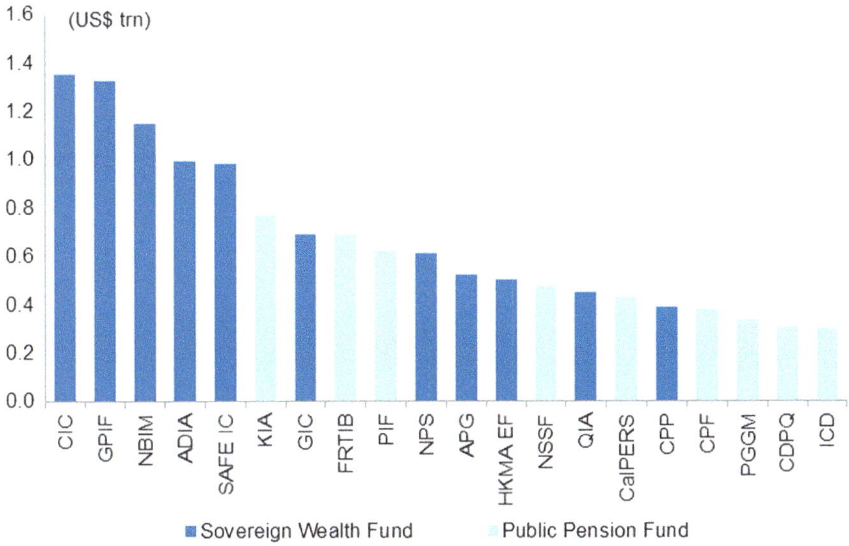

Fig. 2 Top 20 state-owned investors of the world in 2022. *Source* Global SWF. 2023 annual report. *Note* Global SWF regards GIC as a PPF and NSSF as an SWF; here adjustments are made based on China's realities. The top twenty sovereign asset owners are, in order: China Investment Corporation, Government Pension Investment Fund (Japan), Norges Bank Investment Management, Abu Dhabi Investment Authority, State Administration of Foreign Exchange Investment Company, Kuwait Investment Authority, GIC Private Limited (Singapore), Federal Retirement Thrift Investment Board, Public Investment Fund (Saudi Arabian), National Pension Service (Korea), Algemene Pensioen Groep (Netherland), Hong Kong Monetary Authority Exchange Fund, National Social Security Fund (China), Qatar Investment Authority, California Public Employees' Retirement System, Canada Pension Plan, Central Provident Fund of Singapore, Stichting Pensioenfonds Zorg en Welzijn (PGGM in Netherland), Quebec Deposit and Investment Fund (Canada), and Investment Corporation of Dubai. Global SWF categorizes GIC as PPF and NSSF as SWF; we have made adjustments according to the situation in China

(NBIM, US $1.1 trn; NBIM manages the Norwegian Government Pension Fund Global) [4].

2.1 Trends in Sustainable Investing

As the paradigm of ESG investments evolved, according to a survey by Schroders that covered 770 global institutional investors, including state-owned investors, a growing number of investors are seeking to measure, manage, and deliver impact. The survey indicates that institutional investors are particularly focused on four key areas: energy transition investments, pathways to net-zero emissions, ownership-based influence, and investment performance, as well as challenges related to "greenwashing" [5]. The survey also shows that 59% of investors believe that making tangible progress

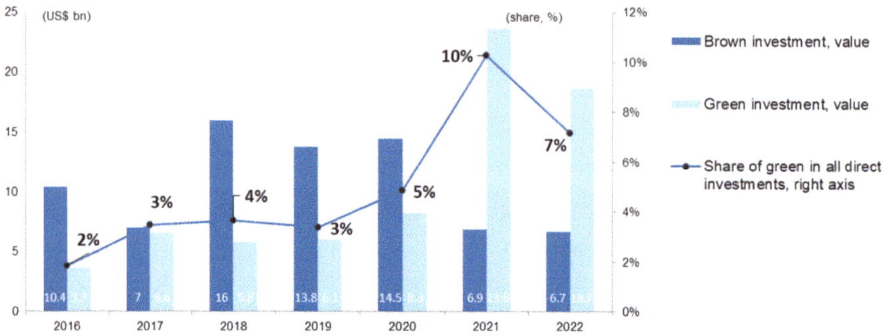

Fig. 3 Trends of investments in renewable projects and fossil fuels by state-owned investors. *Source* Global SWF. 2023 annual report

in the real world is the most important component of a proactive ownership strategy. The six most crucial areas are (i) governance and oversight, (ii) human rights, (iii) climate, (iv) human capital management, (v) inclusivity and diversity (such as gender equality), and (vi) natural capital and biodiversity. As per the estimate of the Global Sustainable Investment Alliance (GSIA), the AuM of funds engaged in ESG-related investments amounts to US $35.3 trn, accounting for approximately 36% of the global AuM, indicating significant room for growth [6].

In project investments, renewables offer comparatively strong, stable, and long-term "bond-like" returns that align with the long-term capital of state-owned investors and present lower stranded risks, providing institutional investors with opportunities for diversification. State-owned investors have reduced their investments in fossil fuels by more than half compared to 2018. In 2022, although investments in renewable projects edged down, the figure remained approximately three times that of fossil fuel investments (Fig. 3).

2.2 *Motivations for Sustainable Investing*

During the early stages of sustainable investing, the traditional view was that fiduciary duty conflicted with ESG considerations. Some pension funds focused on maximizing financial returns in the short term and overlooked factors pertaining to sustainability, the environment, and social impact.

State-owned investors engage in sustainable investing for various reasons and are motivated and constrained by a range of policies (policy factors will be further discussed in Part IV). Through investigations, we believe that sovereign asset managers are driven by the following three motivations for engaging in sustainable investing.

First, amidst the long-term global trend of carbon neutrality, sovereign asset managers seek long-term financial returns by capitalizing on profitable growth opportunities in green sectors.

Second, sovereign asset managers engage in sustainable investing to better mitigate and manage climate and environmental risks, which is the flip side of the first motivation.

Third, sustainable investing is seen as part of responsible investing at the corporate level, which is especially important for state-owned investors with public attributes.

3 Sustainable Investment Practices of State-Owned Investors

State-owned investors around the world have built up sustainable investment practices. Based on the annual reports and ESG reports released by public pension funds and sovereign wealth funds as well as interviews, surveys, and roundtable discussions, we have summarized seven practices of state-owned investors at the corporate and investment levels. At the corporate level, leading state-owned investors include low-carbon transition in their strategic goals, manage the carbon footprint of their portfolio, and engage with international organizations. At the investment level, they account for ESG factors when screening and evaluating asset management firms, take on a stewardship role, employ investment strategies such as ESG integration and negative screening, and engage in sustainability themed investing.

3.1 The Corporate Level

3.1.1 Incorporating Low-Carbon Transition in Strategic Goals

An increasing number of state-owned investors are recognizing the severity of climate change and elevating low-carbon transition to a strategic priority, driving capital markets towards a more sustainable path of development. Some state-owned investment institutions have revised their investment policies to prioritize climate change and environmental factors, and set medium-to-long-term net-zero investment targets in line with the *Paris Agreement*. However, setting net-zero emissions targets alone is insufficient. Asset owners need to use existing baselines, data, and other tools to project future emissions and report them comprehensively, including portfolio targets, sectoral targets, financing targets, and ownership targets.

Some state-owned investors have adjusted their internal organizational structures to support sustainable investment decision-making based on the requirements of ESG investing, such as establishing a sustainability committee or ESG committee. From an organizational perspective, having board members familiar with sustainable investing is crucial for integrating climate and environmental factors into the decision-making process.

3.1.2 Managing the Carbon Footprint of Portfolios

State-owned investors can enhance their understanding of the investment risks and opportunities related to climate change, and respond to stakeholder concerns by accounting for carbon emissions and comparing them with global benchmarks. This, in turn, improves their business reputation and helps them make more sustainable investment decisions. In recent years, nearly one-third of pension funds have been tracking and disclosing carbon emissions in their portfolios, using common indicators such as carbon footprint, carbon emissions, and carbon intensity [7].

In terms of accounting methods, the Partnership for Carbon Accounting Financials (PCAF) developed specific carbon accounting standards for the financial industry, building upon the Greenhouse Gas Protocol (GHG Protocol). The standard, titled the *Global GHG Accounting and Reporting Standard for the Financial Industry*, sets allocation factors and calculation methods for different asset classes, allowing financial institutions to calculate their investment-related carbon emissions. They can then report these emissions based on international disclosure frameworks such as the Task Force on Climate-related Financial Disclosures (TCFD). As of June 2023, a total of 401 institutions around the world have joined PCAF, with an AuM of US $92.1 trn, covering asset management companies, commercial banks, investment banks, insurance companies, and other institutions.[2]

Regarding the scope of accounting and reporting, due to data availability and calculation methods, most state-owned investors only include direct emissions (Scope 1) related to their own activities and indirect emissions (Scope 2) associated with electricity use, primarily limited to equity assets [8]. However, Scope 3 emissions, measure emissions throughout the entire supply chain rather than just within the state-owned investor itself. Addressing Scope 3 emissions is crucial for implementing effective emission reduction measures [9]. In this regard, the PCAF is also exploring ways to improve the availability and quality of statistics concerning Scope 3 emissions. Some pension funds at the forefront of ESG practices, such as GPIF, have expanded the scope of their GHG calculations since 2020 and now include indirect emissions from the sale of products and services (Scope 3), and indirect upstream emissions from the procurement of products and services (Scope 3 upstream).

When it comes to the comparability of carbon emissions data, or domestic stocks, GPIF compares its portfolio's carbon emissions with the benchmark carbon emissions of the TOPIX index [10]. For foreign stocks, it compares them with the MSCI ACWI ex Japan index [11]. As for bonds, GPIF compares the emission figures with foreign bonds, thereby clarifying the level of its carbon emissions relative to similar assets both domestically and internationally. The GIC also emphasizes that it would compare disclosed information with statistics reported by companies in the same industry or region, as well as with its own data, to ensure valid quantitative analyses, consistent definitions, and verifiable results.

[2] PCAF website. https://carbonaccountingfinancials.com/financial-institutions-taking-action. Acquisition time: June 12, 2023.

Table 1 Sector-specific carbon intensity targets of CalPERS

Cluster	Sector	2025 target	2030 target	Sectoral carbon performance measure
Energy	Electricity utilities	0.288	0.138	Carbon intensity of electricity generation (metric tonnes of CO_2 per MWh)
	Oil and gas	51.52	40.95	Carbon intensity of primary energy supply (gCO_2e/Mj)
Transport	Automobiles	68	40	New vehicle carbon emissions per kilometer (grams of CO_2 per kilometre)
	Airlines	1071	616	Carbon emission per revenue tonne kilometer (gCO_2/RTK)
	Shipping	5.63	4.31	Carbon emissions per tonne kilometer (gCO_2t/t km)
Industrials and materials	Cement	0.43	0.373	Carbon intensity of cementitious product (tonnes of CO_2 per tonne of cementitious product)
	Diversified mining	49.79	41.54	Carbon emissions per tonne of copper equivalent (tonne CO_2e/tonne of steel)
	Steel	1.046	0.815	Carbon emissions per tonne of copper equivalent (tonne CO_2e/tonne of steel)
	Aluminium	4.004	3.069	Carbon intensity of aluminium production (tCO_2e/t aluminium)
	Pulp and paper	0.427	0.353	Carbon intensity of pulp, paper, and paperboard production (tonnes of CO_2 per tonnes of product)

From the perspective of measurement and targets, CalPERS established comprehensive sector targeted carbon intensity targets based on the Transition Pathway Initiative (TPI) developed by the International Energy Agency (IEA). By setting carbon intensity targets for each industry for 2025 and 2030, as well as carbon performance indicators for high-emitting sectors (Table 1), CalPERS evaluates the decarbonization progress of its portfolio and assesses the carbon performance of companies and investors to determine their alignment with the 1.5 °C pathway. Investors, including the Climate Action 100+, are extensively utilizing these measures to provide information for state-owned investors in investment decision-making and facilitate joint action with shareholders of high-emitting companies.

3.1.3 Participating Actively in International Processes and Initiatives

Given the limited capacity and scope of state-owned investors, they can participate in sustainable development alliances or institutions/organizations targeting asset owners or broader financial institutions to access information and resources that facilitate shareholder engagement. That will also help reduce time and costs, and expand their influence over investee companies. At the moment, multiple types of global green finance organizations focus on goals including responsible and sustainable investing, information disclosure, and net-zero commitments. State-owned investors play a particularly key role in realizing those goals.

With regard to responsible and sustainable investing, the UN introduced the Principles for Responsible Investment (PRI) in 2006. As of the end of 2022, 5296 institutions have signed and joined the PRI, including 722 asset owners (Fig. 4). The PRI proposed "incorporating ESG issues into investment analysis and decision-making processes" as one of the six investment principles, encouraging investors to improve returns and better manage risks through responsible investing. The PRI also recognizes the important role investors play in advancing DEI (diversity, equality, and inclusive) efforts for all groups in society, including women, people of color, indigenous communities, and others. After the *Paris Agreement* was signed, in light of the unique advantages of sovereign wealth funds in promoting long-term value creation and sustainable market outcomes, the One Planet Sovereign Wealth Fund Working Group was established during the One Planet Summit in 2017. It aims to integrate climate-related financial risks and opportunities into the management of large, long-term asset pools.

Concerning regulation and disclosure, the Financial Stability Board of the Bank for International Settlements (BIS) established the TCFD in 2015 to encourage consistent climate-related financial disclosures by companies and enhance comparability. To achieve this goal, the TCFD developed 11 disclosure recommendations covering governance, strategy, risk management, and metrics and targets, aiming to improve transparency in disclosing climate-related risks and opportunities to investors, lenders, and insurance underwriters. At the moment, over 3800 organizations, including over 1500 financial institutions, have become supporters of TCFD recommendations, with US $217 trn in assets [12]. Similarly, the Sustainability Accounting Standards Board (SASB) identified subsets of ESG issues most relevant

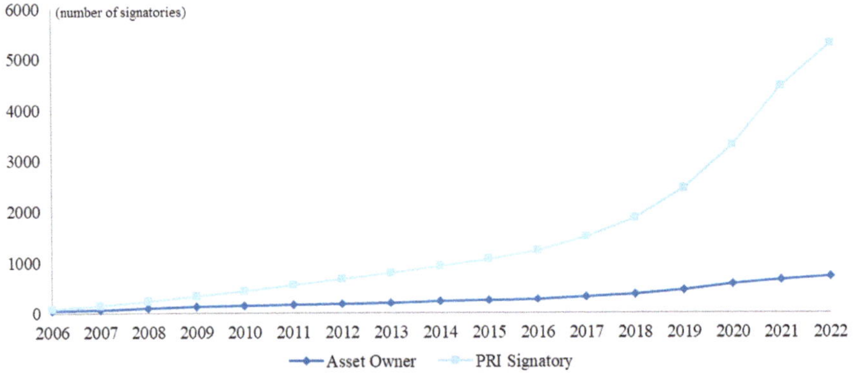

Fig. 4 Number of PRI signatories and asset owners rapidly growing. *Source* PRI. https://www.unpri.org/signatories/signatory-resources/signatory-directory

to financial performance and enterprise value in 77 industries. This enables companies to provide industry-based sustainability disclosures about risks and opportunities that affect enterprise value. The ISSB and SASB are also aligning their standards to avoid duplication.

International organizations focusing on information disclosure typically target a broader range of asset owners and institutional investors, including state-owned investors. Many large sovereign funds and pension funds actively support and sign up for these organizations' disclosure standards for reporting. At the same time, state-owned investors exert their investor influence by pressuring investee companies to enhance their ESG disclosure. Between 2020 and 2022 alone, TCFD supporters more than doubled [13], organizations disclosing through CDP grew by 79%.[3] Investor pressure is one of the key factors driving the growth of sustainability disclosure requirements within regulatory frameworks.

Table 2 summarizes the current status of the leading state-owned investors' participation in representative international organizations, with nine of the top 20 having signed and joined the PRI. It is evident that pension funds such as APG, CalPERS, and NBIM are actively involved in international initiatives related to responsible investing, net-zero targets, and disclosure. In comparison, state-owned investors from South Korea, Singapore, and the Middle East are not fully involved.

3.2 *The Investment Level*

One of the key differences between state-owned investors and conventional asset managers lies in the scale of their delegated investments. Investments by state-owned investors can be categorized into delegated and direct investments. Delegated investments are realized through investment funds managed by investment advisors or fund managers. State-owned investors typically externalize specific geographic regions or asset categories for strategic purposes to yield stable returns through external managers with sector expertise. Direct investments, on the other hand, are made by internal investment managers within state-owned investors either in secondary market equity funds or through direct investments in specific asset classes. According to a joint survey by the Bank of New York Mellon and the Official Monetary and Financial Institutions Forum (OMFIF), public pension funds tend to have around 40% of their assets managed internally and allocate 60% of their assets to third-party managers for investments [14].

[3] https://www.cdp.net/en/companies/companies-scores.

Table 2 Overview of major state-owned investors' participation in representative international initiatives

	Abbreviation	Country	UN PRI	TCFD	Climate action 100+	SASB	CDP	UN NZAOA
Government Pension Investment Fund	GPIF	Japan	✓	✓	✓			
Norges Bank Investment Management	NBIM	Norway	✓	✓		✓	✓	
Abu Dhabi Investment Authority	ADIA	Abu Dhabi						
Kuwait Investment Authority	KIA	Kuwait						
Government of Singapore Investment Corporation	GIC	Singapore		✓	✓		✓	
Federal Retirement Thrift Investment Board	FRTIB	U.S.						
Public Investment Fund	PIF	Saudi Arabia						
National Pension Service	NPS	South Korea	✓			✓		
Algemene Pensioen Groep	APG	Netherlands	✓	✓	✓	✓	✓	
Qatar Investment Authority	QIA	Qatar						
California Public Employees' Retirement System	CalPERS	U.S.	✓	✓	✓	✓	✓	✓
Canada Pension Plan Investments	CPP	Canada	✓	✓		✓	✓	
Central Provident Fund	CPF	Singapore						
Pensioenfonds voor Gezondheidszorg Geestelijke en Maatschappelijke	PGGM	Netherlands	✓	✓	✓	✓	✓	

(continued)

Table 2 (continued)

	Abbreviation	Country	UN PRI	TCFD	Climate action 100+	SASB	CDP	UN NZAOA
La Caisse de Dépôt et Placement du Québec	CDPQ	Canada	✓	✓	✓	✓	✓	✓
Investment Corporation of Dubai	ICD	Dubai						

3.2.1 Accounting for ESG Factors When Screening and Evaluating Investment Companies

When screening and evaluating investment companies, some state-owned investors incorporate ESG factors into their selection criteria and performance assessments. In terms of pre-investment screening, the National Pension Service (NPS) of South Korea has established an additional rating system since November 2020 that considers responsible investing factors when selecting external asset managers [15]. In terms of post-investment evaluation, GPIF regards ESG integration as a critical factor in evaluating external asset managers.

3.2.2 Exercising Stewardship (Active Ownership)

There is no unified definition for the term "stewardship". As defined by the latest report of the UK-China Green Finance Taskforce, stewardship, also known as active ownership, refers to investors utilizing their scale and influence to fulfill their fiduciary duties, and actively leveraging shareholder rights to engage in sustainable investment governance of investee companies [10]. Corporate engagement plays a vital role as a measure for state-owned investors to participate and effecting change in the real economy. Voting participation and involvement in corporate governance are two complementary approaches, particularly in cooperation with other shareholders, as they can effectively influence corporate behaviors.

First, participation mechanisms for stewardship involve holding constructive dialogue with portfolio companies through voting and other means to them to comply with the *Paris Agreement*. Leading institutions select the most relevant issues directly impacting the company's ESG performance and stakeholder interests for extensive discussions. Second, asset managers are encouraged to incorporate non-financial factors, including ESG considerations, into investment decision-making. Furthermore, investee companies are encouraged to consider how ESG risks and opportunities will impact their long-term value creation capabilities. Finally, in the process of exercising active ownership, broader social issues beyond climate and environmental concerns are also considered. The pursuit of gender equality is often embedded in stewardship to ensure that women have a voice in sustainable investing field and that funds account for the impact on women throughout the investment process.

3.2.3 Using ESG Integration and Negative Screening Strategies

According to the classification standards of the Global Sustainable Investment Alliance (GSIA), sustainable investment strategies can be put into seven categories based on the degree of active or passive involvement. These categories are norms-based screening, positive screening, negative screening, sustainability-themed investing, ESG integration, impact investing, and shareholder action. In terms of the investment scale, ESG integration and negative screening are currently the most prominent responsible investment strategies globally (Fig. 5).

Negative screening is an early and relatively straightforward sustainable investment strategy. It involves excluding companies, industries, or regions from portfolios based on established principles (often associated with ethics) or ESG scores. For example, since 2012, the GPFG has gradually divested itself from companies with high financial risks arising from carbon-intensive business models [16]. In 2019, the Norwegian Parliament passed a proposal to divest itself from the fossil fuel industry, leading to the exclusion of 95% of coal mining and 80% of coal-fired power generation companies from the GPFG. In 2021, the fund divested itself from four coal companies with high and difficult-to-quantify coal risk exposure [17]. Similarly, CalPERS divested itself from 14 thermal coal companies in 2017 following unsatisfactory outcomes of shareholder action. In the Netherlands, PFZW has mandated that from 2024 onwards, it will only invest in fossil fuel companies that fully comply with the *Paris Agreement*, with short- and medium-term targets [18].

It is noteworthy that negative screening may potentially lead to regional and sector biases. For example, global portfolios may be biased towards European companies with higher ESG scores while reducing asset allocations to companies in developing

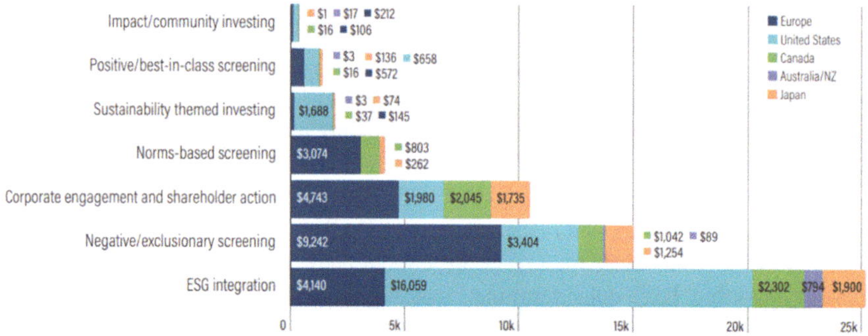

Fig. 5 Assets of sustainable investing by strategies and regions in 2020 (unit: US $ trn). *Source* GSIA. Global sustainable investment review 2020

countries [19]. They may also favor industries such as information technology and healthcare while minimizing exposure to the energy sector. In practice, investors have adopted a more consistent stance toward negative screening, and many state-owned investors find it to be suboptimal.

While negative screening is a common approach, its substantive impact on climate and environmental issues may be limited, making it more suitable as part of risk and return assessment. With the gradual improvement in the quality of ESG data, investors can now conduct more detailed and sophisticated screening, and filter out companies that do not meet their criteria or demonstrate below-average scores concerning specific ESG factors. The blunt divestment strategy of negative screening is being replaced by ESG integration.

ESG integration is a strategy through which state-owned investors systematically incorporate ESG factors into investment decision-making and portfolio building across a range of asset classes through quantitative models or qualitative analysis.[4]

3.2.4 Making Sustainability-Themed Investing

State-owned investors can also directly invest in assets related to sustainable development, with the aim of achieving positive social and environmental impacts, and obtaining corresponding returns. It is important to note that although sustainability-themed investments have the most direct contribution to the environment and climate, state-owned investors cannot simply invest based on the climate and environmental contributions of investment targets. Sustainability-themed investments by state-owned investors are largely distributed in renewables. In specific investment practices, the infrastructure portfolio of CalPERS allocated US $4.76 bn to renewables, energy-efficient infrastructure, sustainability certifications, and carbon-neutral assets, accounting for 51% of the portfolio's net asset value. Some state-owned investors established sustainable investment funds to promote thematic investing.

4 Fostering a Sound Policy Environment for Sustainable Investing by State-Owned Investors

At the policy level, countries worldwide have issued laws and regulations related to sustainable investing and ESG to support sustainable investing from all angles. According to PRI statistics, since 2000, there have been more than 800 new and revised mandatory and voluntary policies related to sustainable investing, with 225 policies issued in 2021.[5] Policies that can support sustainable investing by state-owned investors generally fall into at least four categories: (1) low-carbon transition policies, (2) green financial system development, (3) investment rules, and

---

[4] USISIF. ESG Incorporation. https://www.ussif.org/esg.

[5] https://www.unpri.org/policy/regulation-database.

(4) stewardship codes. Specifically, the first two types of policies are universally applicable, covering financial institutions that are not state-owned investors. Low-carbon transition policies facilitate the development of green industries and provide quality sustainable investment projects for sustainable investing. The creation of a green financial system enables the discovery and transformation of green values, reduces the search cost for investors, and creates more financial support for green development. Investment rules and stewardship codes directly target state-owned investors.

4.1 Macro Policies Supporting Low-Carbon Transitions

In a world where more and more countries have set net-zero targets, national policies supporting the low-carbon transition are becoming increasingly diverse. Fiscal policies such as subsidies for solar power and new energy vehicles, as well as structural monetary policies like carbon reduction support tools and special refinancing loans for the utilization of clean and efficient coals, have created a favorable environment for the development of low-carbon industries. With these policies, investors have also found more suitable investment targets. For example, the U.S. *Inflation Reduction Act* (IRA), the *European Green Deal*, and the EU *Green Deal Industrial Plan* all aim to further empower low-carbon industries.

The U.S. *Inflation Reduction Act* aims to combat inflation by lowering energy costs. The act plans to invest US $369 bn in climate change and energy security, including tax credits for U.S. clean energy manufacturing companies, subsidies for electric vehicle purchases, and grants for smart agriculture, among other initiatives. Widely regarded as the largest climate investment act in U.S. history, the bill could potentially help the country reduce GHG emissions by over 40% from the 2005 level by 2030, according to a preliminary assessment by the U.S. Department of Energy (DOE). According to estimates by the U.S. Committee for a Responsible Federal Budget, a total of US $750 bn in energy and climate incentives will be required between 2023 and 2042. However, imposing a 15% minimum corporate tax on companies with profits exceeding US $1 bn alone will allow the government to raise US $850 bn in tax revenue, fully offsetting the incentives for green industries [19].

The EU's statement on the *Green Deal Industrial Plan* indicates that 37% of the EUR725 bn allocated to the Next Generation EU recovery fund will be dedicated to the green transition [20]. While complementing the *European Green Deal* and the REPowerEU plan, the Plan also lays a solid foundation for subsequent legislation related to the green industry. From January to June 2022, the Next Generation EU program issued EUR120 bn in bonds, including EUR28 bn in green bonds.

These macro-level supportive policies for green projects help enhance the expected cash flows of projects, thereby attracting more direct and indirect investments. Meanwhile, the continuous expansion of the scale of industrial projects fosters economies of scale, further reducing the costs of the green transition and generating positive feedback. However, it is worth noting that such industrial and trade policies in the United States and Europe also have an impact on global supply chains, which is out of the scope of this report.

4.2 Policies on Green Finance and Green Financial Ecosystem

At the moment, in regions such as the EU and China that started early in the development of green finance, a green financial system enabling low-carbon development has initially taken shape, covering five pillars: (1) standard systems including definitions and taxonomies, (2) financial institution oversight and disclosure requirements, (3) incentive and constraint mechanisms, (4) product and market systems, and (5) international cooperation.[6] The following paragraphs briefly elaborate on each of these pillars.

First, the accurate definition of green terms and the establishment of taxonomies and standard systems have played a role in standardizing green financial practices and clarifying the scope of support.

Second, disclosure mechanisms have enhanced the transparency of green financial practices and ensured effective regulatory oversight. The ISSB, which is under the IFRS Foundation, has published, released, and implemented the final version of the IFRS Sustainability Disclosure Standards (ISDS).

Third, incentive and constraint mechanisms help mobilize institutions to engage in green finance and ensure orderly green financial practices. According to disclosures by the Sustainable Banking and Finance Network (SBFN), as of September 2021, 282 policy documents, including laws, regulations, and industry norms, related to national sustainable finance frameworks, had been issued by 43 emerging market member countries, including China.

Fourth, the product and market systems are the direct channels through which green finance supports the real economy, and a diverse range of financial products can broaden the scope of potential investment targets for state-owned investors. In recent years, the issuance of green bonds, social responsibility bonds, sustainability-linked bonds (loans), and transition bonds has grown rapidly. According to the Climate Bonds Initiative (CBI), as of September 30, 2022, the cumulative issuance of global Green, Social, Sustainability, Sustainability-Linked, and Transition Bonds (GSS + bonds) reached US $3.5 trn [20].

Fifth, international cooperation mechanisms enhance the international recognition of green finance standards and products, while improving market participation. Various multilateral platforms and cooperative mechanisms, such as the G20 Sustainable Finance Working Group, the United Nations Environment Programme Finance

---

[6] Chen Yulu, Deputy Governor of the Central Bank: The three major functions and five pillars of green finance have been initially formed. http://finance.sina.com.cn/money/bank/bank_hydt/2022-05-12/doc-imcwipii9471798.shtml?cref=cj.

Initiative (UNEP FI), the Network of Central Banks and Supervisors for Greening the Financial System (NGFS), and the International Platform on Sustainable Finance (IPSF), have jointly promoted international exchanges in green finance and deepened global cooperation.

4.3 Investment Rules

Unlike the previous two categories of policy support, investment rules are specific requirements targeting specific investors and directly constrain the investment behavior of state-owned investors. Over the past ten years, regulatory policies for sustainable investing by state-owned investors have seen rapid progress, and a relatively comprehensive policy framework has taken shape.

4.3.1 National/Local Laws and Regulations

In recent years, regulators in many countries and regions have enacted mandatory provisions through legislation or regulatory amendments to enforce sustainable investing by state-owned investors. Developed countries such as the U.S. and European countries boast mature capital markets and are the birthplaces of responsible investing. They enforce the inclusion of ESG factors in investment decision-making and impose certain disclosure requirements, which is the case in places like the EU, the UK, and France.

In 2014, the European Union introduced the *Non-financial Reporting Directive* through legislation, which mandates companies with over 500 employees to disclose information on the environment and other aspects. In 2016, the EU specifically addressed investment requirements for pension funds through the IORP II Directive, which requires pension funds of a certain scale to consider ESG issues and disclose how they incorporate these risks into their investment policy statements.

As one of the European markets with leading progress in sustainable investing, the UK introduced amendments to the Local Government Pension Scheme (Management and Investment of Funds) Regulations in 1999, requiring each managing institution to prepare, maintain, and publish a written statement of the investment principles for pension funds, which must include social, environmental, or ethical considerations. In 2005, the UK Department for Work and Pensions first included incorporated environmental, social, and ethical considerations in pension safeguard regulations, namely the *Occupational Pension Schemes (Investment and Disclosure) Regulations* and the *Statement of Investment Principles.* Since 2014, ESG policy regulations in the UK have been revised approximately every two years. The Financial Reporting Council, Law Commission, and London Stock Exchange have played a crucial role in this process [21]. In particular, the 2018 amendment to the *Occupational Pension*

Schemes (Investment and Disclosure) Regulations[7] extended the fiduciary duty to include ESG considerations and climate change-related disclosures.

France has consistently been at the forefront of sustainable practices in the EU and beyond, particularly in terms of policy initiatives and mandatory regulations for investors. Article 225 of *Grenelle II*, adopted in 2010, stipulates that listed companies, companies with an annual balance sheet total or turnover exceeding EUR100 mn, and companies with an average of 500 permanent employees are obligated to disclose certain social and environmental information in their annual management reports. The law requires explanations for any undisclosed information, whereas previously only listed companies were subject to such disclosure requirements. Article 224 of the law also mandates that public funds must mention how they account for ESG objectives in their investment policies through their annual reports and documents [22]. In 2015, France enacted the famous *Law on Energy Transition for Green Growth* to lead global climate governance legislation. The law officially brought asset managers and institutional investors under regulatory supervision. Article L533-22-1 of the law specifies that portfolio management companies must disclose their policies regarding the incorporation of ESG quality standards into their investment strategies to beneficiaries and the public, and explanations must be provided for any non-disclosure.

In the United States, early ESG regulatory policies primarily focused on listed companies, but in recent years, states like California and Illinois have introduced more legislation targeting asset owners and sustainable investments. A notable example is the legislation in California. In 2015, the state passed *Senate Bill 185*, which prohibits its two major retirement funds, the CalPERS and the CalSTRS, from making new or additional investments in thermal coal companies. The legislation requires the funds to divest from all fossil fuel assets and gradually transition to clean energy by July 1, 2017. Passed in 2019, *Senate Bill 964* mandates the two funds to disclose financial information related to climate risks in their publicly traded portfolios, alignment with climate objectives, and other relevant information.

In addition to European countries and the U.S., in 2020, the Ministry of Health, Labor, and Welfare of Japan revised the Basic Policy on Reserves (BPR), requiring the Japanese government pension funds under its supervision to incorporate ESG factors into investment decision-making. South Korea's *National Pension Act*, amended in 2015, also requires the NPS to consider ESG issues in its investment decision-making process or provide reasons for not considering them. In South Africa, following the revision of Section 28 of the *Pension Funds Act* in 2011, pension funds were required to establish investment procedures related to fund conditions and regulations, taking into account factors concerning long-term returns that include ESG considerations.

[7] UK. The Pension Protection Fund (Pensionable Service) and Occupational Pension Schemes (Investment and Disclosure) (Amendment and Modification) Regulations 2018. https://www.legislation.gov.uk/uksi/2018/988/regulation/4/made.

4.3.2 Regulations and Guidelines Issued by Treasuries or Financial Regulators

The Norwegian Ministry of Finance plays a crucial role in governing the sustainable investment practices of the GPFG. In the *Management mandate for the Government Pension Fund Global,* the ministry states that a good long-term return depends on sustainable economic, environmental, and social development. Furthermore, the mandate also requires a thorough due diligence review of the unlisted real estate and unlisted renewable energy infrastructure portfolios, including the assessment of risks associated with health, safety, environmental, corporate governance, and social factors [23]. In 2014, the ministry issued the Guidelines for Observation and Exclusion of Companies from the Government Pension Fund Global, which outlined the criteria for the observation and exclusion of GPFG companies [24]. From the perspective of products, the guidelines require the GPFG to observe or exclude companies that: (1) derive 30% or more of their revenue from thermal coal, (2) base 30% or more of their activities on thermal coal, (3) extract more than 20 mn metric tons of thermal coal annually, or (4) operate a power generation capacity of over 10,000 MW from thermal coal. In terms of business conduct, the guidelines require observation or exclusion of companies causing severe environmental damage or exhibiting unacceptable GHG emissions at the company level. As of December 31, 2021, a total of 104 companies have been excluded for violating the product criteria, and 48 companies have been excluded for violating the conduct criteria. In April 2019, the ministry approved the GPFG to engage in unlisted renewable energy investments and lifted the limit for its thematic investments related to the environment from NOK60 bn to NOK120 bn [25].

The U.S. Department of Labor (DOL) is an executive agency under the federal government responsible for national employment, wages, benefits, labor conditions, and employment training. It plays a role in shaping pension fund investment regulations. In 2016, the DOL issued *Interpretive Bulletin 2016-01*, which allowed for the inclusion of ESG factors in investment policy statements or the integration of tools, metrics, and analyses related to ESG to assess investment risks and returns. However, it did not impose mandatory requirements [26]. Subsequently, in 2018, the department issued *Field Assistance Bulletin 2018-01*. It states that to the extent ESG factors involve business risks or opportunities that are properly treated as economic considerations themselves in evaluating alternative investments, the weight given to those factors should be appropriate to the risk and return profiles relative to other relevant economic factors. However, it also pointed out that fiduciaries under the Employee Retirement Income Security Act (ERISA)[8] must always prioritize economic interests when providing retirement benefits and should not excessively rely on the correlation between ESG factors and financial returns. Therefore, some argue that *Field Assistance Bulletin No. 2018-01*, while intended to provide further clarification, may create

[8] The Employee Retirement Income Security Act of 1974 (ERISA) is a federal law that sets minimum standards for most voluntary retirement and health plans in the private sector in order to protect individuals in those plans.

confusion between encouraging and discouraging ESG investments. In November 2022, the DOL issued sustainable investment regulations allowing pension fund managers to consider ESG factors in investment decision-making. However, the regulation was overturned by the U.S. Senate this March on the grounds that consideration of ESG factors could potentially harm pension fund returns. President Biden later vetoed the Senate's decision. Hence, the latest sustainable investment rules remain effective.

4.3.3 Norms and Recommendations from Industry Associations

Industry associations and other self-regulatory organizations inherently play a role in enabling supervision, fairness, self-discipline, and coordination as leaders and promoters of sustainable investment principles. Switzerland has one of the most comprehensive pension systems in the world, with a robust system of self-regulatory organizations, covering the Swiss Association for Responsible Investments (SVVK-ASIR), the Asset Management Association Switzerland (AMAS), and the Association of Swiss Pension Funds (ASIP). These industry associations typically oversee and guide their members' business activities, issue professional codes of conduct and recommendations, and provide guidance on green and sustainable investing.

The SVVK-ASIR offers its members services related to responsible investing, including portfolio screening and monitoring based on specific standards and exclusion recommendations. In 2019, the association published the *Engagement and Exclusion Process*. The document outlined how investors should assess companies' violations, choose whether to engage with a company, and set engagement goals, and the complete process and decision-making logic leading to exclusion and re-inclusion [27].

Encompassing over 900 pension funds, the ASIP represents about two thirds of the persons insured by occupational pension funds and holds approximately CHF650 bn in pension assets. In 2022, it published the *ESG Guidelines for Swiss Pension Funds* and issued the *ESG Reporting Standard for Pension Funds*.

4.4 Stewardship Code

Many countries have adopted regulations on how state-owned investors can better fulfill their stewardship duties and explicitly defined their responsibilities based on investor rights and influence. These regulations directly address the issue of stewardship for state-owned investors. Examples include the *Swiss Code of Best Practice for Corporate Governance*, the *UK Stewardship Code*, and Japan's *Guidelines on the Duties of Pension Fund Managers*. The emphasis on stewardship duty varies across countries and is closely related to the characteristics of their financial systems. For instance, the U.S. boasts a well-established corporate system and places greater emphasis on voting rights in stewardship.

ESG stewardship has become a key channel for institutional investors to directly drive companies to capture transition opportunities. In China, ESG stewardship remains in an early stage, but it aligns well with major goals such as carbon neutrality. In terms of issuers, there are three representative and globally influential stewardship models, which are stewardship codes formulated by: (1) domestic regulatory or quasi-regulatory bodies, (2) industry organizations, and (3) third-party organizations (as shown in Table 3) [28]. The following paragraphs will focus on the stewardship codes adopted by countries and regions that best represent the three models.

Table 3 Typical stewardship codes

Issuer type	Country/region	Name	Year published	Issuer
Regulator	UK	The UK Stewardship Code	2010	Financial Reporting Council
	Japan	Principles for Responsible Institutional Investors	2014	Financial Services Agency
Industry Organization	U.S.	Stewardship Framework for Institutional Investors	2017	Investor Stewardship Group
	EU	EFAMA Stewardship Code	2011	EFAMA
	The Netherlands	Dutch Stewardship Code	2011	Eumedion
	Switzerland	Guidelines for Institutional Investors	2013	Association of Swiss Pension Fund Providers
Third Party	South Korea	Korea Stewardship Code	2016	Korea Corporate Governance Service
	Singapore	Singapore Stewardship Principles for Responsible Investors	2016	Stewardship Asia Centre

Source Asset Management Association of China, ZD Proxy. Institutional Investors' Participation in the Corporate Governance of Listed Companies: A Review of Overseas Regulations and Leading Practices [M]. Beijing: China Financial and Economic Publishing House, Nov 2021

4.4.1 National/Local Stewardship Codes: Case Study of Stewardship Practices in the UK and Japan

The first type of stewardship code includes rules and guidelines formulated by domestic regulatory or quasi-regulatory bodies. This approach is adopted by many countries, with adjustments made to accommodate the actual needs of their respective capital markets. The UK and Japan are the most typical examples of this category of stewardship codes.

Drawing lessons from the 2008 financial crisis, the FRC introduced the first version of the *Stewardship Code* for institutional investors in 2010, the world's first stewardship code, laying the groundwork for the establishment of such guidelines in other countries. *The UK Stewardship Code* adopts a voluntary, non-mandatory model of implementation, which has gradually proven effective and influenced countries and regions worldwide. *The UK Stewardship Code 2020* stipulates that pension funds, insurance companies, fund management institutions, and other financial service providers must publicly disclose their long-term approach to maintaining and enhancing investment value. They are also required to consider the ESG performance of investee companies [29]. In terms of stewardship codes for pension funds, the UK introduced the Occupational Pension Schemes Regulations through the Department for Work and Pensions (DWP). According to the Occupational Pension Schemes (Investment and Disclosure) (Amendment) Regulations 2019 [25], asset owners are required to explain how they exercise their investment rights and monitor investee companies, as well as their voting behavior.

4.4.2 Stewardship Codes by Industry Organizations: U.S. and the Netherlands

The second category of stewardship codes, issued by industry organizations, primarily manifests as mandatory rules on stewardship established by industry organizations such as investor associations, although the ESG-related regulations in this category are mostly voluntary.

4.4.3 Stewardship Codes by Third Parties: South Korea and Singapore

The third category of stewardship code is developed by third parties, mostly independent organizations in the field of corporate governance, with South Korea and Singapore as representative countries. South Korea introduced the draft document of the *Korea Stewardship Code* in 2016 through the Financial Services Commission to regulate institutional investors' fulfillment of investment responsibilities and encourage contribution to reducing GHG emissions. The NPS adopted the code starting in 2018, and as of August 31, 2022, the number of participating institutions has jumped to 193 [30]. Singapore, on the other hand, saw the release of the revised version of the *Singapore Stewardship Principles for Responsible Investors* by the Stewardship Asia Centre, established by Temasek, in 2022. In response to the latest market developments, this revised version incorporates ESG principles into investment decision-making. While compliance with these principles remains voluntary, the revised version encourages signatory companies to submit evidence of stewardship governance to the Secretariat of the Steering Committee annually. The revision has received support from the Monetary Authority of Singapore (MAS) and the Singapore Exchange (SGX).[9]

5 Practices and Policy Context of Sustainable Investing by Chinese State-Owned Investors

According to the annual reports released by the Global SWF from 2021 to 2023, the total assets managed by Chinese state-owned investors reached US $3.5 trn at the end of 2022, approximately 20% of the country's GDP in 2022, with an average annual growth rate of about 9.7% over the past three years. In particular, state-owned financial enterprises such as China Investment Corporation (hereinafter referred to as "CIC") constitute a major component of sovereign wealth funds. The total assets of CIC reached US $1.4 trn in 2022, with a CAGR of about 13.6% from 2020 to 2022.

5.1 Practices at the Institutional Level

5.1.1 Sovereign Wealth Fund: CIC

In 2022, the scale of assets managed by sovereign wealth funds decreased for the first time in history, dropping from US $11.5 trn in 2021 to US $10.6 trn in 2022. During the same period, the assets of CIC grew by 11%, overtaking the GPFG as the world's biggest sovereign wealth fund.

In 2021, based on its own experiences and peer practices, CIC issued the *Sustainable Investment Policy* [31]. To address global climate change and contribute to the realization of carbon peaking and carbon neutrality goals, CIC issued the *Guidelines on Attaining Carbon Peak and Carbon Neutrality Goals and Practicing Sustainable Investing* (hereinafter referred to as "*Guidelines*") in May 2022. The *Guidelines* presented the company's roadmap for achieving carbon neutrality and reducing portfolio emissions over the next five years and beyond.

[9] https://www.aprea.asia/knowledge-hub/featured-insights/esg/stewardship-asia-centre-launches-new-edition-of-singapore-stewardship-principles-for-responsible-investors.

5.1.2 Public Pension Funds: NSSF

The investment philosophy of the NCSSF originates from its mission to ensure fund security and maintain and increase the value of assets. As specified in the *Regulations on the National Social Security Fund*, the NCSSF "should prudently manage and operate the National Social Security Fund, ensure security, seek returns, and focus on long-term results". In recent years, the NCSSF has made active ESG efforts. In 2022, the NCSSF issued the *Guidelines of the National Council for Social Security Fund for Industrial Investment* [28] to clarify the medium- to long-term development plan for industrial investment In June 2022, the Ministry of Finance and other four departments issued the *Measures for the Management of the Budget Performance of the Social Insurance Fund*, pointing out that the "performance indicators of the social security fund mainly involve economic performance, social performance, sustainable performance, satisfaction, etc.". In addition to the economic performance, sustainable development is also explicitly incorporated in the performance indicators.

5.2 *Status Quo of Policy Systems*

China has initially established a macro policy system that supports low-carbon transformation. These efforts have further improved the "1 + N" policy system for carbon peaking and carbon neutrality.

However, compared to the U.S. *Inflation Reduction Act* and the EU's *Green Deal Industrial Plan*, China's existing decarbonization policies lack explicit and quantitative requirements for funding support. For example, China's Ministry of Finance issued the *Opinions on Financial Support for Carbon Peaking and Carbon Neutrality* in 2022.[10] It outlines the key targets and areas of financial support, but remain at the macro level and lack detailed explanations of financial support, without setting clear targets for the amount or proportion of funds to be allocated.

Currently, China's green financial products mainly consist of green loans and green bonds, with green loans accounting for more than 80% of all green financial products as of the end of 2022.[11] However, the development of funds, insurance, and carbon financial products has been slow.

While China has introduced a wide range of national low-carbon and green finance policies, the country has almost yet to release few corresponding investment rules, and the effectiveness and constraint of the relevant documents differ from those adopted by developed countries.

A comprehensive stewardship code applicable to all industries has yet to be established. Based on the development of mature capital markets, as the market matured,

---

[10] Ministry of Finance, 2022. http://www.gov.cn/zhengce/zhengceku/2022-05/31/content_5693162.htm.

[11] The Study Group estimated based on the loan data released by the People's Bank of China and the scale of other green financial products in China Responsible Investment Annual Report.

institutional investors have become more involved in the corporate governance of listed companies. In this respect, practices in other countries and regions provide valuable and ample references for China's capital market.

In conclusion, China faces major challenges in its low-carbon transition. Bank credit alone is insufficient to meet the funding needs of RMB139 trn for achieving carbon neutrality [32], which requires broader participation from financial institutions. Meanwhile, the pursuit of high-quality development fully aligns with the broader concept of economic, environmental, and social development defined by sustainable investing, with clear demands on channeling more funds into the environmental and climate sectors. However, as China's green financial system is still developing, it has yet to provide sustainable investment rules and a comprehensive stewardship code applicable to all industries. At the current stage, asset owners are still exploring ESG practices, which demand policy support and official guidance.

6 Policy Recommendations

To facilitate the realization of carbon neutrality and carbon peaking goals, China's green financial system needs to expand from its current model of green loans and green bonds issued by commercial banks to include systemic support from a broader range of financial institutions. This report, supported by CCICED, provides recommendations for policymakers, regulatory authorities, and state-owned investors based on sustainable investment practices of state-owned investors worldwide and examines the status quo and flaws of sustainable investment in China.

6.1 Recommendations for Policymakers and Regulatory Authorities

Recommendation 1: First, efforts could be made to improve the policy framework for sustainable investing, establish effective incentive and constraint mechanisms, develop sound frameworks and mechanisms for green finance, and consistently optimize the low-carbon transition policy system.

First, in terms of incentives and constraints, regulators could encourage state-owned investors to allocate a certain percentage of their funds to sustainable investment and financing, allowing pilot demonstration funds to incorporate ecological and environmental values into performance evaluation systems. This would increase flexibility in investment return requirements, and encourage innovative utilization of risk-sharing tools. At the same time, regulators could require state-owned investors to develop clear sustainable investment principles to gradually reduce the environmental and climate impacts of their operations and their portfolios. The developed investment principles could establish clear strategic objectives and organizational safeguards, and set out clear requirements for working arrangements such as carbon emissions verification and disclosure. Regulators could also promote a fair transition to address the impact on socially vulnerable groups. This could be achieved through supportive policies that offer more effective community services, reemployment programs, and training and unemployment benefits. While addressing climate and environmental issues, regulators could also promote inclusive social development, with a focus on matters such as livelihood, employment, and gender equality at the community level.

Second, regulators could issue a stewardship code to encourage institutional investors, including state-owned investors, to exercise active ownership and press asset management institutions to make sustainable investments. A stewardship code will also allow authorities to regulate institutional investors' stewardship practices in at least four areas: (1) establish and disclose stewardship policies on sustainable investing, (2) supervise and engage with investee companies to enable sustainable investment practices, (3) disclose voting principles and measures adopted for climate and environmental issues, and (4) report stewardship duties to clients and beneficiaries. Also we recommend including asset owners, asset management institutions, and relevant service providers in the stewardship code to unify the behaviors of key decision-makers in the capital chain and serve the long-term interests of beneficiaries.

Third, a green financial system with sound mechanisms could be established to expand the implications of green finance and provide effective incentives and constraints for non-bank financial institutions. Efforts could be made to unify taxonomies, with prompt updates that include the latest green technologies. Authorities could also introduce mandatory climate and environmental disclosure requirements to promote convergence with international rules, such as the IFRS® Sustainability Disclosure Standards of the ISSB. Viable innovation of financial products could be encouraged to enrich investment targets, such as SLBs, transition bonds, green insurance, and REITs that focus on renewable energy infrastructure. We also advise regulators to prompt institutional investors to incorporate green investing into their evaluation systems, channeling private capital into frontier carbon neutrality technologies and facilitating the green transition of companies and industrial chains. When building a green financial system, state-owned investors and other financial institutions need to work together to form an ecosystem of sustainable investment. This covers four areas. First, strategic synergy, that is, convergence in strategic goals. The second is process coordination, which covers the whole life cycle from project development, investment decision-making, risk control to exit. Third is product synergy, in which financial institutions can better understand the investment characteristics of state-owned investors, and provide them with more choices by developing rich financial products.

Fourth, continued efforts are needed to optimize policy support for the low-carbon transition and increase fiscal and monetary support for innovative green and low-carbon technologies, achieving a combination of efficient markets and well-functioning government. Despite such efforts, the core objective of state-owned investors is still to yield high returns for beneficiaries and achieve appreciation. Therefore, channeling more funds into the green sector fundamentally requires enhancing the risk-adjusted returns of green projects. From this perspective, the key to promoting sustainable investing by institutional investors is to appropriately intervene in the real economy to improve the correlation between sustainability performance and financial performance. A range of policy tools would be required to achieve this. (1) Establish a return-on-investment mechanism to improve predictable cash flow such as increasing public investments in clean energy infrastructure and R&D, providing subsidies, tax exemptions, and fiscal interest subsidies for low-carbon projects. (2) Lower the financing costs of green projects, such as by establishing refinancing tools. (3) Improve market mechanisms for monetizing green value (e.g., electricity markets, carbon markets, state certified emission reduction, green power certificates, and other trading platforms) and to send clear policy signals to investors. (4) Build a risk-sharing mechanism among governments, financial institutions and investors, and use tools such as the first-loss layer, guarantee and insurance to reduce the risks of green projects. (5) Encourage diversified capital market participation and improve market exit mechanisms. These interventions in the real economy can strengthen the correlation between sustainability performance and financial performance, allowing investors to guide capital toward the realization of sustainable goals in a manner consistent with their financial objectives and obligations.

Recommendation 2: State-owned investors could be encouraged to actively participate in multilateral cooperation mechanisms and initiatives in areas of international consensus. As a result, they can play a greater role in the development of international rules and standards concerning taxonomies, information disclosure, transition finance, and climate risk management.

First, policymakers could encourage state-owned investors to actively participate in multilateral cooperation mechanisms and initiatives in areas of international consensus and advocate for rules and propositions beneficial to China through statistics and case studies, promoting consensus in relevant fields. China has promoted and engaged in the development of a series of multilateral cooperation mechanisms, including the UNEP FI, the SBFN, the G20 Sustainable Finance Working Group, and the NGFS, which laid the groundwork for further participation in multilateral mechanisms. For example, the G20 Green Finance Study Group, established through joint efforts by China and the UK in 2016, was upgraded to a working group in 2021 and developed the G20 Sustainable Finance Roadmap. This move consolidated the consensus among countries on financial support for the green transition and provided a basis for systematically advancing sustainable finance worldwide. In the future, policymakers can guide state-owned investors to selectively join the formulation of international guidelines or global initiatives and actively engage in core processes and key organizations, such as the PRI and ISSB. Such organizations play a crucial role in setting standards, regulating disclosures, promoting transition finance, and enhancing climate risk management, with extensive participation from state-owned investors worldwide.

Second, policymakers can promote cooperation in green finance and transitional finance at both multilateral and bilateral levels, further unify taxonomies, enhance information disclosure, improve the policy system for green finance, and pave the way for state-owned investors to make sustainable investments. Specifically, China and

Europe could develop a transitional finance taxonomy based on the *Common Ground Taxonomy*. They could also promote the adoption of the *Common Ground Taxonomy* or its adaptation as the basis for developing national taxonomies through platforms like the IPSF. China and Europe could keep improving the taxonomy while addressing practical challenges, such as providing a clearer definition of "do no significant harm" to avoid divergent interpretations. Additionally, the EU has strengthened its standards for determining green investing. The latest Proposal for a Directive on Green Claims requires environmental claims to be independently verified by third parties and supported by scientific evidence to reduce "greenwashing" risks. In this regard, China can collaborate with the EU and adopt leading practices in calculating, managing, and disclosing carbon footprints.

Third, state-owned investors can be guided to play a greater role in actively leading the development of more pragmatic international sustainable investment and financing activities to explore the establishment or co-establishment of a green alliance or a global investment fund together with state-owned investors from around the world. This can facilitate exchanges among domestic and international investors on responsible investing. Platforms established through organizations like China Sustainable Investment Forum (ChinaSIF) and Norwegian Sustainable Investment Forum (NorSIF) can invite sustainable investing leads from sovereign wealth funds to share their practices. The annual ChinaSIF forum also serves as a channel for popularizing the concept of responsible investing.

6.2 Recommendations for State-Owned Investors

Recommendation 1: Climate and environmental factors could be incorporated into key areas such as corporate governance and investment decision-making, and consistent improvements of the sustainable investment framework are required.

First, state-owned investors could recognize low-carbon transition as a strategic goal, adjust the organizational structure, and establish decision-making processes that better account for climate and environmental factors. Sustainability factors could be systematically considered at the institutional level, which involves the establishment of low-carbon transition roadmaps, sustainable investment strategies, sustainable investment teams, and investment frameworks. It is important to note that sustainable investment policies must align with the obligation to generate returns. Failure to achieve this consistency may lead to political and regulatory risks, reputational risks, and litigation risks associated with an excessive emphasis on sustainable investing, particularly narrowly-defined ESG investment strategies. In practice, sustainable investment strategies require organizational structure and talent support. For instance, the Chief Sustainability Officer (CSO) of the CPPIB provides professional advice for environmental and climate-related proxy voting by its board; sovereign wealth funds in Norway have had an environmental investment strategy and a team dedicated to the strategy since a very early stage, which allowed them to accumulate extensive experience. As the strategies for environment and climate became more integrated, these fund managers were integrated into the entire investment team, facilitating the spread of environmental and climate-related expertise within the organization.

Second, the carbon footprint of portfolios could be managed by employing scientifically sound measurement methods. Specifically, efforts could be made to build carbon footprint calculation models by sector, collect and organize sector-specific data on carbon emissions intensity, and determine the applicable weighting methods. State-owned investors can also compare their portfolio's carbon footprint with international benchmarks and actively disclose the benchmarking results in annual reports or ESG reports. Climate stress tests for specific portfolios could be carried out to improve the capabilities for managing climate-related physical risks and transition risks. In addition to calculating and disclosing their own carbon footprint, state-owned investors can also actively influence their respective asset management institutions to calculate and disclose carbon footprints. In order to create comparable data sources, state-owned investors could recommend that outsourced asset managers adopt international standards for disclosure, while refraining from intervening in specific disclosure practices to avoid adding excessive administrative costs to the asset managers.

Third, state-owned investors can participate in relevant international initiatives and adopt leading international practices. Moreover, involvement in standard-setting institutions and international organizations focusing on the research of key agenda allows state-owned investors to contribute to the formulation of international rules and stay ahead of peers.

Fourth, during the investment process, climate and environmental factors could be considered as appropriate when screening and evaluating institutional investors and fund managers. State-owned investors could, for example, include climate and environmental indicators in compensation incentive mechanisms. They may also build a sustainable index system for investment institutions and fund managers, systematically evaluate their sustainable investment capabilities during the screening process, and monitor their performance on an ongoing basis.

Fifth, efforts could be made to exercise stewardship based on active ownership. To begin with, state-owned investors could establish effective governance structures, institutional rules, and decision-making processes to support and guarantee the implementation of stewardship practices. Additionally, they may build a self-assessment system for sustainable investing (or an ESG scoring system) or rely on third-party services to evaluate the sustainable governance of asset management institutions and investee companies (both listed and non-listed entities) and identify substantial issues. Furthermore, based on the assessment results, state-owned investors could urge asset management institutions and fund managers to fully incorporate climate and environmental factors into investment decision making through voting, meetings, or written notifications, with a focus on substantial issues identified at key companies. Finally, state-owned investors could also continue to monitor and access the engagement outcomes and optimize engagement strategies to achieve the low-carbon transition of investee companies.

Sixth, we recommend establishing sustainable investment guidelines for various asset classes and building a sustainable investment benchmark system to play a guiding role in investment across all asset categories. In addition to broad-based indexes, customized ESG benchmark indexes could be developed. In equity investments, the adoption of ESG integration and negative screening strategies could be scrutinized, and a company's ESG performance, particularly environmental factors, could be incorporated into investment decision-making. For example, based on a pre-determined scoring system, when a target company's ESG score falls within the bottom 20%, state-owned investors could require written justifications for its inclusion in the portfolio. Asset managers could incorporate ESG factors into their overall risk-return assessments, instead of simply assessing the volume of ESG investments. State-owned investors could also carefully consider the adoption of negative screening strategies. Even if a sovereign asset owner divests itself from fossil fuels, any excess returns may quickly be offset by other funds, which diminishes the practical significance of addressing climate change. Globally, many state-owned investors also believe that pressing companies to engage in low-carbon transition is preferable to blunt divestment strategies. Of course, if positive outcomes cannot be achieved through active engagement over a period of time, "voting with your feet" can still have an impact, as it demonstrates the resolve of state-owned investors.

Seventh, state-owned investors could proactively engage in sustainability themed investing. For instance, they may invest more in green technologies, renewables, and green supply chains. Projects in these areas come with long investment cycles and relatively stable returns, which align with the long-term capital of state-owned investors and present lower stranded risks. State-owned investors could also develop sound risk-sharing mechanisms based on the overall conditions of their portfolio, with appropriate capital allocation. Additionally, state-owned investors could join hands with peers or other institutional investors to establish a global investment fund focusing on the green sector to share experiences and risks. With the cost reduction of photovoltaic, wind power and other costs and the enhancement of competitive advantages, a large amount of capital has entered the renewables sector. Under this background, sovereign asset owners can consider setting up impact investment sub-funds to guide funds into relatively immature fields, such as ecological environmental protection, soil and air pollution control, biodiversity protection and other fields.

References

1. Climate Policy Initiative. (2021). *Global landscape of climate finance 2021.*
2. International Energy Agency. (2023). *Credible pathways to 1.5°C: Four pillars for action in the 2020s.*
3. International Renewable Energy Agency. (2020). *Renewable energy finance: Institutional capital.* IRENA.
4. Global SWF. (2023). *2023 annual report: "State-owned investors in a multipolar world".*
5. Schroders. (2022). *Sustainability: Institutional investor study 2022.*
6. Global Sustainable Investment Alliance. (2020). *Global sustainable investment review.* GSIA.
7. UN-Convened Net-Zero Asset Owner Alliance. (2022). *Advancing delivery on decarbonization targets: The second progress report of the net-zero asset owner alliance.* NZAOA.
8. The Institutional Investors Group on Climate Change. (2022). *Annual report 2022.* IIGCC.

9. BNY Mellon. (2022). *The evolution of public asset owners.*
10. National Pension Fund. (2021). *2021 annual report.*
11. Norges Bank Investment Management. *Climate change: Expectations of companies.*
12. Government Pension Investment Fund. (2022). *2021 ESG REPORT*. GPIF.
13. Norges Bank Investment Management. (2021). *Responsible investment government pension fund global 2021.*
14. Asset Management Association of China. (2021). *Institutional investors' road to "carbon neutrality": Practical experience of overseas pension funds and sovereign wealth funds.*
15. Alessandrini, F., et al. (2020). *Optimal strategies for ESG portfolios.*
16. California Public Employees' Retirement System. (2015). *Global governance principles.*
17. GSIA. (2020). *Global sustainable investment review 2020.*
18. Climate Bonds Initiative. (2023, March 19). *A Q3 2022 overview of the global sustainable bond market.* https://www.climatebonds.net/files/reports/cbi_susdebtsum_highlq32022cn.pdf
19. The Ministry of Finance. Norway. *Guidelines for observation and exclusion of companies from the government pension fund global.* https://www.regjeringen.no/contentassets/9d68c55c272c41e99f0bf45d24397d8c/guidelines-for-observation-and-exclusion-of-companies-from-the-gpfg-19.11.2021.pdf.pdf
20. The Asset Management Association Switzerland. Portrait. https://www.am-switzerland.ch/en/ueber-uns/portraet
21. SFAMA&SSF. *Sustainable asset management: Key messages and recommendations of SFAMA and SSF.* https://www.sustainablefinance.ch/upload/cms/user/EN_2020_06_16_SFAMA_SSF_key_messages_and_recommendations_final.pdf
22. ASIP. ESG-Wegleitung für Schweizer Pensionskassen. https://www.asip.ch/media/filer_public/30/ec/30ec7b97-8107-4e65-b3e9-55a4b528102b/asip_esg-wegleitung.pdf
23. Asset Management Association of China, Ziding Shareholder Service Agency. (2021). *Institutional investors participating in listed company governance overseas regulations and practices compilation.* China Finance and Economics Publishing House.
24. Financial Reporting Council. (2020). *UK stewardship code: Guidance for investors.* FRC.
25. The Secretary of State. (2019). *The occupational pension schemes (investment and disclosure) (amendment) regulations 2019.* London: Parliament.
26. Financial Services Agency of Japan. (2020). *Code of due diligence for Japanese companies.* Financial Services Agency of Japan.
27. China Investment Corporation. (2023, August 03). *Opinions on fulfilling dual carbon goals and sustainable investment actions.* http://www.china-inv.cn/china_inv/Media/2022-05/1002031.shtml
28. National Council for Social Security Fund. (2023, August 03). *Guidelines for industrial investment of national council for social security fund.* http://www.ssf.gov.cn/portal/rootfiles/2022/09/27/1665925788891218-1665925788914601.pdf
29. China Banking Association. (2009). *Guidelines on corporate social responsibility of Chinese banking financial institutions.* https://www.china-cba.net/Index/show/catid/14/id/734.html
30. Social Impact Investment Alliance. (2022). Thematic Insights I Proxy Voting Season for 2022 General Meeting of Shareholders from the ESG Perspective (Part 1)[EB/OL].(2022-12-28)[2022-03-11]. https://xueqiu.com/1145329483/238712805
31. China Investment Corporation. (2021). *Sustainable Investment Policy [EB/OL].* [2023-03-08]. http://www.chinainv.cn/china_inv/Media/2021-11/1002006.shtml
32. CICC Global Institute, CICC Research Department. (2021). *Carbon Neutrality Economics: Macro and Industry Trends under New Constraints [M].* CITIC.

Chapter 10
Sustainable Development Innovation Mechanism Boosted by the Belt and Road Initiative

1 Foreword

Marking the 10th anniversary of the Belt and Road Initiative (BRI), the year of 2023 ushers in a new phase and major opportunities for the development and cooperation of green BRI. Meanwhile, as climate change, energy security, eco-environment conservation and other global challenges are intertwining with each other, more institutional innovations for BRI are required to drive the process of sustainable development. With global consensus on climate change, coupled with energy security risks caused by geopolitical conflicts, more and more countries are developing their own renewable energy industries to ensure the energy supply in a more low-carbon, safe and sustainable way. Most of the BRI participating countries are from the developing world.

On September 21, 2021, at the General Debate of the 76th Session of the United Nations General Assembly, Chinese President Xi Jinping announced that China will step up support for other developing countries in developing green and low-carbon energy and will not build new coal-fired power projects abroad. Renewable energy has become a key area in building the Green Silk Road. Based on our previous studies, this report will focus on BRI's renewable energy cooperation, to sort out the best practices and cooperation demands for sustainable financing for green and low-carbon development, and map out the overall cooperation mechanism for green BRI. Through these efforts, this study proposes relevant policy recommendations for the innovation mechanism of green BRI.

Green Empowerment and High Quality Development,
https://doi.org/10.1007/978-981-96-4218-2_10

2 Best Practices and Needs for BRI Sustainable Financing for Green and Low-Carbon Development

2.1 International Cooperation Mechanism for BRI's Green and Low-Carbon Development

As an important builder, participant and leader in global ecological conservation, China has established stable international cooperation mechanisms with the BRI participating countries in ecological conservation and green and low-carbon development. Relying on dialogues and exchanges conducted across multiple levels, China has maintained stable and pragmatic bilateral ties with major BRI participating countries and regional organizations, which are multi-level and broad in scope. At the same time, the multilateral mechanism, as the best solution proposed by China for international cooperation to tackle global climate change, introduces subjects, perspectives, and voices into the construction of the green BRI. Such an approach has provided a platform for communication and cooperation between China and the BRI participating countries.

2.1.1 Bilateral Cooperation for Steady and Efficient Development

China currently maintains stable and pragmatic bilateral ties with major BRI participating countries and regional organizations, which are multi-level and broad in scope. This mutual support will ensure the constant development of eco-environmental cooperation. The two sides have established a mechanism for regular meetings between prime ministers and ministers, such as the China-ASEAN Environmental Cooperation Forum and the ASEAN + 3 Environment Ministers Meeting, to provide direct channels for policy coordination and information exchange. China has also set up environmental cooperation centers with key countries in the region, such as the China-ASEAN Environmental Cooperation Center (CAEC), the China-SCO Environmental Cooperation Center (CSEC), the China-Africa Environmental Cooperation Center (CAECC), to carry out multi-sectoral events concerning ecological conservation. Through project cooperation, capacity building and other activities, the China-CEEC 16 + 1 Cooperation Mechanism and the China-Pacific Island Countries Economic Development and Cooperation.

2.1.2 Multilateral Cooperation to Promote Diversified Exchanges on Green and Low-Carbon Areas

The multilateral cooperation mechanism established by China and BRI participating countries is being renewed day by day. It provides cooperation platforms covering all sectors and broad themes, with a special focus on green investment and financing, green technology and other targeted fields. Such a mechanism has introduced diversified voices into low-carbon development. The Belt and Road Initiative International Green Development Coalition (BRIGC), proposed by President Xi Jinping, has been the first international environmental social organization within the green BRI framework. BRIGC has attracted more than 150 partners from over 40 countries to promote practical cooperation in areas of policy coordination, exchanges and dialogues, capacity building, and environmental technology in building the Belt and Road. BRIGC has become an international cooperation platform on environment.

2.2 *Status-Quo of BRI Green Energy Investment*

2.2.1 Green Energy Investment and Financing

In 2022, global investment in low-carbon energy reached the same level as fossil fuels for the first time, at USD 1.1 trillion. The majority of this investment was directed to renewable energy (USD 495 billion), closely followed by electrified transport (USD 466 billion). Despite this progress, investment needs to increase significantly to limit warming to Paris-compatible temperatures, particularly in low- and middle-income countries along the BRI participating countries. However, these countries often face insufficient domestic investment and thus rely on international financial support.

Chinese financial institutions already play a pivotal role in providing cross-border green energy investment. Given the increasing demand of countries for green investment and the diminishing role of multilateral development banks (MDBs), China can leverage its experience and reputation to manifest itself as a global sustainable leader by ramping up support for countries to transition to low-carbon power, transportation and heating systems. This is in line with efforts to green the BRI while bringing economic benefits for China's renewable energy industries, as discussed in part two of the report.

2.2.2 China's Overseas Green Energy Investment

The BRI has become a widely-welcomed international public product and cooperation platform. From 2000 to 2021, China provided USD 235 billion in development finance for overseas energy projects, which is more than MDBs have provided in the same period. In 2022, financing and investment totaled USD 67.8 billion, with more than a third—USD 24.1 billion—going to the power, transportation, and heating sectors.

Globally, the relatively low supply of Chinese finance for renewable energy compared to traditional energy sources is related to high perceived risk and limited demand for solar and wind, which can partly be explained by high financing costs due to risk perception. However, with a changing international policy landscape, growing awareness of climate change and drastic decreases in solar costs, recipient countries increasingly demand a supply of green finance and investment.

In terms of renewable energy, China has become a key player on international investment in recent years, signing agreements with over 100 countries to develop low-carbon electricity generation. The share of renewable energy financing and investment in total BRI-related financial commitments has been steadily increasing. In 2022, Chinese financial institutions committed around $6 billion to the renewable energy sector, though the numbers vary greatly across different sources.

Given BRI's focus on improving connectivity through infrastructure investment, it is also important to note that China's development finance institutions have provided $15 billion in loans for grid construction and upgrades and electricity transmission and distribution infrastructure.

Aside from renewable energy generation, China has also increased investment in battery storage facilities. In 2022, projects were implemented in Hungary, Germany and the U.S. Investment figures for other sectors such as green transportation and green heating are not available, and these sectors could be targeted in future work. However, data on solar and wind projects are available and are analyzed in the next section.

2.2.3 Chinese Overseas Financing for Solar and Wind

China's support for overseas solar and wind projects takes various forms. Historically, China mainly served as an equipment supplier and engineering contractor. In recent years, however, Chinese companies have increasingly provided FDI, including greenfield investment and M&A.

China's support for overseas renewable energy projects has increased significantly in recent years, totaling 25.3 GW. The increase began around 2015 for wind and in 2018 for solar. Of the total, 6.2 GW is financed through loans, split between commercial banks (5.5 GW) and the two policy banks (0.7 GW), while 22.7 GW is equity financed by individual companies. This results in an overlap of 3.5 GW between the two forms of financing, representing projects co-financed by FDI and loans. Solar projects account for 12.1 GW, while wind projects account for 13.1 GW.

In contrast to the financial commitments of Chinese institutions, MDBs financed renewable energy projects with a total capacity of 39.1 GW, split almost evenly between wind and solar energy. MDBs have thus financed more renewable energy capacity than Chinese institutions. The difference is about 7.4 GW more for solar capacity and 6.4 GW more for wind capacity.

MDBs have supported around 241 GW of energy capacity worldwide (in operating and in the pipeline as of 2020). However, their total financial commitment to the power sector peaked in 2010. Chinese financing, on the other hand, has increased significantly since 2010. Chinese institutions are now among the top financiers of power projects, with 151 GW across operational and pipeline projects.

The regional distribution of financial commitments shows similar patterns for both Chinese financial institutions and MDBs, although the total amount of investment varies significantly between regions. Most financial transactions by both China and MDBs are directed towards solar and wind energy projects in the Middle East and Africa, while the least amount is directed towards North America. MDBs are much more involved in Asia–Pacific, while Chinese institutions and firms have supported more renewable energy capacity than MDBs in Europe and South and Central America.

The share of Chinese investment in solar and wind projects varies greatly by region. In the Middle East and Africa, solar capacity is 14 times greater than wind capacity due to large solar projects in the United Arab Emirates, which is in line with the region's abundant solar radiation. In all other regions, the capacity of wind projects financed by Chinese institutions is greater than that of solar projects.

Looking at regional investment over time, cumulative solar and wind capacity with Chinese financial support in Asia–Pacific and Europe saw an almost linear increase starting from around 2010. Projects in the Middle East and Africa, as well as in South America and Central America, saw a rapid, exponential increase in 2015. All regions witnessed a drop in capacity additions from 2020 onwards. Based on existing financial commitments, additional capacity is expected in 2024, with most capacity added in the Middle East, Africa and South America and Central America.

2.2.4 Future Investment and Finance Needs

To achieve net-zero emissions globally by 2050, annual global investment in green energy needs to increase from $1.38 trillion in 2022 to around $4 trillion. This will require a 30% increase in planned energy investment by all countries as well as a shift from fossil to green energy sources. Thus, there is currently a large investment gap particularly for low-income countries, many of which are part of the BRI.

An alternative way to quantify the investment gap is to compare current investment levels with the amount needed for countries to fulfill their Nationally Determined Contributions (NDCs) to the Paris Agreement on climate change. All developing countries combined require around USD 1 trillion (671 GW) in renewable energy investment, and for BRI participating countries it is around USD 469 million.

This green investment gap presents a unique opportunity for China to manifest its role as a global sustainable leader. Chinese financial institutions can provide green investment to meet the growing demand from countries. Additionally, China can proactively develop low-carbon infrastructure to meet its stated commitments. Such green investments could reduce emissions by hundreds of millions of tons of carbon dioxide.

To support other countries in their low-carbon development, China can finance research and development (R&D), equipment, and infrastructure. Urgent capital supply is needed for green transportation, sustainable buildings and renewable energy technology. Large upfront investments are particularly crucial to accelerate the deployment of low-carbon technologies such as solar and wind. To achieve net zero by 2050, global investment in solar needs to increase from USD 115 billion to USD 237 billion per year until 2050, while (onshore) wind investment should increase from USD 98 billion to USD 389 billion.

It is worth noting that not all countries require the same level of Chinese investment. Regions such as Africa and Southeast Asia are expected to see a significant increase in their electricity supply from renewable technologies far beyond their current installed power capacity. To meet projections, these regions need to increase annual mitigation investment flows by up to 16 times by 2030, compared to a 2–3 times increase needed in Europe.

However, sustainable investment in these regions is insufficient due to underdeveloped financial markets and high financing costs, especially for solar and wind power. Underinvestment creates a climate investment trap: as climate change increases the perceived risks to investors, it raises the barriers to sustainable investment even further. As a result, low- and middle-income BRI participating countries in these regions heavily depend on Chinese finance and investment. Comparing future needs with current investment patterns, China could increase its engagement in Africa and Southeast Asia, while ensuring that it also resumes investment in Latin America. This would enable these regions to pursue low-carbon development pathways through Chinese green energy investment.

2.2.5 Management Needs of BRI Green Energy Investment and Financing Projects

The success of China's efforts to green the BRI depends not only on the amount of investment, but also on the effectiveness and longevity of each project. The previous Green BRI SPS emphasized the need for a "whole lifecycle" approach to renewable energy project development. These best practices can ensure the successful implementation of projects abroad by managing each phase of the project.

Policymakers and other entities have multiple options to improve compliance. China is actively implementing the green BRI guidelines for outbound investors and aligning with best practices. Empirical evidence also demonstrates that co-financing for China's overseas lending and development finance projects is associated with better project completion rates and environmental outcomes, suggesting that cooperation can improve practices [1].

2.3 Innovative Financing Mechanisms to Promote Green and Low-Carbon Development of BRI Energy Projects

The Green BRI Special Policy Study Report in 2022 outlined three innovative financing mechanisms for green and low-carbon development of BRI energy projects. First, it suggested supporting Chinese contractors and investors to participate in renewable energy projects abroad through blending grants and loans, setting up special funds for grants or loans and establishing a BRI Public–Private Partnership (PPP) Project Development Facility. Second, Chinese financial institutions can expand their portfolios by improving their capacity to provide comprehensive services, engage in project preparation and introduce country-specific financing schemes. Third, it suggested supporting renewable energy investors with long-term equity stakes in PPP projects. Options to do so include a range of diplomatic efforts, such as intergovernmental cooperation and economic instruments, including an investment security mechanism.

Despite the innovative mechanisms outlined in last year's Green BRI SPS, many developing countries along the Belt and Road face major challenges in financing renewable energy development projects. This section elaborates on the specific barriers for BRI participating countries in renewable energy project financing and explores in detail how Chinese financial institutions could provide innovative financing.

2.3.1 Renewable Energy Project Development Lifecycle

This section provides an overview of the different phases of the renewable energy project development lifecycle, the standard scope of work or tasks to be completed in each phase, and the expected outcome in terms of renewable energy deployment. It builds on the "whole lifecycle" approach expounded above. This information is a prerequisite for discussing the barriers to renewable energy development in Belt and Road countries, and solutions to overcome them.

The general project development lifecycle encompasses eight phases: creating an enabling environment; conceptualizing the project; conducting a pre-feasibility study; assessing bankable feasibility; financial close; construction; operations and maintenance; and decommissioning and repurposing. The previous Green BRI SPS emphasized the need for a "whole lifecycle" approach to renewable energy project development. The "whole lifecycle" approach involves a comprehensive and inclusive process of project selection, design, implementation, and closure. Project preparation includes consultation with government and community stakeholders to identify the type of project that is most likely to be successful and address local priorities, such as reducing air pollution by transitioning away from coal-fired power plants. Next, project design considers the environmental and social impacts of project development, including land and water use, on affected community members. As implementation begins, project managers can cultivate partnerships with local suppliers for both direct and indirect inputs.

2.3.2 Challenges for Financing Green and Low-Carbon BRI Energy Projects in Developing Countries

Based on this understanding of the project development lifecycle, there are clear challenges that Chinese financiers may face when seeking to support green and low-carbon energy projects in BRI participating countries, especially in the early stages of project development.

First, there may not be a sufficiently enabling environment. Successful project development starts with the creation of an enabling environment by the host government. Appropriate policies, such as renewable energy targets and a transparent framework for the procurement of the renewable energy projects, are key, as they provide potential investors with the assurance that the projects will be procured according to existing legislation or procurement frameworks.

Second, there may be limited technical capacity in the host country to carry out key steps such as environmental assessments, financial modeling, market studies and project information memoranda. This is relevant for projects based on newly commercialized and new-to-market low-carbon technologies, including renewable energy projects in most developing economies.

Third, early-stage processes may be prohibitively expensive. A high-level review or assessment of a project's environmental impact, grid connectivity, resource potential, energy costs and electricity tariffs can require significant capital outlay, particularly for a small start-up or new entrant.

Fourth, there is a lack of pre-feasibility funding for early-stage project development. In the early stages of the project development lifecycle, the project is not yet de-risked and there is little interest from potential investors. There is limited seed capital available for project scoping, site acquisition and initial design, which are pre-requisites for project preparation funding.

Fifth, despite the availability of funds for pre-feasibility studies, project developers may have stringent requirements for access to these funds. Most funds require the developer to contribute a portion of the required funding, commonly referred to as having "skin in the game."

Sixth, there is often a lack of related infrastructure to support project development. This means that although the site and resources may be ideal for a solar or wind power project, the developer may not be able to transmit the power if the grid connection near the project site is not on the utility's list of priority projects.

Finally, developers may not be able to secure off-take arrangements. In order to complete the pre-feasibility study, the developer must secure indicative off-take letters of interest. This process normally requires that the developer to have a strong balance sheet and several years of relevant experience to demonstrate that the developer has a track record. Thus, power infrastructure projects, including renewable energy projects, cannot make a profit through traditional methods such as billing. Developing countries urgently need to explore a profit model suitable for their national conditions.

Despite a general lack of support for the early phases of the renewable energy project pipeline, there are some existing pre-feasibility funds and project preparation facilities for renewable energy projects. In the context of a green and low-carbon Belt and Road, there are some examples of how MDBs have helped overcome some of the aforementioned barriers in the early stages of renewable energy project pipelines, making the projects more suitable for financing by China's FDIs and other financiers.

3 Renewable Energy Development Policy, Demand and Model Innovation for Belt and Road Countries

In this part, we looked into China's renewable energy industry, which has made remarkable achievements from scratch and caught up with the rest of the world over the past 40 years of reform and opening-up. Therefore, China's renewable energy development is of great reference for BRI participating countries in their developing stages.

This report compares the greenhouse gas reduction and renewable energy targets and related policies in major BRI participating countries. It also studies the current practices and future demands of their renewable energy industries. By combining these findings with China's policy and implementation experience in renewable energy, this chapter summarizes and puts forth insights about renewable energy growth and green low-carbon transformation in BRI participating countries, as well as practices of some major countries and regions in launching green energy cooperation.

3.1 Renewable Energy Development Policy Objectives, Practice and Demand in Key Belt and Road Regions

As of June, 2023, China has signed more than 200 documents with 152 countries and 32 international organizations, including 52 African countries (regions), 40 Asian countries (regions), 27 European countries (regions), 11 Countries (regions) in Oceania, 9 South American countries (regions), and 13 North American countries (regions). Based on indicators such as electricity access for the entire population, access for the rural population, the proportion of renewable energy and China's investment in them, this study selects the countries with great potential and urgency for renewable energy development from the BRI participating countries. This session focus on issues of renewable energy development targets, policies and practical experiences, as well as the pressures and challenges, and with reference to China's policy experience and successful cases. This session discusses the situations in BRI participating countries, analyzes the issues that need to be addressed in these regions, and the role that China, especially China's centrally-administered state-owned enterprises, needs to play.

3.1.1 Renewable Energy Development Policy Objectives, Practice, and Demand in Southeast Asia

Renewable Energy Policy Objectives of Major Southeast Asian Countries

Southeast Asian countries have always been important trade and international partners of China, with great potential for cooperation in renewable energy development. Southeast Asia boasts great renewable energy resources but is highly vulnerable to climate change, which has led to an increased focus on renewable energy development in recent years. With low-cost labor and land, Southeast Asia can attract global renewable industries where it has established strong industrial advantages in terms of both equipment manufacturing and deployment. In the updated nationally determined contributions (NDCs), major Southeast Asian countries set emissions targets with renewable energy targets as important emissions reduction instruments. However, IRENA's data finds that there remains a considerable gap between the current renewable energy capacity and the predefined targets.

Renewable Energy Practice in Major Southeast Asian Countries

Southeast Asian countries have put in place a variety of supportive policies and incentives for renewable energy development and clean energy transition. Based on different national conditions and the maturity of the renewable energy market, these countries have widely implemented various incentives. Renewable energy development targets have been widely used in Southeast Asian countries to indicate their overall renewable energy development goals. Consumption measures such as feed-in tariffs, self-consumption programs, competitive bidding/auctions, and the introduction of licensing mechanisms and technical standards have been mainly designed to increase renewable energy consumption. Fiscal policies such as tax incentives, preferential loans and capital subsidies have increased the returns from renewable energy projects and provided incentives for investment in these projects. As an emerging energy incentive tool, the mechanism for the "Green Electricity Certificate"[1] is in its experimentation and exploration stage.

The feed-in tariff (FiT) is the most sophisticated and critical driver for the largest installed capacity in Southeast Asian countries. The system has been adopted by countries such as Thailand, Indonesia, Vietnam, Malaysia and the Philippines in the early stages of their renewable energy generation markets. With changes in the local renewable energy generation technologies and costs as well as incentive targets, these countries have fine-tuned FiTs to satisfy the needs of different stages. Market-based trading of renewable energy is facilitated by measures such as bidding/auctions for feed-in tariffs.

[1] An electronic ID card for green electricity, which is the confirmation and attribute proof of non-hydro renewable energy generation and the only proof of green electricity consumption.

These measures vary from country to country. Many Southeast Asian countries have also streamlined power generation approval, market registration and settlement procedures of measures like feed-in tariffs. In addition, the investment and financing policies introduced by these countries have mobilized financial instruments such as public finances, development banks and related bonds and funds to support renewable energy. These policies represent innovative practices to promote industrial development through financial means. Despite the relatively small investment base for renewable energy in Southeast Asian countries, public finances and development banks still account for a large share and there will be more investment in the future.

Renewable Energy Development Demand of Major Countries in Southeast Asia

In recent years, Southeast Asian countries are putting more and more attention to renewable energy development, but many problems remain unaddressed due to the late start and the economic environment.

Heavy reliance on fossil energy and difficulties in achieving clean development goals. Southeast Asia boasts rich coal resources. The local energy supply is dominated by fossil energy while renewable energy meets less than 15% of demand. Most countries are already behind schedule in renewable energy development, except for Thailand and Vietnam, where solar capacity has been installed as planned. In this context, it is difficult for Southeast Asian countries to achieve their clean energy targets.

Less developed mechanism. Renewable energy projects require a large amount of land, but most Southeast Asian countries lack transparent land use permitting procedures. It is complicated to acquire, retain, and transfer land use rights. On top of that, there are issues like the period and costs of land purchase. As a result, renewable energy projects are paused by procedural delays and cost overruns. FiTs change frequently, so the return on investment is low. Investment has the potential to be more efficient. These institutional issues need to be addressed for renewable energy growth in Southeast Asian countries.

Insufficient transmission infrastructure. Renewable energy power generation and grid connection are universal problems. Areas that are resource-rich in renewable energies generally have limited consumption capacity. Transmitting renewable energy power relies heavily on grid transmission and regulation capacity. However, the grids of most Southeast Asian countries are less developed. Transnational and cross-border cooperation is still in its infancy and cannot transmit clean renewable energy power to areas with great power demands, stifling renewable energy growth.

Difficulties in investment and financing. In terms of policies and regulations, some Southeast Asian countries have poor regulations, underdeveloped capital markets and relatively high commercial risks. Renewable energy projects are small in scale with poor local financial markets and insufficient protection for project refinancing, exit and investment, which makes them less attractive to investors in the private sector.

3.1.2 Renewable Energy Development Policy Objectives, Practice and Demand in Central Asia

Renewable Energy Policy Objectives of Major Central Asian Countries

Central Asian countries—Kazakhstan, Uzbekistan, Turkmenistan, Kyrgyzstan, and Tajikistan—are rich in oil and gas and have been heavily relying on these traditional fossil energy resources. The five countries have been the major destinations of China's outbound energy investments. However, to achieve carbon neutrality, ensure energy security and transform the economic structure, Central Asian countries are facing pressure to transform their traditional energy mix based on hydrocarbon energy.

The five Central Asian countries are also rich in renewable energy resources such as wind, solar and hydro power. Although there are no quantitative renewable energy targets in the NDC document, developing alternative energy resources is the primary choice for greenhouse gas emission reduction in the region, according to each country's relevant action plans, laws, and regulations.

Renewable Energy Practice in Major Central Asian Countries

Differences in the endowment of fossil energy resources in Central Asia have led to different renewable energy planning and development. In the west, Turkmenistan is rich in hydrocarbon resources. These countries are more flexible and adaptable in their energy transitions and thus have greater potential.

In addition, various supportive measures are the foundation for renewable energy development in Central Asian countries. Kazakhstan has introduced a bidding mechanism for renewable energy; Kyrgyzstan has issued a presidential decree to carry out all-out research and promotion of dominant energy and established an energy development center; and Turkmenistan founded a specialized research center for solar energy in 2014 and promoted an energy conservation program in 2020 to develop solar and wind power generation programs in each province.

Renewable Energy Development Demand of Major Central Asian Countries

The five Central Asian countries differ greatly in terms of resources and energy, and this uneven distribution of water and oil remains prominent. Issues arising from the energy integration in Central Asia during the Soviet era compounded with the uneven distribution of water and oil resources resulted in disputes and differences in energy supply and demand among the five Central Asian countries that are now facing great challenges in energy independence. These are problems that Central Asian countries need to resolve with new mechanisms and institutional design as they transition towards a low-carbon energy system.

Fewer geopolitical conflicts and regional security are political and economic preconditions for renewable energy development. After the United States withdrew its troops from Afghanistan, Central Asian countries need to invest more energy and financial resources to maintain the regional balance of power. Therefore, Central Asian countries need to maintain regional stability and peace to provide a favorable business environment for renewable energy such as photovoltaic power and hydropower.

3.1.3 Renewable Energy Policy Objectives, Practice, and Demand in Africa

Renewable Energy Policy Objectives of Major African Countries

Africa is rich in natural resources as Northern African countries like Libya, Nigeria, Algeria, Angola, Sudan, Egypt, and Chad enjoy abundant oil and gas reserves while the vast continent is abundant in solar and wind energy resources. However, Africa's economic development is extremely uneven. Northern African countries are more developed economically thanks to their oil and gas exports and technological advantages. Therefore, renewable energy has become a unanimous choice for both North and South Africa. Like the uneven economic development, there is a huge gap between Northern Africa and Southern Africa in renewable energy policy objectives. Northern Africa is more ambitious as it has a stronger economic foundation while Southern Africa set lower targets, most of them being demonstration projects, as their priority is to tackle the energy access deficit.

Renewable Energy Practice in Major African Countries

To realize the renewable energy targets, African countries adopted a series of policy measures which are as follows.

Eliminate fossil fuel subsidies. By the end of 2020, African countries like Egypt, Ethiopia, Ghana, Morocco, Rwanda, and Togo have committed to or already started removing fossil fuel subsidies, which has since relieved governments from huge fiscal burdens and rid the countries of dependence on fossil fuels, reducing political corruption and social injustice.

Introduce carbon pricing. The economically more developed South Africa introduced carbon pricing, which covers over 41% of energy-related carbon emissions, reduces negative externalities of burning fossil fuels, and brings new opportunities for renewable energy development.

Reduce fossil fuel investments. Egypt and South Africa have made commitments to abandon coal, and the international community has also promised to stop funding coal plants, including those in Africa.

Boost renewable energy investments with fiscal policies and safeguard mechanisms for energy consumption. Kenya has reduced import tariffs on raw materials used in the production of solar equipment and implemented a zero-tariff policy and zero VAT rate on imported solar products to enhance the competitiveness of the local solar industry.

Renewable Energy Demand of Major African Countries

Africa still faces many challenges in renewable energy development.

A weak industrial base. Africa's renewable energy industry lacks a strong industrial base and supporting sectors. Therefore, many supporting facilities are imported from overseas, result in increasing project costs and decreasing project Return On Investment (ROI).

Insufficient grid infrastructure. Renewable energy-generated power is delivered through transmission and distribution (T&D) networks. However, except for North Africa and Southern Africa, there are very few national lines and they mainly serve capitals and major cities. Without T&D lines, power generated by renewable energy cannot be delivered which is unhelpful to further renewable energy development.

Incoherent and inconsistent renewable energy policies in most African countries. Most countries in Africa do not have detailed energy planning and policies that are coherent and consistent, and this has greatly discouraged investment and entrepreneurship in the renewable energy industry.

Talent shortages. Africa suffers greatly from talent shortages both in renewable energy and equipment maintenance, which significantly limit the sustainable development of renewable energy. Africa needs to address talent shortages by improving general education and eliminating discrimination and prejudice in vocational education.

Lack of funding. Renewable energy development requires substantial investment, which is currently not in place because government funding in most African countries are insufficient and foreign investments are not enough due to the unfavorable investment environment in Africa. Therefore, African countries need to revitalize foreign and local investment, increase funding for research and foreign experts, and invest more in production to boost renewable energy development.

Inadequate marketization. African governments monopolize electricity production for its high tax revenue. As a result, some renewable energy projects cannot operate smoothly and efficiently.

3.1.4 Renewable Energy Policy Objectives, Practice, and Demand in Latin America

Renewable Energy Policy Objectives of Major Latin American Countries

Many Latin American countries have set renewable energy targets, especially long-term ones with clear targets in energy production, consumption, and installed capacity. Long-term, reliable policies are essential to attract developers to invest in the renewable energy industry, which also shows great commitment to renewable energy development from the government side. Although most Latin American countries did not specify their renewable energy targets in their NDCs, most of them have set phased targets and plans in their domestic policies to guide renewable energy development according to their potential, technology readiness level, investment expectations, and political will.

Renewable Energy Practice Basis in Major Latin American Countries

Most Latin American countries have set phased targets and plans in their domestic policies to guide renewable energy development according to their potential, technology readiness level, investment expectations, and political will. In order to realize the renewable energy development targets, some Latin American countries, including Argentina, Colombia, Chile, Honduras, and Mexico, have enacted renewable energy laws. They have also used policy tools such as auctions, feed-in tariffs, renewable energy quotas, net metering, and fiscal incentives to support renewable energy in terms of quantity and price.

Auctions. Latin America has seen great interest in auction schemes which are great drivers for renewable energy development. Currently, 13 Latin American countries have deployed renewable energy auctions but focus on different aspects.

Feed-in tariffs. Subsidizing low-income residents and power utilities with feed-in tariffs puts a huge fiscal burden on governments. Therefore, only Argentina, Dominica, Honduras, Panama, Uruguay, Nicaragua, Brazil, and Ecuador have made such attempts as opposed to other Latin American countries.

Renewable energy quotas. Renewable energy quotas define the minimum shares of renewable energy sources in the energy mix of power utilities and are often implemented together with "Green Electricity Certificates."

Net metering. Net metering is a billing mechanism that allows utility customers to install their own renewable energy generating systems and export that power to the grid to reduce their future electric bills. Brazil primarily targets consumers who generate less than 1 MW for retail, while Costa Rica has opened net metering to individual consumers whose credits are equal to their annual electricity consumption.

Fiscal incentives. Most Latin American countries offer tax breaks, such as exemption from VAT, income tax, fuel tax, and import and export tax. Five countries, including Argentina, Colombia, and Mexico, have also implemented accelerated depreciation policies to encourage renewable energy investments. In addition, many Latin American countries have established public funds to support the sustainable development of renewable energy.

Renewable Energy Demand of Major Latin American Countries

Despite rapid growth, renewable energy development in Latin America still faces many barriers, including technical barriers, market barriers, and social barriers.

Technical barriers. Major technical barriers to renewable energy development in Latin American countries are insufficient information and integration barriers.

Market barriers. Market barriers include barriers to entry, transaction costs, contractual risks, subsidies to fossil fuels, lack of financing instruments, and political and economic instability. Electricity markets in Latin America vary greatly in their openness to private or independent power producers and are not friendly to international and local investors. The markets and project scales for renewable energy in Latin America are relatively small, which can result in higher transaction costs and less appeal to investors. Independent power producers and the electricity sector face greater uncertainty and risk because utility contracts are often not legally enforced. Renewable energy projects often have payback periods that exceed the scheduled repayment period of debt financing and do not meet the high equity ratio requirements for equity financing. Therefore, it is difficult to obtain financing through these means, and few competitive financial products are available from the public and private sectors.

Social barriers. Social barriers include public misconceptions, complacency, "NIMBY (not in my back yard)" effects and lack of human resources. Fossil fuels dominate energy consumption in Latin American countries, and some countries are accustomed to maintaining this status quo; residents oppose the construction of large-scale renewable energy projects in their neighborhoods out of concern for people's health and negative impacts on the environmental quality and economic development; people without adequate qualifications and training may fail to deliver in management and application of renewable energy projects, and even bring negative influence to renewable energy policy design.

3.2 Policy Evolution, Innovation Models, and the Experience of China's Renewable Energy Development

Since the beginning of this century, China's promotion policies for renewable energy have gradually taken root and matured, forming a renewable energy policy tool system with the *Renewable Energy Law of the People's Republic of China* as the core, which includes policies and measures such as overall targets, fiscal and financial policies, fixed electricity rates, and guaranteed purchase. This has provided strong support for the steady development of renewable energy in China.

3.2.1 Evolution of China's Renewable Energy Development Policy

China currently holds a leading position in solar and wind power manufacturing worldwide. The development of the solar and wind energy industries is attributed to China's constantly strengthening industrial strength and expanding domestic energy demand. The solar and wind energy industries in China exhibit different growth patterns. This section highlights the evolution of policy concerning solar and wind power generation in three distinct stages, namely early technology development, manufacturing scale-up, and industrial globalization.

Early Stage of Technology Development

In 1995, the *Electricity Law of the People's Republic of China* officially encouraged and supported the use of renewable energy including solar and wind energy for power generation. In terms of solar energy, in 2000, the Western Development Strategy set the agenda for rural electrification; in 2002, the Brightness Program allocated 2.6 billion yuan for solar production and installation, while the Delivering Power to Villages project achieved nearly 20 MW of installed PV capacity. In 2001, the *10th Five-Year Plan* made an explicit request for the annual capacity of solar photovoltaic cells, and in 2003, the *Catalogue of High-tech Products in China* encouraged foreign investment in solar energy, offering tax incentives to potential market entrants.

The wind energy industry in China has been jointly promoted by the central and local governments. The first 200-kW wind turbine generator was successfully developed in China with funding from the central government.

In September 1997, the *Mid- and Long-term Development Plan for Renewable Energy* was released, setting the guiding target that 10% of all electricity should be generated from renewable sources by 2010. The year 2000 marked a turning point in demand-driven wind power deployment, with the National Bond Wind Power program providing financing for 80-MW of domestically produced wind turbine generators. Driven by external expertise and R&D financing, the early planning laid the foundation for wind power development.

Stage of Scaling Up

Like the early stage of development, the growth of leading local players in wind and solar has benefited from different factors. In the solar sector, the introduction of local production incentives and early demand-side measures between 2005 and 2008 facilitated the scale development of the photovoltaic industry. Driven by the policy, several PV industry leaders completed their initial public offerings in the global market.

In 2001, Goldwind obtained a license for a 750-kW turbine from REpower based in Switzerland; in 2004, it designed a 1.2-MW wind turbine collaborating with the German company Vensys. Several small European companies established joint ventures with Chinese companies, including Nantong CASC Wanyuan Acciona, XEMC Darwin Co., Ltd., REpower North (China) Co., Ltd., Harbin Hafei-Winwind and AVANTIS YINHE. In 2006, China proposed to enhance its independent innovation capability, and independent innovation as a national strategy was given an important place in the *11th Five-Year Plan*. In 2006, China proposed to enhance its capacity for independent innovation, and independent innovation became a national strategy occupying an important position in the *11th Five-Year Plan.*

Stage of Globalization

After 2008, the government prioritized demand-led growth in the wind and solar sectors, focusing on improving the efficiency of existing leading companies and actively implementing policies to boost domestic demand and drive solar component consumption. In 2009, the implementation of the Solar Rooftop Program and the Golden Sun Demonstration Project laid the groundwork for the development of a nationally unified benchmark feed-in tariff for solar PV power in 2011. In 2013, the Ministry of Industry and Information Technology (MIIT) set performance standards and capacity requirements, for example, mandating that manufacturers invest at least 3% of their revenues in R&D. By 2015, China's domestic installed capacity of solar had reached 43,500 MW.

The wind energy Industry also actively deployed demand-led incentives. After 2008, the Chinese government increased the incentive for domestic companies to invest through tax reductions and exemptions.

3.2.2 Policy Tool for Renewable Energy Development in China

After decades of exploration, learning and innovation, China's renewable energy promotion policies have gradually become systematic and mature, providing a solid institutional guarantee for the development of the renewable energy industry. With stable guidance from the Renewable Energy Target Policy, and support from financial, taxation, and fixed pricing policies as well as guaranteed purchase policies, China's renewable energy industry has made significant progress: its renewable energy installed capacity, including wind and solar power, ranks first in the world, making it the "No. 1 country of renewable energy"; the level of renewable energy technology and equipment has been significantly improved, with key components basically achieving domestic production and the number of newly added patents ranking high in the world, and a complete internationally leading industrial chain has been built; the progress of China's renewable energy, including improvements in technology and market expansion resulting in a reduction in costs, has greatly lowered the threshold for the development and utilization of renewable energy, contributing significantly to the vigorous development of renewable energy worldwide.

Renewable Energy Target Policy

The Renewable Energy Target Policy refers to the policy of setting specific strategic goals and plans for the development and utilization of renewable energy resources, covering the development goals, construction layout, key tasks and innovative development methods concerning various aspects of the renewable energy industry such as power generation equipment production, infrastructure construction, power generation plans and grid-connected utilization for a certain period of time in the future. It is the guiding policy for the development of renewable energy in China.

Fiscal and Financial Policy

Fiscal and financial policies refer to the policies that support the development of renewable energy through fiscal subsidies, tax incentives, capital subsidies, loan support, etc. These policies aim to reduce the production cost, lower the threshold for investment and financing in the renewable energy industry, and alleviate competitive disadvantages concerning technology and cost in the early stages of renewable energy development, thus promoting the rapid growth of the renewable energy industry, especially those small and medium-sized enterprises in the market.

Feed-in Tariff Policy

The fixed power tariff policy is a regulation or scheme for feed-in tariffs or related adjustments for different renewable energy technologies and power generation projects within a certain period. It also contains policies on renewable power price formation mechanisms, tariff reform programs, grid transmission and distribution, etc.

Guaranteed Purchase Policy

The guaranteed purchase policy refers to relevant measures and methods for grid connection, dispatching support, and consumption promotion of renewable power, aiming at increasing the proportion of renewable energy in total energy consumption and ensuring the fulfillment of tasks and commitments proposed in target-planning policies.

3.2.3 Integrated Development and Utilization Mode of Renewable Energy

In the practice of renewable energy development, the Chinese government actively promotes an integrated mode of RE development and utilization, applying in diversified and innovative scenarios, such as "photovoltaic plus portfolio" (PV + X) to support the solar PV industry and modern agriculture, forestry, animal husbandry, fishery, control of desertification, building integration, green hydrogen manufacturing and even rural vitalization, so to improve the comprehensive efficiency of land use, boost local residents' well-being, and help achieve the goal of carbon peak and neutrality.

Most of the BRI participating countries are developing countries. While realizing green transformation, it is necessary to take into account the fairness of the transformation and the protection of the vulnerable, to facilitate economic development, poverty reduction in rural areas, increase green employment and enhance people's livelihoods. Some typical "PV + X" modes widened the space for further application of renewable energy, extending the paths for profits. Mode replicable, the "PV + X" application has certain reference significance for the green transformation of the Belt and Road countries.

PV + Agriculture

Many regions in China have carried out pilots and operations of "PV + agriculture" projects that were applied in new scenarios, laying photovoltaic panels in rice fields, terraced fields, greenhouses and other scenarios for power generation. Relevant projects need to pay attention to light and land resources at the spot, coordinate the cleaning of panels and farm water, and design solar systems according to the conditions of crops to ensure that the yields are immune from impacts. In addition, there were also fish farming projects explored, which combines PV greenhouses with fishery and vegetables-growing technology. Excrement and residual feeds produced in the fish farming process are transformed into nutrients needed for vegetable growth, thus turning waste in water into treasure. The water for fish farming is absorbed and purified by the plant roots, and then the water is reinjected into the ponds, delivering a recycling of water resources.

PV + Livestock Breeding

"PV + livestock breeding" is also a commonly seen application. In areas that enjoy flat, open and unsheltered terrains and sound irradiation, some projects took the space of greenhouses and animal houses, livestock shed roofs, and between greenhouses and sheds to install PV sets and raise large livestock like cattle and sheep. And there are some poultry breeding projects, as poultries such as geese are fed with forage, helping remove weeds for PV plants.

PV + Desertification Control

Deserted areas often enjoy sufficient sunlight and flat terrains, suitable for photovoltaic power generation. However, facing natural challenges such as sandstorms and water shortage, it is of greater difficulty to take care of PV equipment. What China achieved in desertification control was recognized by the entire world, in which the deeds of Saihanba Forest Farm has won the title of "Guardian of the

Earth", the highest honor in the field of environmental conservation of the United Nations. China's desert has continued to decrease for 20 consecutive years, with constantly reduced level, making a historic turnaround from "sand forcing humans to retreat" to "trees forcing sand to retreat". On this basis, China is exploring the establishment of PV power plants in desertified areas to promote the synergy between photovoltaics and sand control. Shading light and shielding wind, PV panels help reduce the evaporation of soil moisture, and effectively keep down the wind speed.

3.2.4 Multi-effect Synergy of China's Innovative Renewable Energy Development Model

Renewable energy is a green, low-carbon energy source that plays an important role in China's "multi-wheel drive" energy supply system, and it is crucial for ensuring energy security, improving the energy structure, protecting the ecological environment, coping with climate change, and achieving sustainable economic and social development.

Renewable energy has yielded fruitful results in benefiting and serving the people, contributing green power to the building of a moderately prosperous society in all respects. Over the past decade, China has made solid progress in extending power grids to areas without electricity. It has completed the upgrading of power grids in 6 Provinces and Regions, and completed the upgrading ahead of schedule in many remote villages. It has also effectively improved the production and household electricity utilization of over 210 national-level poverty-stricken counties. China has actively implemented independent renewable energy supply projects, providing access to green power for millions of people without electricity.

Significant achievements have been made in reducing pollution and carbon emissions, providing solid support for ecological civilization and green, low-carbon development. In 2021, China's development and utilization of renewable energy reached a scale of 530 million tons of standard coal, equivalent to replacing 1.05 billion tons of raw coal, which was 3.5 times the average annual import of coal in China over the past three years. In the meantime, it reduced carbon dioxide, sulfur dioxide, and nitrogen oxide emissions by approximately 2.07 billion tons, 400,000 tons, and 450,000 tons, respectively.

With growing international cooperation, China is making its contribution to tackling climate change. As the world's largest market for renewable energy and a leading manufacturer of renewable energy equipment, China has continued to deepen international cooperation in this field. Its hydropower business extends to multiple countries and regions worldwide, and the photovoltaic industry supplies over 70% of the components for the global market. The widespread application of renewable energy in China has effectively promoted the decrease in renewable energy costs, further driving the development and utilization of renewable energy all over the world and accelerating the global transition to green energy.

3.3 *Insights from China's Renewable Energy Development Experience to the BRI Participating Countries*

3.3.1 China's Renewable Energy Development Experience

Despite facing numerous challenges, China's framework and path for green and low-carbon transformation provide a reference for global sustainable development. China is willing to use BRI as an opportunity to share its experience with participating countries, help eliminate their reliance on the traditional high-carbon growth models, and encourage them to pursue an innovative, efficient development path with lower emissions and pollution, thus promoting global low-carbon transformation. In the process of "going global", relying on its leading manufacturing capability sufficient capital, and huge potential domestic market, China continuously deepens its global supply chain layout and specialization-based division of labor.

Following the success of domestic wind and solar manufacturing, Chinese companies have expanded their investments in overseas manufacturing. Outbound investment in solar manufacturing is driven both by international tariffs and by the incentive to access global demand centers.

By contrast, the wind power supply chain remains less globalized. Wind turbine generators have higher transportation costs and involve more complex production processes. However, the growing demand among BRI participating countries could present expanding opportunities, especially as components become increasingly modular and lightweight.

The progress of China's renewable energy industry is first and foremost due to its relatively abundant natural resources, and the sound industrial production system. On this basis, a highly strategic and relatively comprehensive supportive policy system has provided a solid guarantee for the starting and prosperity of the renewable energy industry.

Development finance can play an important role in supporting the expansion of renewable energy production and installation in BRI participating countries. Based on China's experience, development finance can bridge the gap between the capacity of local enterprises and the renewable energy goals, giving new market entrants time to expand their business. Manufacturing is highly capital-intensive. Development finance can help local businesses cover high upfront capital costs and get paid back once the facilities are put into operation.

3.3.2 Policy Recommendations for Supporting the Development of Renewable Energy in the BRI Participating Countries

Enhancing quantitative guidance. BRI participating countries need to review and update national energy plans and set up development targets for different types of renewable energy based on research.

Optimizing subsidy policies. BRI participating countries need to improve the on-grid power tariff system that is composed of levelized cost of energy (LCOE) and power price premium to provide market incentives for investors. Meanwhile, with the renewable energy sector gradually entering the stage of commercialized development, the phase out of favorable policies and subsidies should be considered, so that a system integrating government guidance and market mechanisms could be formed.

Streamlining access procedures. BRI participating countries need to establish transparent renewable energy information management systems to optimize the procedures for the application and allocation of subsidies. The system allows governments to keep track of information on power generation and project construction, adjust industrial policies accordingly and improve development rights acquisition system.

Improving the environment for investment. BRI participating countries need to further streamline access procedures, lower the threshold for investors, help local governments to get access to capital from a variety of sources, and attract green bonds and climate funds to reduce the risk of investment.

Improving power accommodation through multiple channels. BRI participating countries need to coordinate power grid planning and power source construction through the innovative application of multi-energy complementation. The "dual coordination" between power source and power transmission and between distributed power generation and power distribution could effectively ensure the safe and stable operation of power grids. Efforts should also be made in promoting cross-border grid interconnection. A connected power grid system enables the optimized allocation of clean energy on a large scale with Ultra High Voltage (UHV) power transmission systems, effectively solve the problem of the mismatch between resources and needs, and lay the groundwork for the large-scale development and cross-border accommodation of renewable energy.

Enhancing international cooperation. BRI participating countries need to strengthen international cooperation in renewable energy, especially in the development of technical standards, demonstration projects for the application of technologies and R&D cooperation centers, to support the development of hydrogen fuel cells, electric vehicles and ships, biofuels and biomass pellet fuels that are used for energy efficient stoves. Joint talent training and capacity building programs should also be carried out to support the development of renewable energy technologies in the region.

3.3.3 Issues for Special Attention in Renewable Energy Cooperation with Some BRI Participating Countries or Regions

International Cooperation with Central Asia

Endowed with huge reserves of energy and natural resources, Central Asia is playing an increasingly important role in the international energy landscape. Meanwhile, the security of Central Asia is affected by international and regional geopolitical landscape.

Besides, it is important to engage in cooperation with third parties to develop renewable energy in Central Asia. Technology R&D and trade cooperation in the field of hydrogen energy should be prioritized to help Central Asian countries to turn oil and natural gas into hydrogen. It is also feasible to help Central Asian countries to reduce the tension between water resources and energy. Through building hydropower stations and reservoirs for seasonal pumped energy storage and developing water electrolysis for hydrogen production, Central Asian countries could realize the dual storage of water resources and energy, fulfilling the demand for both energy and water resources.

International Cooperation with Africa and Latin America

China should (1) assist African and Latin American countries to carry out capacity building and actively engage in cooperation projects initiated by IEA, IRENA and other international organizations; (2) help Africa to formulate long-term plans and measures for the development of renewable energy, including development goals, financial and taxation policies, supporting policies for technology transfer and standards to ensure the continuity and stability of policies; (3) make full use of the "10–100–1000 Initiative" for South–South Cooperation on Climate Change under the Framework of the BRI and other cooperation platforms to train talents for the development of renewable energy and improve the understanding and acceptance of renewable energy in the local area.

To solve the problem of financing in Africa and Latin America, China should work together with African Development Bank, the World Bank and the Special Climate Change Fund to explore innovative means of financing, establish investment and financing platforms, and set up special funds to guarantee financing for the development of renewable energy in African and Latin American countries and bridge the financing gap in most countries.

4 Policy Suggestions on the Innovation Mechanism of BRI to Promote the Process of Sustainable Development

4.1 Strengthen the Innovation of BRI Green Development Cooperation Mechanisms and Promote the Establishment of a Support System for Renewable Energy Projects

Promoting the green and low-carbon development of energy in BRI participating countries is an essential means of supporting their sustainable development, and its core is strengthening renewable energy investment and financing. At present, many

BRI participating countries, especially developing countries, are facing huge financial and technological gaps in their renewable energy, whereas China has precisely the cooperation demands and production capacity advantages in this regard. Thus, the following policy recommendations are proposed.

First, leverage and activate market-oriented cooperation in renewable energy through innovative measures. Technical assistance and capacity building cooperation are needed for the emerging renewable energy markets along the Belt and Road, to train more small-and-medium developers. Special funds for BRI's green development projects are to be set up, providing financial support for the cost of the pre-feasibility study and project preparation stage, to leverage the project development process. The decision-making mechanism for renewable energy in China can be used as a reference to improve the efficiency of fund allocation for green BRI projects and fill the funding gap in developing countries.

Second, China can join hands with the BRI participating countries to set up a database of pre-feasibility financing options accessible to renewable energy project developers to help better understand and adapt existing resources, which will maximize the project financing capacity for BRI participating countries. This database should provide catalogued projects with renewable energy financing portfolio, which may include financing + design, procurement, construction (EPC + F), refinancing by international financial institutions, sovereign wealth funds, BRI renewable energy bonds, international development funds, overseas industrial funds, international syndicated loans and other diversified financing support.

Third, utilize existing international cooperation platforms to promote joint renewable energy investment and formulate regional cooperation strategic policies and action plans. Through these platforms, we should push for the formation of BRI renewable energy cooperation standards, norms and guidelines for international participation and mutual recognition; strengthen digital empowerment for green development; collaborate in providing technical assistance and capacity building for emerging renewable energy markets along the Belt and Road; train more small and medium-sized renewable energy developers; and enhance the capacity of BRI participating countries to receive investment. By promoting diversified cooperation between industry organizations and stakeholders in BRI participating countries, we can establish a professional cooperation network for green transformations.

4.2 Strengthen the Synergy Among BRI's Green Development Cooperation Mechanisms in Various Fields and Promote the Establishment of a Policy Environment Conducive to BRI Green Development Cooperation

At present, most of the BRI participating countries are still on the fast track of both economic development and carbon emission growth. Full and effective environmental and climate cooperation under the BRI framework is the cornerstone of enhancing mutual trust, reducing differences and developing cooperation, and it will make an important contribution to global climate governance. During the last ten years of cooperation under BRI, the Chinese government has established BRI green development cooperation mechanisms and worked with other countries in the areas of environment, energy, green finance, transportation and communications to build the Belt and Road. The aforementioned mechanisms are important driving forces for promoting BRI's green development. In this regard, the following policy recommendations are proposed.

First, strengthen vertical and horizontal coordination in key areas of green development through existing BRI cooperation platforms. Currently, there are independent green BRI mechanisms in different regions and sectors, and information sharing is not sufficient among those mechanisms. With a focus on cross-sectoral cooperation, we should make use of the overall cooperation mechanism for BRI construction and strengthen the communication and exchange frequency in the key areas of green infrastructure, green energy green transportation, green industry, green finance and green technology, coordinating and promoting BRI green development cooperation in various fields. Exchanges and cooperation mechanisms should be further improved at multiple sectors and levels, with policy discussions conducted on a regular basis and an information sharing institution established. More non-state actors are encouraged to be included in the scope of cooperation, to help construct a network with diverse subjects.

Second, improve the policy environment for the development of overseas cooperative green BRI projects. We should promote the reform of the overseas investment approval system; form new financing system favorable to green projects; incorporate renewable energy investment into the enterprise performance tracking system; and relax performance requirements for overseas green energy investments appropriately. In addition, we should establish a BRI climate investment and green credit system and appropriately reduce the financing cost of low-carbon investment projects. A favorable policy environment will encourage financial regulators to adopt green policies to stimulate the development of green finance.

4.3 Implement Innovative BRI Demonstration Projects and Support the Development of Customized Sustainable Development Solutions for BRI Participating Countries

Finding suitable paths for sustainable development and realizing green, low-carbon and sustainable development is the only option for humankind and a challenge that developing countries must overcome. Previously, the Green BRI Special Policy Studies have highlighted the close relationship between the Green Belt and Road and the realization of the United Nations sustainable development agenda. In the future, cooperation on BRI's green development can and should bring important opportunities and solutions for the realization of green, low-carbon and sustainable development in participating countries. To this end, the following policy recommendations are proposed.

First, utilize existing cooperation platforms to coordinate resources from various parties for building green BRI cooperation demonstration projects in renewable energy and other fields. On the basis of previous policy recommendations, we should urge the Chinese government and BRI participating countries to jointly build a number of green BRI cooperation demonstration projects. Combining the efficient cross-sectoral cooperation mechanism with the green investment and financing channels of financial institutions, we can provide full support for the planning, design, financing and implementation of demonstration projects. These demonstration projects will influence BRI participating countries by providing customized greens solutions for developing countries, encouraging the formation of their own corresponding green policies, and promoting the development their green industries.

Second, explore the cooperative demonstration of "PV+" and other innovative application scenarios, as well as profit models of such projects that align with the characteristics of developing countries. On the basis of the aforementioned demonstration projects, we should strengthen the demonstration role of renewable energy projects. We should support the preliminary feasibility study, construction and operation of "PV + X" projects, with pilot projects carried out in the BRI participating countries, such as "PV + Agriculture," "PV + Aquaculture," "PV + Industrial Parks," and the like. In view of the challenges generally faced by energy projects, such as difficulties in collecting electricity charges in developing countries, we need to explore innovative profit modes, such as through PV supporting industries, to further promote renewable energy projects.

References

1. Lu, Y., Springer, C., & Steffen, B. (2023). *Collaborating for sustainable development: The role of cofinancing in shaping outcomes of Chinese lending and overseas development finance projects* (GCI working paper). Boston University Global Development Policy Center. https://www.bu.edu/gdp/2023/04/04/collaborating-for-sustainable-development-the-role-of-cofinancing-in-shaping-outcomes-of-chinese-lending-and-overseas-development-finance-projects/

Part V
Main Reports for the 2023 CCICED AGM

Chapter 11
CCICED Issues Paper Annual General Meeting 2023 Green Innovation

1 Foreword

Established in 1992, the China Council for International Cooperation on Environment and Development (CCICED) is a high-level international advisory body to the Government of China. Along with China's rapid social and economic progress, CCICED has witnessed and taken part in the country's historic shift in development philosophy and model. It opens the door to advanced international experience in sustainable development and connects China with the international community focused on environment and development. CCICED also provides a valuable platform of exchange, enabling the international community to understand China and support the country's engagement with the world.

Since 2002, the International and Chinese Chief Advisors have produced an Issues Paper each year for the use of CCICED Council Members, high-level policymakers, and others during the Annual General Meeting, where research findings and recommendations are discussed.

Green Innovation is the 22nd Issues Paper published by the CCICED. Against a backdrop of the much-needed recovery of the global economy, the paper provides thoughts on how to empower green innovation and promote high-quality development from a market and synergy perspective. The preparation of the Paper is led by CCICED Chief Advisors, Mr. Scott Vaughan, and Mr. Liu Shijin, with contributions from International Chief Advisor Support Group and the Chinese Associates.[1]

Context: The world economic outlook continues to be volatile, facing inflation and geoeconomic fragmentation that is affecting rates of economic growth, unemployment, trade, and foreign direct investment. The World Bank's June 2023 *Global Economic Prospects* report warns of precarious economic prospects ahead, especially for many emerging and developing countries, while World Trade Organization

[1] The ICA supporting group mainly includes Mr. Knut Alfsen and Mr. Dimitri de Boer. The Chinese supporting team mainly includes Mr. Zhang Huiyong and Ms. Mu Quan.

Green Empowerment and High Quality Development,
https://doi.org/10.1007/978-981-96-4218-2_11

(WTO) trade statistics reveal the increased use of export restrictions, as well as a decline in intermediate merchandise trade in 2022.

Following pandemic restrictions, uncertainties in the property sector, weak export demand, and other factors, China faces slower-than-expected rates of economic recovery.

Compounding economic challenges are the mounting costs of extreme weather events linked to climate change. 2023 marked the highest world average temperature ever recorded. China has experienced record heat waves and severe flooding in several regions. Wildfires across Canada blanketed large regions of North America, exposing millions to chronic air pollution and causing billions in economic losses. The *State of the Global Climate 2022* assessment by the World Meteorological Organization (WMO) expects extreme events to worsen in the coming 5 years due to a combination of heat-trapping greenhouse gases (GHGs) and El Niño, affecting the food security of millions.

Multilateral Environmental Commitments: Despite an overall increase in geopolitical fragmentation, there have been several important international breakthroughs. The successful completion of the Kunming-Montreal Global Biodiversity Framework (GBF) at the United Nations Convention on Biological Diversity (UN CBD) Conference of the Parties (COP 15) in December 2022 exceeded most expectations as countries pledged to halt and reverse biodiversity loss, meet "30 × 30" target for nature, and mobilize new and additional financial resources. An important step in GBF implementation is the June 2023 agreement to establish, via the Global Environment Facility (GEF), a new GBF trust fund. Other important steps include the new UN oceans treaty adopted in June 2023 and ongoing negotiations toward a global plastics treaty.

A Carbon Control System: In July 2023, President Xi Jinping called for the transition from the current energy control system to a carbon control system. The elements to enable a comprehensive transition toward a carbon control system include incentives, carbon markets, and green finance to accelerate both supply-led low-carbon technologies and demand-side efficiency measures, working in tandem with regulations and standards to limit carbon pollution. Several jurisdictions have adopted a national climate law to support regulatory coherence and raise the profile of both compliance promotion and effective enforcement. A carbon control regime can also be supported through national and sub-federal carbon budgets, as well as real-time GHG emissions reporting to measure current conditions and scenarios for carbon peaking and carbon neutrality.

From Trade-Offs to Take Off: The expansion of China's installed renewable energy has outpaced planned targets. Estimates suggest that 2023 will see China exceeding 150 GW of new solar power and wind power. This faster-than-expected growth in renewable energy is further evidence that trade-offs between either energy security or low-carbon energy and developments are obsolete.[2]

---

[2] China is the main driver behind this renewable energy breakthrough. In 2022, according to the China Electricity Council (CEC), solar energy increased by 28.1% to 392.6 GW, and wind power

A new paradigm of high-quality green development is underway, creating millions of new jobs and expanding highly competitive markets. International Renewable Energy Agency (IRENA) reports that USD 500 billion was invested in 2022 in photovoltaics and onshore and offshore wind turbines, while sales of electric vehicles (EVs) continue to grow. A recent trade-climate scenario report by the World Economic Forum (WEF) envisions that 15% of global merchandise trade could be made up of low-carbon goods by 2030.

Green Innovation: At the heart of high-quality green development is innovation. There is no single innovation blueprint. No two innovations are identical. Yet an increasingly vital part of green innovation involves synergies between low-carbon green technologies and digitization. A recent review of China's climate policies, led by Professor Nicholas Stern and Professor Min Zhu of the London School of Economics and Political Science, underscores the need for a comprehensive innovation system to implement the dual control targets. Chinese Chief Advisor of CCICED Shijin Liu recently pointed to the dynamic links between low-carbon technologies, digitization, and market-based innovation at the 2023 China Development Forum.

Digitization and climate are connected on numerous levels. As significant energy users, data hubs run by leading companies like Tencent and Microsoft are implementing ambitious net-zero and negative carbon targets, including through the use of green power (including through large-scale green power purchase agreements where they are available). They are investing in new solar and wind-generating capacity, upgrading energy efficiency, and using Direct Air Capture (DAC) technologies. Data hubs and digital infrastructure are also increasing their resilience to climate-extreme weather events like flooding.

However, the most important synergies are those in which digitization is being put to work to identify least-cost decarbonization pathways across many areas and sectors, from emerging smart, low-carbon, and resilient cities to hard-to-abate industrial sectors. Coalitions like C40 Smart Cities are sharing case studies in low-carbon pathways while digital technologies and industrial Internet platforms work within emerging digital twin approaches to identify low-carbon engineering solutions in hard-to-abate steel and other heavy industrial manufacturing processes. Other examples of linking digitization, decarbonization, and sustainability outcomes include finding new ways to inform online consumers of sustainable fashion options; the use of digital blocks to improve sustainable sourcing through better traceability systems for soft commodity value chains like palm oil and soy; and widening applications of digitization to provide granular and real-time environmental and climate monitoring.

increased by 11.2% to 365.4 GW. With plans to install 160 GW of additional capacity in 2023, solar and wind power are poised to more than double the total production capacity of China's total hydropower generation (currently at 413.5 GW). Bloomberg recently estimated that based on current trends, China's rates of new renewable energy could exceed 200 GW in 2024.

2 Unleashing Innovation Through Markets

The private sector is the most effective in moving inventions made in the lab to new products available at scale in the market. Governments have a crucial role to play in fostering both inventions and innovations, as well as being first movers through procurement. A new wave of green industrial policies that include large-scale subsidies for green technologies, like large-scale battery storage systems for electricity grids, are lowering the cost of decarbonization pathways for companies and consumers lucky enough to receive subsidies.

As part of green development innovation, carbon markets are a vital instrument. The World Bank identifies carbon markets in over 70 jurisdictions, including China's national carbon market. Analysis by the Organisation for Economic Co-operation and Development (OECD) concludes that carbon prices-the central feature of carbon markets-are especially effective in decarbonizing power markets, with coal being the most responsive to cross-price supply elasticities linked to carbon pricing.

An important aspect of energy policy and climate policy coordination is finding synergies between the anticipated pricing effects of carbon markets with wider power market reforms. China is currently reforming its national power market, to build a coherent national market that will improve the ability of power companies to interconnect between jurisdictions and integration of renewable energy. Led by the National Development and Reform Commission (NDRC), the first iteration of the reforms will be completed in 2025.

China's national Emissions Trading System (ETS) has the potential to become a key catalyst in accelerating market dynamics and allocative resource efficiencies to encourage innovation in meeting China's dual control goals.

However, to unleash the full market potential of the ETS system, four design adjustments warrant consideration within the wider context of ongoing power market reforms. First, increase the scope of the ETS beyond the current power sector to include other sectors like steel, cement, aluminum, chemicals, and others by which different marginal abatement costs across sectors would increase the efficient allocation of resources. A second step is to move from an intensity-based system to an absolute emissions cap, which would send a clear signal about the supply scarcity of carbon credits.

A particularly effective market-based instrument applied to large, stationary, energy-intensive, and trade-exposed sectors like steel, cement, aluminum, and other sectors is an Output-Based Pricing System (OBPS) or performance-based system, which links sector-based emission factor averages to a graduating carbon price. While there are varying carbon prices across different markets, the International Monetary Fund (IMF) suggested a carbon price of USD 75/tonne by 2030, which is a useful benchmark for the price trajectory of China's ETS over time.

A final feature is clarifying the ETS's roles and responsibilities, including the supporting roles of a quality GHG data system, compliance promotion through training and education, and enforcement.

A Comprehensive System Approach: As noted, China leads the world in renewable energy. It will likely install more solar and wind power in 2023 alone than the entire renewable energy capacity of either the European Union (EU) or the United States (U.S.). Estimates suggest as much as 200 GW of renewable power could be installed in 2024. At the same time, there has been an increase in new fossil fuel-based electricity generation capacity since 2021. While numerous older and inefficient coal plants have been retired, the net effect of recent approvals is an estimated 10% rise in China's coal generation, equivalent to 100 new plants.

In light of the faster-than-planned expansion of renewable energy, coupled with the forthcoming power market reforms, a review of the need to proceed with recently-approved coal-fired electricity generation is warranted since newly installed renewables are on track to supplement the declining output from hydropower.

Flexible Electricity Grid Innovation: A key innovation challenge is to ensure that all segments of the power generation system have innovative features designed to work together to create a modern, holistic, and green power system. This emphasis on an integrated system is a conclusion of both CCICED's 2023 research and the International Energy Agency (IEA) China 2022 report, which recommends the adoption of a flexible power system.

There are numerous examples linking renewable energy with more flexible demand-side models. Innovation in renewable energy production is being matched by innovation in energy delivery and pricing models such as pay-as-you-go, plug-and-play, and others. China's Whole-County Rooftop Solar program enables flexible ownership and leasing arrangements to scale up rooftop solar panels in residential and commercial buildings. The analysis identifies two major flexibility clusters: (i) panels are sold to the property owner, who in turn can sell surplus power back to the project developer, or (ii) the developer retains ownership of the solar panels, providing the property owner electricity at a discount in return. To date, the program is being implemented in over 600 counties across the country.

The key link between clean renewable power production and flexible demand-side consumption is the transmission grid system. China continues to make significant new investments in Ultra-High Voltage (UHV) transmission lines, investing a reported RMB 150 billion (USD 22 billion) in the second half of 2022. New models of electricity grids are quickly emerging that include smart grids, battery-storage banks, distributed power generation, embedded sensors, digital and AI capabilities, applied research in flexible alternating-current transmission systems (FACTS), and other features. Taken together, more innovative transmission systems have been characterized as on the cusp of a wholescale shift in grid technology "from electro-mechanical to electronic and from rigid physics to programmability electricity grid models".

There are a growing number of new-generation electricity grid models. The California Energy Commission plans to install 49,000 MW of battery storage capacity by 2045 as a vital component in supporting carbon neutrality goals.[3] In Germany,

[3] Researchers from the National Renewable Energy Laboratory have shown that for meeting an electricity storage duration of 120 h, the least-cost approach combines hydrogen systems with geologic storage and natural gas with carbon capture.

due in part to permitting backlogs, rather than replacing older transmission lines, investments are underway to upgrade older lines with sensor and digital capacities to allow continuous monitoring that is improving efficiency by up to one-third during colder temperatures.

For good reason, most electricity grid operators are state monopolies: infrastructure and operating costs are high, public planning often requires complex public reviews and approval, and reliability needs to be matched with accessibility and affordability to all communities. However, China could consider opening gird operations to more competitive, market-type dynamics. Examples include provincial-level grid entities operating in Canada (Ontario and Alberta), the U.K. Office of Gas and Electricity Markets (Ofgem), and the designation in Texas of Competitive Renewable Energy Zones (CREZ).[4]

One step in opening grids to market dynamics is for China to introduce more options for households and businesses to purchase green power. Examples include California's Renewable Auctions Mechanism and the United Kingdom's auction system to accelerate investments in renewable energy under its Contracts for Difference scheme. As more companies adopt net-zero targets, large green power purchase agreements led by Amazon, Microsoft, Google, and Walmart exceeded 36 GW in 2022.

Regulations and Compliance Markets Working in Synergy: While carbon markets are an essential tool for achieving carbon peaking and carbon neutrality, they work best in the context of regulations and mandatory standards. Decades of experience in mandatory energy-efficiency standards for household, office, and business appliances has fostered a thriving market for innovative products, such as air conditioners, lighting, refrigeration, and heating systems that employ millions and are a vital part of the world's trading system and billions of households. In saving energy, these standards are also a major cause of avoided GHG emissions: the IEA estimates that mandatory standards have led to annual energy savings equivalent to 15% of total energy use in many jurisdictions while avoiding 700 million tonnes of GHG emissions per year in China, the EU, and the U.S.

Carbon Compliance Markets and Carbon Sequestration: As noted, there are a growing number of jurisdictions with compliance carbon markets, including China's ETS; the EU's ETS; Canada's graduation federal carbon price; and sub-federal systems, such as those in California, Quebec, and the northeastern U.S. states (through the Regional Greenhouse Gas Initiative (RGGI)).

Governments and companies are paying closer attention to the role of carbon offset markets to help meet their decarbonization pathways. For example, China allows companies to offset a maximum of 5% through offsetting. Unlike the current state of voluntary carbon markets, which continue to face significant credibility problems, credits derived from compliance markets are the preferred option.

[4] An important obstacle for energy state-owned enterprises (SOEs) is accessing sufficient private capital to modernize their grids and other operations, a topic examined in a helpful analysis of Association of Southeast Asian Nations energy SOEs.

As domestic carbon markets expand, more work is needed to ensure governance systems are in place across different jurisdictional levels, whereby credits are only issued once they adhere to robust and transparent carbon sequestration methodology based on Intergovernmental Panel on Climate Change (IPCC); use models, inventories, and field testing appropriate for different ecosystems, such as forests[5]; have clear roles and responsibilities for credit approvals; and apply robust and transparent accounting standards.

Special attention is warranted for emerging carbon capture and storage technologies. The current contribution of roughly 60 Carbon Capture, Utilization, and Storage (CCUS) projects to total carbon sequestration is roughly 0.5% compared to the 99.5% that natural carbon sequestration solutions like oceans, forests, peatlands, and other ecosystems provide. However, following the IPCC's report noting the potential role of CCUS, many companies, especially in carbon-intensive sectors, are increasing CCUS investments.

No question that calculating rates of carbon sequestration in a given landscape or seascape is more complex than monitoring GHG emissions from stationary or other sources. Recent mandatory climate risk disclosure and reporting standards, like the EU's Sustainable Finance Disclosure Regulation (SFDR), the U.S. Securities and Exchange Commission's draft Climate Disclosure Rules (its final version is once again delayed), and the International Financial Reporting Standards (IFRS) Foundation's International Sustainability Standards Board (ISSB)'s climate disclosure rules contain frameworks for the disclosure of carbon offsets by companies. It will be important that China's evolving disclosure rules, based on Task Force on Climate-Related Financial Disclosure (TCFD), are comparable and interoperable with those of the international community.

3 Coordination

Green innovation presents enormous opportunities to accelerate the implementation of China's dual carbon goals. At the same time, innovation presents challenges around policy coordination on at least four levels, noted briefly below.

---

[5] Progress has been made in forest inventories at differing scales, which are integrating both satellite-based data with tailored models. One example of national system is the China Forest Resource Inventories (CNFRI), conducted every 5 years. This data has been combined with other inventories, such as the China Forest Ecosystem Inventory System (CFEIS), as well as with model algorithms (linear regression, random forest, and extreme gradient boosting) to estimate the biomass of the subtropical forests in Hunan Province, China. In 2015–2016, Guizhou Province conducted its fourth forestry inventory through the Forest Resource Planning and Design Survey. Data from the survey breaks out over 100 tree stand attributes and site conditions. During the survey, some 3 million stands were recorded, including land-use type, forest land type, plant type, dominant tree species, average age, age class/group, stand volume per hectare, stand area, origin, soil type, community structure, disaster class, health class, and other data points.

Other examples include U.S. Agriculture and Forestry GHG Inventory 1990–2018: estimates of carbon density and carbon pools are presented as carbon dioxide equivalent per hectare.

Development First: An important outcome of China's 2023 Two Sessions was the importance of coordinating the development, social prosperity, economic security, carbon peaking, and carbon neutrality goals.

A central focus of climate mitigation policies across multiple jurisdictions is ensuring jobs and labor markets adjust in ways that ensure stability at the household and community levels. Ongoing programs around just transition in Canada, the EU, France, and elsewhere entail ensuring that the jobs and regions in sunsetting and carbon-intensive industries are supported through public policies to deliver new jobs in sunrise green sectors. Since these shifts are comparable to structural changes associated with trade competition, lessons from trade policy can be useful in the green transition.

So too is a growing body of research that suggests a net gain in employment because of the green transition. The WTO and IEA estimate that over 2 million Chinese jobs are related to green power, while researchers at Oxford point to higher employment, better financial returns, and wider social benefits from the green transition compared to high-carbon and polluting industries.

An emerging tool to contribute to development, sustainability, nature, and climate progress is through the innovative use of enhanced international green credit arrangements. Among the outcomes of the June 2023 New Global Financing Pact, which included Premier Li Qiang, is the creation of a new task force to examine options for the use of new debt instruments like debt-for-climate swaps, a tool the IMF has noted has the potential to help emerging and developing economies in meeting their Paris Nationally Determined Contribution goals as well as addressing rising debt distress levels.

Co-control: A second feature of coordination involves what the 2022 20th National Congress emphasized as "reducing carbon emissions, managing pollution, restoring ecology and promoting growth" in a coordinated manner. The CCICED 2023 Special Policy Study (SPS) on Collaborative Mechanism for Carbon Reduction, Pollution Reduction, Green Expansion and Growth demonstrates the multiple co-benefits of tackling GHGs emissions and criteria air pollutants simultaneously, including substantial public health benefits from reduced long-term exposure to $PM_{2.5}$, smog, and other pollutants.

Climate Action Coordination: A third feature of coordination involves aligning the growing number of policy tools that together make up carbon neutrality goals. The characteristics of nearly all net-zero policy frameworks are remarkably similar and comprise sector-based and wider targets within which carbon markets, green subsidies, research and development (R&D) investments, regulations and mandatory standards, green finance, and operational measures like green procurement and greening government operations are enacted. To illustrate the scope of climate actions, the U.K. climate framework is supported by over 360 indicators to track progress, too many to provide a clear overview of progress. Other jurisdictions like the EU, France, Sweden, Canada, and New Zealand are implementing several dozen decarbonization measures.

The World Resources Institute (WRI) has recently identified good country practices in carbon neutrality, which include crucial coordination governance mechanisms. The U.K. Climate Policy Dashboard is intended to unclutter hundreds of data points to inform decision-makers and the public of key GHG trends. France has adopted measures to screen all budgetary measures through a climate lens.

Such coordination approaches are also important in ensuring gender issues are highlighted and mainstreamed in the design of low-carbon measures, especially within the context of support for shifting labor markets and the need for social safety net policies to help displaced workers and communities. Gender issues remain a strategic priority for CCICED, and a report on gender mainstreaming based on the current 2022–2023 CCICED SPS is forthcoming.

Private–Public Partnerships in Innovation: A final coordination challenge involves effective partnerships between the private sector and government. Yet reviews of innovation policies identify common bottlenecks, from different levels of risk acceptance to complex interactions across innovation chains from start-up manufacturing, finance, consumer markets, and global supply chains. Governments can help clear innovation pathways by uncluttering minor impediments to innovation that together create barriers. Examples include obtaining multi-year R&D funding; adjusting sales tax to incentivize R&D; improving R&D grants within and across funding agencies; reducing first-mover risks, as well as some operational costs like installation; patent applications; labor skill gaps; and community distrust of new technologies.

Conclusion: Green Innovation is a key catalyst for achieving the dual control goals, as well as concurrently tackling pollution and ecosystem stewardship. As climate action engages economy wide in all key sectors and regions, linking a green innovation national system with policy coordination will be key.

Chapter 12
China Council for International Cooperation on Environment and Development 2023 Annual General Meeting Policy Recommendations for the Chinese Government

Maintaining the Strategic Dual-Carbon Determination and Exploring the Innovative Path of Multi-objective Synergy: Accelerating Green and Low-Carbon High-Quality Development

As the world emerges from the COVID-19 pandemic, the global economy is still grappling with volatility, inflation, and geopolitics, with prolonged sluggishness in economic growth, employment, trade, and investment. The World Bank's June 2023 *Global Economic Prospects* report warns of precarious economic prospects ahead. Many emerging and developing countries are expected to bear greater pressure. The world economy faces the dilemma of how to restart and sustain healthy growth.

At the same time, the damage caused by extreme weather events linked to climate change is escalating with each passing day. The world has just experienced the hottest summer in 174 years, with alternating droughts and floods, and frequent wildfires, all of which have posed urgent and severe challenges to public security and health. In today's turbulent world, all countries are facing the challenges of accelerating economic recovery, ensuring energy and food security, and addressing climate change in a coordinated manner. The process of modernization in human society is once again standing at a juncture in history.

The development of green, low-carbon industries, exemplified by renewable energies, has significantly accelerated, emerging as a new driving force to maintain economic growth and push for transformation. The International Renewable Energy Agency reports that USD 500 billion was invested in photovoltaics, onshore wind power generation, and offshore wind turbines in 2022, and sales of electric vehicles (EVs) are experiencing ongoing expansion. A recent trade-climate scenario report by the World Economic Forum envisions that as much as 15% of global merchandise trade could be made up of net-zero goods by 2030. The development of green and low-carbon industries has become an indispensable driver of new growth, prompting countries to reflect on their development philosophies and encourage innovation in development strategies, organizational models, institutions, and mechanisms.

Green Empowerment and High Quality Development,
https://doi.org/10.1007/978-981-96-4218-2_12

After 5 years, China reconvened the National Conference on Ecological and Environmental Protection, where Chinese President Xi Jinping delivered an important speech. CCICED Council Members expressed a strong appreciation of China's confidence and determination to firmly advance ecological civilization and modernize the harmonious coexistence between humanity and nature. They believe that this commitment has injected greater assurance and positive momentum to the sustainable development of not only China but also the global community.

Based on the research outcomes of the CCICED Special Policy Studies (SPSs) and the discussions at the 2023 Annual General Meeting, Council Members recommend that China should maintain its strategic determination, take green and low-carbon development as the endogenous driving force, and promote high-quality development with multi-objective synergy in a coordinated manner. Promote synergies between carbon and pollution reduction and take an integrated approach to addressing energy, supply chains and food security to ensure a gradual, systematic, and controlled progression toward the carbon peaking goal. Accelerate the digital and green upgrading and transformation of traditional industries through digital technological innovation and support the high-quality development of industries and cities through digitization. Establish a green financial system to support low-carbon transformation, coordinate carbon and pollution reduction with a focus on the transportation sector and accelerate the development of a new type of power system. Improve the legal safeguards for addressing climate change. Integrate the sustainable blue economy as a national strategic goal and incorporate climate adaptation capacity assessment into river basin planning to establish a green, low-carbon, and resilient spatial pattern covering mountains and oceans. Promote open cooperation to improve the policy environment for overseas green cooperation projects and to integrate green and sustainable criteria into global supply chains. Build a green Belt and Road Initiative (BRI) and share opportunities for low-carbon transformation. Give play to the role of the COP 15 presidency and work together to implement the Kunming–Montreal Global Biodiversity Framework (GBF) to achieve harmonious coexistence between humanity and nature. Align low-carbon green development with poverty alleviation, job growth, youth engagement, and gender equity.

Specific recommendations are as follows:

A. Maintain strategic determination and firmly implement the "dual-carbon" goal.

   1. Establish a roadmap for institutional transformation and promote the transition from "dual control of energy consumption" to "dual control of carbon emissions." During the mid-to-late stages of the 14th Five-Year Plan, initiate pilot projects for dual control of carbon emissions in select provinces, cities, and key industries. In the early stage of the 15th Five-Year Plan, test the dual control of carbon emissions at the national level, with carbon intensity as a binding indicator and total carbon emissions as a predictive indicator. Beyond 2030, refine the comprehensive carbon reduction system with a primary focus on total carbon emissions control.
   2. Establish a framework law for climate change response or carbon neutrality promotion and identify opportunities to embed carbon control measures

into sector-specific laws in fields such as transportation, construction, and urban development. Formulate action plans at the provincial and municipal levels to implement the "1 + N" policy system. This involves improving the management mechanism in terms of target setting, data and analytical technologies, public participation, continuous monitoring and assessment, and dynamic adjustments. Climate change litigation should be incorporated within the scope of environmental public interest litigation and a preventive environmental public interest litigation system should be established. Set up judicial guidelines for climate change cases, with special attention given to short-lived climate pollutants like methane.

3. Promote the whole-chain application of green innovation and expedite the development of a new low-carbon power system. Match the reliable supply of renewable energy with energy delivery, pricing mechanisms, and a more competitive market environment. Implement nationwide economic dispatch of the power system to reduce renewable energy curtailment. Develop new and more ambitious goals for clean energy and energy storage. Encourage energy storage and demand-side management resources via market mechanisms for a balanced power system. Accelerate the technology research and development and policy preparations for Vehicle-to-Grid (V2G) systems for EVs to supply power back to the grid. Make full use of spatial planning to optimize renewable energy systems that do not undermine biodiversity conservation areas, high-value agricultural areas, and residential areas.
4. Incorporate energy security, asset-stranding risks, and social equity into the top-level design of the energy system transformation toward decarbonization. Ensure that dispatchable power generation and energy storage meet peak demand, especially during extreme weather periods when wind and solar power generation sharply declines to prevent widespread power outages. Accelerate the flexibility retrofit of certain coal-fired power plants to enhance their adaptability to variable wind and solar power with high penetration levels and meet peak demand requirements. Some coal-fired power plants can be converted to biomass power generation, with consideration of integrating carbon capture and storage technology to reduce asset stranding and social risks. Expand the scale of renewable energy production in urban areas, support skills training, and increase green employment opportunities from the energy transition.
5. Continue to improve the policy package for promoting new energy heavy-duty trucks (HDTs), including setting industry standards, implementing fiscal incentives like vehicle purchase tax exemptions and non-fiscal incentives like priority road rights, and specifying procurement requirements for new energy vehicles in commercial fleets. Accelerate the development of new energy HDT infrastructure, such as charging and battery-swapping stations. Set long-term sales share targets for new energy HDTs, aiming for 45% by 2030, 75% by 2035, and 100% by 2040. Introduce a "dual-credit" policy for new energy HDTs and off-grid energy storage systems for road-charging electricity.

B. Continue to optimize the industrial structure and promote carbon reduction and pollution reduction synergies.

6. Continue to adjust and optimize the structure of energy, industry, transportation, land use and more. Accelerate the shift from end-to-source treatment and promote cross-sectoral synergies in carbon and pollution reduction for enhanced efficiency. Accelerate source emissions reduction, process control, end-of-pipe treatment, and comprehensive utilization in the industrial sector to facilitate green development across the entire production cycle. Establish an efficient, standardized recycling system for sorting, recycling, and reuse of metals from industrial waste. Increase efforts to optimize and adjust the transportation structure, promoting the conversion of highways to railways and waterways, while paying attention to the development of sustainable fuels and other decarbonization technologies related to aviation and shipping. Enhance the quality of green, low-carbon and climate-resilient development in urban and rural construction and take multiple measures to increase the proportion of green buildings and improve energy efficiency.
7. Strengthen synergies in the fields of pollution prevention and control across air, water, soil, and solid waste. Continue to deepen the nationwide battle to prevent and control pollution. Collectively advance deep air pollution reduction, energy saving, and carbon reduction in key industries. Set up a synergized target and evaluation system for environmental quality, pollution control, and greenhouse gas emissions reduction with a focus on the coordinated control of fine particulate matter and ozone. Advance integrated freshwater management that coordinates the management of water resources, water environment, water ecology, and resilience. Strengthen the synergistic control of soil pollution management and encourage green and low-carbon soil remediation. Promote concerted action on solid waste pollution prevention and control and strengthen the development of "Zero Waste Cities." Promote the integration of nature-based solutions (NbS) in conjunction with conventional engineering approaches.
8. Address the challenges related to climate change, biodiversity loss, and food security in an integrated manner through land-use transformation. Optimize agricultural support policies and apply natural capital and ecological accounting to support sustainable farming, fisheries, and forestry. Promote the widespread application of agricultural digitization and smart technologies to facilitate the transition of agricultural production toward green, low-carbon, and regenerative agricultural management to ensure food security and ecological services. Incorporate environmental and health dimensions in the definition of food security. Optimize China's dietary guidelines to provide scientific guidance for food policy formulation and food security evaluation. Adjust food security policies through fiscal incentives to optimize the supply of nutritious and healthy food.

C. Promote the high-quality development of industries and cities through coordinated digital and green transformations.

9. Promote the low-carbon development of existing digital infrastructure, such as data centres, industrial Internet, 5G, etc., and build energy-efficient and climate-friendly digital infrastructures. Conduct annual energy consumption assessments of key national computing and data centres, implement energy-efficiency audits, and establish zero-carbon data centres. Establish a public data centre directory to record key indicators related to data centre operations, including electricity efficiency, renewable energy factor, cooling efficiency ratio, and water usage effectiveness, and to track the carbon emissions associated with the hardware, software, and cloud services of digital facilities. Optimize industrial policies and support the application of renewables in the digital economy, establish a coherent evaluation system for digital and green development, and establish incentive mechanisms for green and low-carbon development.
10. Centred on core indicators like carbon productivity, energy efficiency, water consumption, and material usage/consumption, establish a system of green and low-carbon production metrics. Strengthen continuous carbon monitoring through digitization to identify priority areas for emission reduction. Develop a carbon asset management system for key manufacturing sectors and gradually promote the disclosure of corporate climate-related information. Utilizing the supply chain as a framework, mobilize upstream and downstream companies to track carbon emissions data and product carbon footprint.
11. Optimize the energy supply structure for enterprise production and expand green power trading and new energy power supplies. Introduce time-of-use electricity pricing signals to encourage industrial energy saving during peak demand. Encourage enterprises to expedite the technological upgrading of pollution and carbon reduction and adopt green and low-carbon technologies. Extend the lifespan of information and communication products through eco-design and recycling, gradually phasing out energy-intensive equipment. Carry out digital low-carbon production pilot programs in industries such as steel and metal.
12. Promote a system for measuring and assessing sustainable urban development through digitization, integrating multiple dimensions, such as spatial planning, industry, housing, transportation, management services, etc., and carry out ongoing assessments of smart and sustainable cities.
13. Enhance the climate adaptation capacity with digital technologies, formulate a special meteorological digitization plan, improve the capacity for multi-source meteorological data collection and transmission, and standardize multi-source data integration and security management. Increase climate modelling, simulation, and climate risk assessment, enhance the capacity of weather forecasting and disaster monitoring, and develop diversified smart weather service products.
14. Ensure the digital competence training and rights of key groups, establish a digital competence training and evaluation system for governments and civil servants at all levels, and foster transferable digital competence.

Develop mechanisms to ensure benefits, the rights to be informed, and participatory rights for groups such as women, the elderly, and individuals with disabilities in the context of digital and green development, promoting inclusive and universal digital development.

D. Enhance the green financial system to support green and low-carbon transformation.

15. Shape diverse, green climate investment and financing mechanisms with a comprehensive range of incentives, including taxation, pricing, compensation, and procurement. Accelerate the formulation of categorized directories, rules and standards for transition finance, enhance information disclosure of risks related to climate, environment, and biodiversity loss, and regulate the environmental, social and governance (ESG) investment market. Maintain consistency between domestic green classification standards and international standards, expand disclosure scope, with a focus on complying with international financial reporting standards related to ESG set by the International Sustainability Standards Board (ISSB), and prepare for the upcoming biodiversity risk disclosure standards.
16. Emphasize the green and low-carbon investment potential of sovereign wealth funds and social security funds as the main components of sovereign assets. Encourage sovereign asset owners to conduct sustainable investment and financing pilot demonstrations. Consider incorporating climate, ecological and environmental value into performance assessment systems and provide flexible support for applying investment return assessment and risk-sharing tools. Establish sustainable investment principles for sovereign asset owners, including clear strategic objectives and organizational safeguards. Encourage sovereign asset owners to engage in more exchanges and cooperation on sustainable investment and financing with international partners.
17. Actively participate in multilateral financial cooperation and reform of the international financial systems. Strengthen the alignment with and mutual recognition of international rules and standards related to climate, nature, and sustainable development. Effectively prevent the risks of stranded assets and greenwashing. Raise green standards for overseas investment and financing and improve disclosure, compliance, and accountability mechanisms for financial institutions.

E. Build a sustainable blue economy through land–ocean integration and build resilient river basins.

18. Make the sustainable blue economy a key strategic national development goal and an integral part of the national "dual-carbon" goals. Establish a sustainability-oriented ocean economic accounting and statistical framework to calculate the carbon dioxide emissions from the marine industry

and the contributions to decarbonization through NbS. Develop corresponding monitoring methods. Strengthen the assessment and prediction of the impact of climate change on the oceans and global fisheries.

19. Create a blue finance framework to enhance financial support for a sustainable blue economy. Strengthen coordination and funding for international scientific research cooperation on a sustainable blue economy and marine carbon reductions.
20. Advance the comprehensive development of offshore wind energy, tidal energy, solar energy, hydrogen energy, and other renewable energy sources for electricity generation. Reduce carbon emissions from fishing vessels and ports. Initiate decarbonization plans for maritime operations and for aquaculture and fisheries management. Conduct scientific land-sea-space planning, identify optimal layouts for photovoltaic and wind power, promote multifunctional vertical development and compound utilization, and enhance spatial utilization efficiency.
21. Improve the multi-level Integrated Ocean Management (IOM) system from central to local levels. Develop site selection and implementation standards for marine-related construction projects to protect ocean and coastal ecosystems. Enforce strict control over plastic usage in marine industries and develop comprehensive plans to reduce plastic pollution, including effective extended producer responsibility standards, capacity building, and public education.
22. Assess climate risks in coastal areas (such as the Greater Bay Area). Update urban building codes for cities along rivers and oceans and increase investments in climate adaptation for assets like infrastructure, housing, and industries, to cope with the risk of sea level rise.
23. Under the framework of the Yangtze River Protection Law and other river basin protection laws, develop vertical action plans and horizontal collaborative agreements. Building upon the existing government collaboration mechanisms, such as the National Yangtze River Basin Coordination Mechanism and Local Coordination Mechanism, establish a cross-departmental and cross-administration regional collaborative framework involving multiple stakeholders such as governments, enterprises, the public, and other entities.
24. Expedite the formulation of river basin development plans and territorial spatial plans. Establish a comprehensive assessment mechanism for river basins, systematically evaluating the long-term pressure and short-term impacts of climate change. Incorporate climate adaptation capacity assessment into policy-making and decision-making processes for construction projects. Promote NbS and encourage the establishment of water funds to support pilot projects on sustainable hydropower.
25. Identify key steps to further tackle plastic pollution. Develop an action plan to implement the global plastics treaty after its adoption. Consider launching a series of pilot projects designed to reuse, reduce, and recycle plastics.

F. Maintain green and open development, build sustainable supply chains, and contribute to global low-carbon transformation.

26. Establish a new type of cooperative relationship between importing and exporting jurisdictions and companies to optimize the layout of the global industrial and supply chains and collectively ensure the supply of critical minerals, materials, and components in green and low-carbon industries. Establish BRI green innovation partnerships. Create cross-departmental coordination mechanisms for a resilient and sustainable development of the industrial and supply chains.
27. In the multilateral trade cooperation mechanisms in which China participates, conduct constructive dialogues and pilots to forge green consensus and explore the establishment of green, zero-deforestation, and nature-positive trade standards and certification systems. Establish transparent and traceable technological and policy frameworks, incorporate green soft commodity import and export measures into bilateral and multilateral trade agreements, and integrate certification systems across different stages of the value chain. Ensure that all imported soft commodities are legally sourced in their country of origin and explore opportunities for greening commodity chains through South–South cooperation.
28. Promote the reform of the overseas investment approval system. Implement a comprehensive and coherent green BRI project pipeline system. Scale up solar and wind energy BRI projects, strengthen green technology transfer cooperation, and reduce fossil fuel-based power generation. Integrate green energy investments into the corporate performance assessment system, and appropriately relax performance requirements for overseas green energy investments. Establish a BRI climate financing and green credit system to reduce the financing costs for low-carbon investment projects. Enhance the green energy investment information service system and establish overseas investment risk assessment and early warning mechanisms.
29. Collaborate with BRI participating countries on innovative projects demonstrations. Establish pre-feasibility research and development funds and a database of financing options for green development projects. Actively provide renewable energy financing portfolios to the projects in the database. Strengthen dialogues and exchanges through multilateral cooperation platforms such as the Belt and Road Initiative International Green Development Coalition (BRIGC). Utilize the third Belt and Road Forum for International Cooperation (BRF III) as an opportunity to introduce an international cooperation initiative for green and low-carbon development. Coordinate resources from all stakeholders to facilitate demonstrative cooperation in innovative application scenarios, such as "photovoltaics+," and explore business models for green cooperation projects that suit the characteristics of developing countries.

G. From agreements to synergies in the implementation of the Kunming–Montreal Global Biodiversity Framework.

30. Swiftly update the National Biodiversity Strategy and Action Plan (NBSAP) alongside corresponding policy measures and roadmaps in line with the Kunming–Montreal Global Biodiversity Framework. As the Presidency of COP 15, China should continue to communicate with parties to the convention, observer states, and other stakeholders to promote cutting-edge biodiversity conservation initiatives. Take early action to achieve the "30 × 30" and all the other GBF targets to achieve rapid early results. Boost confidence in the implementation of the Kunming–Montreal Global Biodiversity Framework.
31. Establish a biodiversity expert group to facilitate engagement, coordination, and implementation at the national and international levels. Develop global standards to encourage enterprises to integrate biodiversity conservation into their development strategies, ensuring their activities yield nature-positive outcomes. Large enterprises should pay attention to the impact of their activities on nature and enhance risk disclosure. Encourage, guide, and assist businesses in participating in biodiversity conservation and implementation through platforms like the China Business and Biodiversity Partnership. For challenging specific goals, encourage the development, promotion, and application of methods and tools, and utilize incentives to help achieve the goals. Develop a youth nature education program and initiate corresponding agricultural, forestry, and fisheries practice pilots that align with the Kunming–Montreal Global Biodiversity Framework.
32. Call for and welcome contributions from all signatories to support the Kunming Biodiversity Fund, integrating and coordinating different sources of international financing to support biodiversity conservation in developing countries. Reallocate direct transfer payments that are harmful to biodiversity to optimize the impact of existing funds. Support partner countries to establish a hybrid financing model, formulate national-level financing plans, and mobilize and coordinate funds from government agencies, private sectors, philanthropic organizations, multilateral development banks, voluntary carbon markets, and other relevant stakeholders.
33. Ensure the applicability of green finance classification standards to the biodiversity conservation financing goals in the Kunming–Montreal Global Biodiversity Framework. Gradually implement internationally aligned and mandatory biodiversity disclosure standards at the market level. Develop systematic and comprehensive methodologies to assess the ecological and environmental impacts of subsidy policies, launching pilots in agriculture, forestry, and fisheries.
34. Collaborate with economic sectors to advocate integrated and sustainable land-use practices and promote mainstreaming biodiversity conservation. Reassess and optimize land use based on ecosystem service functions, incorporating science-based climate and nature objectives into decision making and operations. Taking agriculture as a starting point, identify pathways and methods to achieve the sustainable use action goals of the Kunming–Montreal Global Biodiversity Framework. Implement pilot

projects on regenerative agriculture and conservation-oriented farming and promptly summarize the experience gained.

Chapter 13
Progress on Environment and Development Policies in China and Impact of CCICED's Policy Recommendations (2023)

About the Report

As a high-level policy advisory body approved by the Chinese government, the China Council for International Cooperation on Environment and Development (CCICED) is mainly tasked with studying and proposing policy recommendations on major issues of environment and development. As the highest form of policy consultation, CCICED's Annual General Meeting (AGM) invites Chinese and international members, invited advisors and experts from home and abroad to have policy discussions on major environmental and development issues based on CCICED's SPS reports, focusing on urgent and long-term domestic issues while responding to major concerns of the international community and building consensus on ideas. On this basis, annual policy recommendations will be formed and submitted to the State Council and relevant departments of the central government.

This report reviews the progress of China's environment and development policies since 2022. This is the 16th report presented by CCICED Chief Advisors' Expert Support Group and the Chinese Associates.[1]

Foreword

The past year has been extraordinary and momentous. In the face of an intricate international environment and the arduous task of domestic reform, development and stability, the Chinese Government has adhered to the principle of seeking progress while maintaining stability, planned and organized the various undertakings of socialist modernization, and actively created a new situation of high-quality development. At the historical intersection of the two centenary goals, the 20th CPC National Congress was successfully convened, depicting a grand blueprint for advancing the great rejuvenation of the Chinese nation by means of Chinese modernization in an all-round way. The report to the 20th CPC National Congress clearly sets out the historical direction of environment and development, and makes the harmony

[1] The Chinese supporting team mainly includes Mr. Zhang Huiyong and Ms. Tang Huaqing.

Green Empowerment and High Quality Development,
https://doi.org/10.1007/978-981-96-4218-2_13

between humanity and nature one of the important features and essential requirements of Chinese modernization, reflecting the Chinese government's responsibility to promote sustainable development and its historical courage in exploring a new development path.

At the major historical turning point for sustainable development in China, CCICED has given full play to its role as a major platform for international cooperation on environment and development, and made a large number of policy recommendations for implementing the "dual carbon" goals and promoting ecological progress by pooling the ideas of top-notched experts at home and abroad. The policy recommendations have been either consulted or adopted by the Chinese government, giving a strong impetus to China's sustainable development.

1 Environmental and Development Planning

1.1 Strengthening the Institutional Foundation for Ecological Civilization and Building the Modernization Featured by Harmony Between Humanity and Nature

Since the 18th CPC National Congress, China has deepened the reform in systems and mechanisms for ecological civilization in all respects, emphasized the "ecological and environmental protection with the strictest system and the tightest rule of law". During the 14th Five-Year Plan (FYP) period, China's ecological civilization has entered a critical period in which efforts shall be made to promote synergies in reducing pollution and cutting carbon emissions, foster the transition to green economic and social development in all respects, and realize quantitative to qualitative improvements in ecological and environmental quality with carbon reduction as a key strategic direction.

With regard to the realization of the "dual-carbon" goals, the Chinese Government has established a 1 + N policy system, namely, a top-level design document: *Working Guidance for Carbon Dioxide Peaking and Carbon Neutrality in Full and Faithful Implementation of the New Development Philosophy*; and "N", which includes the *Action Plan for Carbon Dioxide Peaking Before 2030* (hereinafter referred to as the "Action Plan"), implementation plans for carbon peaking in the fields of energy, industry, transportation, and urban and rural development, as well as support plans in terms of scientific and technological support, energy supply, carbon sequestration capacity, fiscal, financial and pricing policies, standards and measurement system, and supervision and assessment.

The report to the 20th CPC National Congress puts forward the ambitious goal of "promoting the modernization of harmony between humanity and nature", marking that the ecological civilization has entered a new stage of development. The principle that "lucid waters and lush mountains are invaluable assets" has been put into practice in various places, and ideological concepts such as tree planting and greenery protection, garbage sorting, water and electricity conservation, and the "Clean Your Plate" campaign have been deeply rooted in people's hearts and minds.

1.2 Accelerating Green Transition in All Respects Driven by the Outline of the 14th Five-Year Plan and Long-Range Objectives Through the Year 2035

Green development is the most distinctive feature of development in contemporary China. The *Outline of the 14th Five-Year Plan and Long-range Objectives Through the Year 2035* begins by stating the need to foster green development and harmony between humanity and nature. By 2025, new progress will be achieved in ecological conservation, with continued improvement of the ecological environment. By 2035, the ecological environment will be fundamentally improved, and the goal of building a beautiful China will be basically realized. In the first year of the 14th FYP period, a new journey to build a beautiful China is underway.

During the 13th FYP period, green development was written into the country's five-year plan for the first time. The Outline of the 14th FYP elaborates on accelerating the transition to a model of green development in four aspects, namely, resource utilization efficiency, utilization system, green economy, and policy system, and institutional innovation is particularly important for the realization of these goals. The national ETS was officially launched on July 16, 2021, leveraging market-based mechanisms to stimulate technological innovations by enterprises and reduce carbon emissions intensity. In addition, pollution prevention and control actions will be carried out in depth, to basically eliminate heavily polluted weather, the inferior Class V state-controlled sections and urban black and smelly water bodies.

1.3 Improving Ecological and Environmental Governance Propelled by Green Urbanization

In October 2021, the General Office of the CPC Central Committee and the General Office of the State Council issued the *Opinions on Promoting Green Development in Urban and Rural Areas*. In accordance with the decisions and arrangements made by the CPC Central Committee and the State Council, we should base ourselves on the new development stage, implement the new development concept, develop a new development pattern, adopt the people-centered approach, give priority to ecology, conservation and protection, adhere to the systematic concept, integrate development and security, simultaneously promote material progress and ecological progress, implement the goal and task of carbon peaking and carbon neutrality, promote urban renewal and rural development, accelerate the transformation of urban and rural development patterns, and advance the transition to green economic and social development in all respects, thus laying a solid foundation for comprehensively building a modern socialist country.

1.4 Pursuing High-Quality Development Fueled by Major River Basin Development Plans

The planning and implementation of the major national strategies for coordinated regional development in the new era, including "Development of the Yangtze River Economic Belt (YREB)" and the "Ecological Protection and High-Quality Development of the Yellow River Basin", have stimulated the relevant provinces and regions of the two river basins to thoroughly implement the new development concept for the green development, with significant achievements made.

1. Achieving initial results in green and high-quality development of the Yangtze River Basin

The Yangtze River is the mother river of the Chinese nation and an important support for the development of the Chinese nation. In September 2022, MEE, the National Development and Reform Commission (NDRC) and 15 other departments jointly issued the *Action Plan to Further Advance the Ecological and Environmental Protection and Restoration of the Yangtze River Basin*, focusing on solving outstanding ecological and environmental problems in the Yangtze River protection, and solidly advancing the task of protecting and restoring Yangtze River.

The pattern of green and high-quality development in the Yangtze River Basin has taken initial shape. Firstly, a "1 + N" development planning system with the *Outline of the Yangtze River Economic Belt Development Plan* as the programmatic document has been established, the *Yangtze River Protection Law* has been introduced and implemented.

A "ten-year fishing ban" has been implemented in the key waters of the Yangtze River Basin, with 11,000 fishing boats and 231,000 fishermen returning to shore. As a result, the Yangtze River has been able to recuperate and the condition of its biological resources has gradually improved. The Yangtze River Basin has established a sound inter-provincial joint prevention and control mechanism covering upstream and downstream, left and right banks, main and branch streams of the Yangtze River; successively set up 5 inter-provincial ecological protection compensation mechanisms; and explored inter-provincial and municipal trading of carbon emission rights, pollutant discharge rights, water rights, and energy rights. The provinces and municipalities along the Yangtze River have cultivated a number of competitive and influential industrial clusters through innovations.

2. Depicting the blueprint for high-quality development of the Yellow River Basin

Since the 18th CPC National Congress, the CPC Central Committee with Comrade Xi Jinping as the core has proposed that the protection of the Yellow River is a major plan for the great rejuvenation of the Chinese nation, and the ecological protection and high-quality development of the Yellow River Basin is a major national strategy. At present, China's ecological civilization is advancing in an all-round way, and the principle of "lucid waters and lush mountains are invaluable assets" has been deeply rooted in people's hearts, and the people along the Yellow River have an even stronger desire to pursue lush mountains, lucid water, blue skies and clean land. On October 30, 2022, the 37th Session of the 13th NPC Standing Committee passed the *Yellow River Protection Law*, providing a strong legal guarantee for ecological and environmental protection and high-quality development in the Yellow River Basin.

The *Yellow River Basin Ecological and Environmental Protection Plan* issued by MEE and other departments in June 2022, is a special plan to implement the "1 + N + X" requirements of the *Outline of the Yellow River Basin Ecological Protection and High-quality Development Plan*, which is of great importance in advancing the ecological protection and high-quality development of the Yellow River Basin. In August 2022, MEE and other 11 departments jointly issued the *Action Plan for Yellow River Ecological Protection and Governance*, which takes maintaining the ecological security of the Yellow River as the goal and improving the quality of the ecological environment as the core. To protect the Yellow River, we must rely on the system and the rule of law to protect the mother river. 2022 The *Yellow River Protection Law of the People's Republic of China* was promulgated in October 2022, providing comprehensive regulations on water conservation, soil and water conservation, estuarine regulation, and ecological flow.

In terms of ecological protection and management of the Yellow River, as of 2022, sections with surface water quality from Class I to Class III in the Yellow River Basin accounted for 87.5%, the number of days with good air quality in cities at the prefecture level and above in the river basin accounted for 80.3%, and the ecological economy, including specialized agriculture and animal husbandry and clean energy, was developing healthily in various places along the Yellow River.

1.5 CCICED Policy Recommendations

For ecological civilization, CCICED recommended in 2022 that China should unremittingly hold firm in its strategic determination towards Ecological Civilization, by prioritizing and stabilizing expectations for a green, low-carbon transition, and move from securing short-term economic, energy, food and other security to unleashing win–win short and long-term green economic stimulus, innovation, low-carbon growth in which science links short, interim and longer-term green development targets and timetables. In this way, China will open a new green chapter for high-quality development.

For green urbanization, CCICED recommended in 2019 that the 14th FYP should formulate an urbanization strategy based on ecological civilization. The strategy should move away from the quantity-based model to a quality-based model where green urbanization becomes a key driver of China's high-quality economic development. CCICED recommended in 2020 and 2021 that it will be vital to advance urban green transition in line with the principles of green prosperity, low-carbon, intensive and circular development, equity and inclusiveness, as well as security and health; China should intensify the transformation to green, low-carbon urban infrastructure, improve rural and county green development, and adhere to the "one pole and multiple wings" rural integrated development model dominated by green development and supported by diversified development.

For green development of major river basins, CCICED recommended in 2022 integrated climate-resilient management for low-carbon and resilient river basins shall be strengthened. To be specific, it is advised to improve climate resilience in the integrated management of important river basins; implement the requirements of the *Yangtze River Protection Law* and develop a cooperative governance mechanism based on large-scale spatial planning and the co-management of pollution, ecosystem protection, low-carbon development and climate adaptation. Detailed climate vulnerability assessments should be made throughout the Yangtze River basin, from upstream to downstream areas, major tributaries, key urban and rural agglomerations, river coastlines, estuarine deltas, flood storage areas, and agriculture and natural ecological zones. A risk early warning system should be implemented at the basin level for extreme climate-related weather events, with special attention to flooding, wildfires, drought and heat waves.

2 Governance and Rule of Law

2.1 Supporting Ecological and Environmental Governance from the Judicial Level

A well-established legal system requires a strong judicial system to enforce it, while standardized and strict judicial action is an indispensable part of ecological and environmental governance, as well as an important means to support the effective implementation of systems and regulations.

In July 2022, the State Council's *Legislative Work Plan for 2022* proposed to request the NPC Standing Committee to deliberate the draft *Energy Law* and the draft amendment to the *Mineral Resources Law.*

In November 2022, MEE published the *Measures for the Administration of Lists of Major Entities under Environmental Regulation* (hereinafter referred to as the "Measures"), which came into effect on January 1, 2023, replacing the *Provisions on the Administration of the List of Key Pollutant Discharging Entities (Trial)*, which had been in use for five years.

In May 2023, MEE issued the newly revised *Measures on Administrative Penalties for the Ecological Environment*, to ensure that the ecology and environment enforcement team carries out enforcement activities in a strictly standardized manner and in accordance with the law, and it came into force on July 1, 2023.

Since 2022, local ecology and environment departments at all levels have continued to implement the *Guiding Opinions on Optimizing Enforcement Methods for Ecological and Environmental Protection and Improving Enforcement Effectiveness*, implemented a package of policies and measures to stabilize the economy, and made optimizing environmental supervision methods one of the five key measures to support the smooth operation of the economy in the field of ecological and environmental protection.

2.2 Introducing Pollutant Discharge Permits

The *Work Plan for Improving the Quality and Efficiency Through Pollutant Discharge Licensing of the Ministry of Ecology and Environment (2022–2024)* proposes to effectively improve the issuance quality of pollutant discharge permits, strengthen the quality inspection of pollutant discharge permits, enhance the entity responsibility, reinforce the joint supervision, and develop a closed-loop quality management mode of pollutant discharge permits featured by "source control, process management, ex-post supervision", so as to improve the efficiency of post-permit supervision, enhance the effectiveness of the core system, and give full play to the role of the "one permit-based" management.

On March 29, 2022, the MEE General Office released the *Guiding Opinions on Strengthening the Enforcement and Supervision of Pollutant Discharge Licensing*, which requires that by the end of 2023, key industries shall implement the checklist-based enforcement and inspection of pollutant discharge licensing, the daily management of pollutant discharge licensing, environmental monitoring, and enforcement and supervision shall be effectively linked, and the enforcement and supervision system for stationary pollution sources, with pollutant discharge licensing at its core, shall basically take shape.

On April 2, 2022, the MEE General Office issued the *Implementation Plan for Environmental Impact Assessment and Pollutant Discharge Licensing during the 14th Five-Year Plan Period*, to further consolidate the core system of pollutant discharge licensing.

The quality of pollutant discharge permits is the lifeline of "permit-based pollutant discharge", "permit-based supervision" and "public oversight". In order to improve the technical support system for pollutant discharge licensing, standardize the technical methods for quality inspection of pollutant discharge permits nationwide, and harmonize the standards for determining the quality of pollutant discharge permits, MEE issued the *Technical Specification for Quality Inspection of Pollutant Discharge Permit* (HJ1299-2023) in June 2023.

2.3 Further Improving Environmental Laws and Placing Climate Legislation on the Agenda

The report of the 20th CPC National Congress clearly proposes to "work actively and prudently toward the goals of reaching peak carbon emissions and carbon neutrality". On June 1, 2022, China's first law dedicated to protecting wetlands, the *Wetland Protection Law*, came into effect. As the first law specializing in the protection of wetlands in China. In June 5, 2022, the *Law of the People's Republic of China on Prevention and Control of Noise Pollution* came into force. On October 30, 2022, the *Yellow River Protection Law of the People's Republic of China* was formally adopted, which came into force on April 1, 2023.

In addition, in 2022, China has accelerated the legislation related to addressing climate change, and carried out thematic demonstration on the integration of GHG control into the EIA of construction projects, and studied the proposed amendment to the *Environmental Impact Assessment Law*; revised and issued the *Technical Guidelines for Planning Environmental Impact Assessment: Industrial Parks*; actively pushed forward the legislative process of the *Provisional Regulations on the Administration of Carbon Emission Trading*, and endeavored to improve the legislative safeguards for the national carbon trading market; guided and stimulated the formulation of relevant local laws and regulations at the local level; stepped up the revision of the *Regulations on the Administration of Ozone-Depleting Substances* to include HFCs and other ozone-depleting substances with greenhouse effect in the environmental protection control system.

2.4 Deepening the Green Financial System

Based on China's experience in the rapid development of green finance, a harmonized policy framework and incentive mechanism will be further improved in the future. Under the guidance of "accelerating the transition to a model of green development" as specified in the report to the 20th CPC National Congress, it is expected that the transition finance-related system and the green directory will be improved to support the low-carbon transition plans of high-carbon enterprises. At the same time, the development of criteria for defining green activities can prevent "greenwashing" and effectively channel funds into truly green industries.

In February 2022, the People's Bank of China (PBOC), together with the State Administration for Market Regulation (SAMR), the China Banking and Insurance Regulatory Commission (CBIRC) and the China Securities Regulatory Commission (CSRC), jointly issued the *Development Plan for Financial Standardization during the 14th Five-Year Plan Period*. In 2022, 49 out of 54 listed banks in China had made environmental and climate information disclosure in the form of social responsibility reports, sustainability reports or more explicit ESG reports.

In April 2022, CSRC published the *financial industry standard-Carbon Financial Products*. On a certain basis, carbon financial products were divided and detailed provisions were made. In the course of development, various innovative financial instruments for carbon sinks have been emerging in China. In May 2022, CBIRC issued the *Plan for Standardization of China's Insurance Industry during the 14th Five-Year Plan Period*, which calls for the establishment of index standards for green insurance statistics, green capital utilization, and green insurance assessment, so as to contribute to the building of a green financial system.

MEE, NDRC and seven other departments jointly issued the *Circular on the Pilot Work on Climate Investment and Financing* in December 2021, guiding the pilot places to actively participate in the building of the national carbon trading market, and to study and promote the development and alignment of carbon financial products.

In October 2022, SAMR and eight other departments jointly issued the *Implementation Plan for Establishing a Sound Standards and Measurement System for Carbon Peaking and Carbon Neutrality*. In March 2023, MEE issued the *Letter on Openly Soliciting Proposals on Methodologies for GHG Voluntary Emission Reduction Projects* (HBBH [2023] No. 95). In July 2023, MEE prepared the *Measures for Administration of GHG Voluntary Emission Reduction Trading (Trial) (Draft for Comments)*.

2.5 *Incorporating Carbon Emissions from Key Industries into the EIA System*

On May 31, 2021, MEE issued the *Guiding Opinions on Strengthening Prevention and Control from the Source in Energy-Intensive and High-Emission Projects for Ecological and Environmental Protection* (HP [2021] No. 45).On July 27, 2021, MEE issued the *Notice on Pilot Carbon Emission Environmental Impact Assessment for Construction Projects in Key Industries* (HP [2021] No. 346).

On December 2, 2022, MEE issued the *Notice on Printing and Distributing the Principles for Approval of Environmental Impact Assessment Documents for Construction Projects in Four Industries: Iron and Steel/Coking, Modern Coal Chemical Industry, Petrochemical Industry, and Thermal Power* (HP [2022] No. 31). It adds the requirement for GHG emissions in Article 6 related to the approval of EIAs for the four industries: "incorporating GHG emissions into the environmental impact assessment for construction projects, accounting for GHG emissions from construction projects, achieving synergies from reducing pollution and cutting carbon emissions, and advancing the demonstration and application of innovative carbon reduction technologies".

2.6 Developing a Sound Environmental Credit System

In March 2022, the General Office of the CPC Central Committee and the General Office of the State Council issued the *Opinions on Advancing the High-quality Development of the Construction of the Social Credit System in Furtherance of the Shaping of a New Development Pattern* (hereinafter referred to as the "Opinions"), which improves the credit system for ecological and environmental protection. The *Opinions* stipulates that credit evaluation shall be fully implemented in the fields of environmental protection and soil and water conservation, and the sharing and application of credit evaluation results shall be strengthened.

2.7 Fostering Green and Low-Carbon Lifestyles

The report to the 20th CPC National Congress specifies that a green and low-carbon economy and society are crucial to high-quality development. In January 2022, NDRC and other ministries issued the *Implementation Plan for Promoting Green Consumption*. June 15, 2022 is the 10th "National Low-Carbon Day". MEE and the People's Government of Shandong Province jointly organized the main event of the "National Low-Carbon Day 2022" in Jinan, Shandong Province.

On August 30, 2022, the State Council issued the *Opinions on Supporting Shandong to Further Replace Old Driving Forces with New Ones for Green, Low-Carbon, and High-Quality Development* (GF [2022] No. 18), which requires the in-depth implementation of the "Green and Low-Carbon Initiative for All", and the establishment of incentive mechanisms for green consumption.

The implementation plans for carbon peaking issued by provinces propose to foster green and low-carbon lifestyles. Hainan Province issued the *Implementation Plan for Carbon Peaking in Hainan Province* in August 2022. The *Implementation Plan for Carbon Peaking in Shanghai* proposes to guide citizens to travel in a green and low-carbon manner. The *Implementation Plan for Carbon Peaking in Jiangsu Province* proposes to vigorously advocate simple, moderate, green, low-carbon, civilized and healthy lifestyles, and resolutely curb extravagance, wastefulness and unreasonable consumption.

2.8 CCICED Policy Recommendations

In terms of the rule of law and governance, CCICED recommended in 2022 that efforts should be made to develop a sound governance system for green and low-carbon transition and strengthen innovative and flexible institutional capacity building; prioritize the development of a dedicated climate change law to set the necessary legal basis for China's climate transition, and explore to include dual carbon targets and climate adaptation into the scope of public interest litigation by procurators; create an ongoing working dialogue between financial regulators and relevant government departments, and develop and implement ESG standards. Moreover, it is suggested that China should integrate digitalization with sustainable development and promote green technology innovation and green digital governance, and engage the public in linking digital platforms with low-carbon, green lifestyles.

3 Energy, Environment and Climate

3.1 Making Concerted Efforts to Cut Carbon Emissions, Reduce Pollution, Expand Green Development, and Pursue Economic Growth

The CPC Central Committee and State Council released the *Opinions on Comprehensively Strengthening Ecological and Environmental Protection and Resolutely Fighting the Tough Battle Against Pollution* on April 17, 2018. The *Opinions* emphasizes the importance of reducing pollution, cutting carbon emissions, and expanding green development, and put forward a series of policy measures and targets, including specific requirements for strengthening environmental regulation and pursuing green development.

In June 2022, under the guidance and support of NDRC, the Ministry of Industry and Information Technology (MIIT), and MEE, the China Association of Environmental Protection Industry (CAEPI) issued the *Action Plan for Accelerating High-Quality Development of the Ecological and Environmental Protection Industry and Deepening Pollution Prevention and Control to Support Carbon Peaking and Carbon Neutrality (2021–2030).*

In July 2022, MEE and six other departments jointly issued the *Implementation Plan for Achieving Synergies from Reducing Pollution and Cutting Carbon Emissions*, which, as an important part of the "1 + N" policy system of carbon peaking and carbon neutrality. The concept of "green and low-carbon, energy saving first" has long been integrated into the practice of development across China.

Locally, the Wuhan Municipal Government took the lead in issuing an implementation plan. On November 16, 2022, the Wuhan Municipal Government issued the *Implementation Plan for the Top 10 Actions to Cut Carbon Emissions, Reduce Pollution, Expand Green Development, and Pursue Economic Growth in the Yangtze River Economic Belt in Wuhan.*

At present, the development of green industry has become a major priority for Hangzhou's low-carbon transition. Not long ago, Hangzhou formulated and issued the *Opinions on Implementing the New Development Concept in a Complete, Accurate and Comprehensive Way to Achieve Carbon Peak and Carbon Neutrality*, which provides top-level institutional design for achieving carbon peaking and carbon neutrality in a high-quality way. In fact, Hangzhou has achieved certain development results in the above industrial areas, laying a good industrial foundation for future low-carbon development.

3.2 Continuously Adjusting and Optimizing the Energy Structure

In May 2022, the General Office of the State Council issued the *Implementation Plan on Promoting the High-Quality Development of New Energy in the New Era*, which calls for improving the compensation mechanism for peak-frequency regulation power supply, stepping up flexibility retrofits of coal-fired power units, hydropower expansion, pumped storage and solar thermal power projects, facilitating the rapid development of new energy storage, studying the energy storage cost recovery mechanism, and encouraging the use of solar thermal power as a peaking power source in areas with good light conditions such as the western region.

As for the high-quality development of the new energy industry, power supply construction is also facing new challenges. To this end, the National Energy Administration (NEA) revised and issued the *Regulations on the Administration of Grid-connected Operation of Electric Power* (GNFJGG [2021] No. 60) and the *Measures for the Administration of Auxiliary Services of Electric Power* (GNFJGG [2021] No. 61). In order to further advance the "coordinated three retrofits" in coal-fired power units, NDRC and NEA jointly issued the *Implementation Plan for Retrofitting and Upgrading of Coal-fired Power Units Nationwide* in November 2021, which puts forward flexibility retrofits of 200 million kW during the 14th FYP period. With regard to pumped storage, NEA issued the *Medium- and Long-term Development Plan for Pumped Storage (2021–2035)* in September 2021, NDRC and NEA issued the *Guiding Opinions on Accelerating the Development of New Energy Storage* (FGNYG [2021] No. 1051), the *Implementation Plan for the Development of New Energy Storage During the 14th Five-Year Plan Period* (FGNY [2022] No. 209), and the *Notice on Further Promoting the Participation of New Energy Storage in the Electricity Market and Dispatching Application* (FGBYX [2022] No. 475), to accelerate the large-scale and market-based development of new energy storage.

In January 2022, General Secretary Xi Jinping stressed at the 36th session of the collective study of the CPC Political Bureau that we shall step up efforts to plan and build a new energy supply and consumption system based on large-scale wind and solar power bases, supported by clean, efficient, advanced and energy-saving coal-fired power in their vicinity, and carried by stable, safe and reliable EHV transmission and transformation lines. On May 14, 2022, NDRC and NEA issued the *Notice on the Implementation Plan for Promoting High-Quality Development of New Energy in the New Era*. At the same time, the development and utilization of new energy sources are still subject to such constraints as the lack of adaptability of the power system to large-scale and high-share new energy sources for grid connection and consumption, and obvious constraints on land resources.

It is also suggested that we shall reduce the financing cost of renewable energy (RE) companies and further increase support for RE development in terms of early queuing for initial public offerings (IPOs), targeted lending, equity financing and lowering the reserve ratio.

3.3 Continuing to Promote Energy Conservation and Energy Efficiency Improvement

In February 2022, NDRC and three other departments jointly issued the *Implementation Guidelines for Energy Saving and Carbon Reduction Retrofitting and Upgrading in Key Areas of Energy-Consuming Industries (2022 Edition)* (hereinafter referred to as the "Implementation Guidelines").

On November 17, 2022, NDRC and other departments jointly issued the *Circular on the Release of Advanced Levels of Energy Efficiency, Energy Saving Levels, and Access Levels of Key Energy-Using Products and Equipment (2022 Edition)*, which focuses on key energy-using products and equipment, and makes the relevant arrangements for saving energy and cutting carbon emissions.

In order to mobilize the industry's energy saving, carbon reduction, green and low-carbon transition, and to continuously improve the industry's energy efficiency, the Departments of Industry and Information Technology in many places have publicly solicited opinions on the action plans for energy saving, energy consumption reduction and energy efficiency improvement in petrochemical and chemical industries. The *Several Opinions on Strict Energy Efficiency Constraints to Promote Energy Conservation and Carbon Reduction in Key Areas* issued by NDRC and other departments sets forth that: by 2025, through the implementation of energy saving and carbon reduction actions, the proportion of production capacity in the cement and flat glass industries with energy efficiency reaching the benchmark level will exceed 30%, the overall energy efficiency level of the building materials industry will be significantly improved, the intensity of carbon emissions will be significantly reduced, and the capacity for green and low-carbon development will be obviously enhanced.

In order to ensure the safe and stable power supply, enhance the core competitiveness of the energy industry, NEA issued the *Guiding Opinions for Energy Work 2022* (hereinafter referred to as the "Guiding Opinions") in March 2022. In April 2023, NEA issued the *Several Opinions on Accelerating the Digital and Intelligent Energy Development*.

3.4 *Enhancing Climate Action and Adaptation*

Climate change-induced extreme heat is taking a huge toll in many parts of the world. In June 2022, 17 departments, including MEE, NDRC and the Ministry of Science and Technology (MOST) jointly published the *National Climate Change Adaptation Strategy 2035*. The *Adaptation Strategy 2035* also stipulates that in the urban space, the focus will be on reducing climate risks with respect to populations, social and economic development, and infrastructure, building climate-resilient cities, and improving urban climate risk prevention and control capacity.

In September 2022, MEE issued the *Guidelines for the Preparation of Provincial Action Plans for Adaptation to Climate Change*, which proposes to start the preparation of provincial action plans for climate adaptation as soon as possible, and actively expand international cooperation on climate adaptation. In April 2023, the *Action Plan for Adaptation to Climate Change in Sichuan Province* (hereinafter referred to as the "Action Plan") was released, which is also the first provincial-level action plan for climate adaptation in China. In January 2023, the Department of Ecology and Environment of Zhejiang Province and seven other departments jointly issued the *Implementation Opinions on Promoting Climate Change Investment and Financing in Zhejiang Province.*

3.5 *Steadily Advancing the Building of China's National ETS*

Since its official launch on July 16, 2021, China's national ETS has become the world's largest carbon market in terms of covered emissions. On December 31, 2021, the first compliance cycle was successfully concluded, with a fulfillment rate of 99.5%. As of May 24, 2023, China's national ETS has operated safely for 449 trading days, with a total of 235 million tons of China Emission Allowances (CEAs) traded, involving a total turnover of 10.786 billion yuan, and a total of 61,350 transactions cleared, involving a clearing amount of 21.571 billion yuan.

Overall, it seems that the trading volume in the national ETS is relatively close to the quota shortfall of key emitters. The main purpose of the trading entities is to fulfill their compliance obligations, and the trading volume can basically meet the compliance needs of key emitters. In January, 2023, MEE released the *Report on the First Compliance Cycle of the National Carbon Emission Trading Market*. The framework for the operation of the national ETS has been basically established, the role of the price discovery mechanism has been initially demonstrated, the awareness and capacity of enterprises to reduce emissions have been effectively improved, and the expected targets have been achieved.

In order to effectively upgrade the quality of carbon emission data in the national ETS, improve the long-term mechanism of data quality management, and strengthen the daily supervision of data quality, MEE issued the *Guidelines on Accounting*

Methodology and Reporting of Corporate GHG Emissions: Power Generating Facilities (Draft for Comments) and the *Technical Guidelines for Verification of Corporate GHG Emissions: Power Generating Facilities (Draft for Comments)* in December 2022, to revise the technical specifications for accounting and reporting GHG emissions from power generation facilities. Moreover, MEE prepared technical guidelines for the verification of GHG emissions from power generation facilities.

In June 2023, MEE revised the *Interim Measures for the Administration of GHG Voluntary Emission Reductions Trading* (hereinafter referred to as the "Interim Measures"), and compiled the *Measures for the Administration of GHG Voluntary Emission Reductions Trading (Trial) (Draft for Comment)* (hereinafter referred to as the "Measures"). In addition, MEE has accelerated the preparations for the launch of the voluntary emission reductions (VERs) trading market, and strived for an early launch of the national VERs trading market in 2023.

3.6 CCICED Policy Recommendations

In 2022, CCICED has made a number of valuable policy recommendations on energy, environment and climate. These are detailed below:

Accelerating investment in renewable energy: The broad reform of China's electricity power market towards greater market orientation will strengthen the efficiency of market pricing mechanisms that in turn will attract additional private sector investments in green electricity generation. The current spot market should be expanded, with additional pilot projects that include inter-provincial trading. Increased renewable energy deployment should include land and offshore planning, using best-in-class environmental impact assessments, and respecting the ecological redline and spatial planning that protect ecological systems, including migratory corridors. The further scaling-up of renewable energy should include the early queuing for initial public offering (IPO), targeted loans and equity financing, and lower required reserve ratios. Regional renewable energy pilot projects should focus on correcting poor intra-provincial power consumption and out-ward grid connectivity, the inadequate development of regional power grids, and lagging price transmission mechanisms. Power grids should become more flexible and interconnected, and complemented with additional power storage to better integrate renewable energy sources.

Stabilizing the stock, strictly controlling the increment, and guiding the orderly phase-down of coal power: Efforts should be made to peak coal use by 2025, in order to achieve the peaking of carbon dioxide emissions before 2030. Short- and interim-term planning should be closely aligned with the dual control low-carbon transition, comprised of a short-term shift in coal power from base-load power generation to peak-management power generation; the elimination of outdated coal generating capacity while ensuring reasonable operating hours for high-efficiency and low-emission coal power; modernizing the remaining coal power fleet to further cut criteria air pollutants; paying special attention to cutting methane and other short-lived climate pollutants; paying close attention and leading financial risk disclosure related to coal and other fossil-fuel investments, and adjusting the investors' expectations for action related to stranded asset risks. An open and competitive auction-based mechanism to replace the guaranteed hours and price of coal-fired power generation units should be established, in conjunction with an efficient electricity-price market to provide economic returns for the flexibility of power.

Establishing a multi-objective collaborative mechanism for reducing pollution, cutting carbon emissions, expanding green development and pursuing economic growth: It is recommended to mainstream Nature Based Solutions (NbS); establish a standard Chinese system for NbS that aligns with the 2022 UNEA multilateral definition, and international standards; integrate NbS into existing policies such as the ecological redline, expand the Green Bond Endorsed Projects Catalogue and an updated green finance classification system to include eligible NbS project financing.

Developing a sound governance system for green and low-carbon transition and strengthening innovative and flexible institutional capacity building: It is advised to prioritize the development of a dedicated climate change law to set the necessary legal basis for China's climate transition, and explore to include dual carbon targets and climate adaptation into the scope of public interest litigation by procurators; develop integrated climate data systems and standards to enhance the integrity of the national ETS; and improve the quality of emissions data by building capacity, clarifying the responsibilities of the main emitters and setting penalties.

Systematically assessing risks from green and low-carbon transition and identifying key affected sectors and regions: It is proposed to undertake an ongoing, systemic risk assessment of the low-carbon green transition, paying close attention to inflationary effects of carbon pricing, stranded asset financial risks, price volatility and default risks in high-carbon sectors.

Building a diversified capital investment and financing mechanism: Transition finance should be utilized to facilitate corporate green transitions based on climate, biodiversity, pollution risk disclosure and transition timetables, and to avoid a net increase in fossil energy investments during the transition. The integration of climate, environmental and ecological finance should be encouraged through public–private partnerships and payments for ecosystem services. Moreover, a multi-party cooperation platform should be developed to track and disclose the greenwashing of ESG investments on an annual basis.

4 Pollution Prevention and Control

4.1 Further Advancing Air Pollution Prevention and Control

Combating air pollution is a complex and systematic project that requires unremitting and strenuous efforts. Since the 18th CPC National Congress, China's ambient air quality has improved significantly due to the formulation and implementation of the *Action Plan for the Prevention and Control of Air Pollution* and the *Three-Year Action Plan to Fight for Blue Skies.*

In November 2022, MEE, together with NDRC, the Ministry of Industry and Information Technology (MIIT), the Ministry of Transport (MOT) and ten other departments, jointly formulated the *in-depth fight to eliminate heavily polluted weather, ozone pollution prevention and control of diesel truck pollution treatment plan.* In 2022, the State Council issued the *Comprehensive Work Plan for Energy Conservation and Emission Reduction during the 14th Five-Year Plan Period*, with focus on key regions for air pollution prevention and control, the Pearl River Delta region and Chengdu-Chongqing region.

According to the data released by MEE, in 2022, the three binding indicators of national air quality met the schedule targets; the annual average $PM_{2.5}$ concentration in cities above prefecture level nationwide reached 29 μg, entering the 20+ era for the first time; the number of days with good air quality accounted for 86.5%, and the number of heavily polluted days dropped to 0.9%, with all three indicators meeting the schedule targets during the 14th FYP period.

4.2 Reinforcing Water Pollution Prevention and Control

Since January 2022, many new environmental regulations on domestic water use are being implemented by local governments at all levels. The Interim Measures stipulates that the operation and maintenance entities shall not suspend the operation of rural sewage treatment facilities without authorization.

On November 25, 2021, the 31st Session of the Standing Committee of the 13th Sichuan Provincial People's Congress adopted the *Regulations on Ecological and Environmental Protection of the Jialing River Basin in Sichuan Province*, which came into force on January 1, 2022. On September 9, 2021, the People's Government of Zhejiang Province approved and issued the *Discharge Standard for Water Pollutants from Centralized Rural Domestic Sewage Treatment Facilities*, which came into force from January 1, 2022.

In addition to the policy of domestic water pollution prevention and control, the central government has strongly supported the ecological protection and restoration of river basins in recent years. In May 2023, the CPC Central Committee and the State Council issued the *Outline of the National Water Network Construction Plan,* an important guiding document for the construction of national water networks in the current and future periods.

Shandong Province issued the *Regulations on the Protection of Nansi Lake in Shandong Province*, which came into force since January 1, 2022. The people's governments at or above the county level and the relevant departments receiving the report shall handle according to law and give awards in accordance with the relevant provisions. Hainan Province has pushed ahead with the fight against water pollution, and the "six water co-governance" has continuously enhanced the overall coordination of water control in the province.

4.3 Accomplishing Preliminary Results in Soil Pollution Prevention and Control

In order to deeply implement the spirit of the 20th CPC National Congress, actively implement the relevant requirements of the *Law of the People's Republic of China on Environmental Protection*, the *Law of the People's Republic of China on Prevention and Control of Solid Waste Pollution* and the *Law of the People's Republic of China on Prevention and Control of Soil Pollution*, and give full play to the role of advanced technologies in the prevention and control of solid waste and soil pollution, MEE planned to compile a *National Catalogue of Advanced Pollution Prevention and Control Technologies (in the field of Solid Waste and Soil Pollution Prevention and Control)* in 2022, to recommend technologies for the disposal and recycling of urban and rural domestic waste.

In June 2023, MEE issued a notice on public solicitation of opinions on *the Guiding Opinions on Promoting Green and Low-carbon Soil Pollution Risk Control and Remediation (Draft for Comments)*. For large and complex contaminated sites, risk assessment methods and parameters shall be scientifically selected based on pollutant migration and transformation patterns and effective exposure doses, and remediation and control targets can be reasonably determined to avoid excessive remediation. In 2023, Jiangsu Province has formulated the *Work Plan for the Prevention and Control of Soil, Groundwater and Agricultural and Rural Pollution in Jiangsu Province in 2023*.

4.4 Strengthening Marine Pollution Prevention and Control

On December 27, 2022, the draft amendments to the *Marine Environmental Protection Law* was initially submitted to the 38th session of the 13th NPC Standing Committee for deliberation, with a view to further improving the quality of the marine ecological environment through amendments to the law.

In 2022, China introduced a series of policies for marine environmental protection. In January 2022, in order to implement the decisions and arrangements made by the CPC Central Committee and the State Council on further combating pollution, MEE, together with NDRC, the Ministry of Natural Resources (MNR), the Ministry of Housing and Urban–Rural Development (MOHURD), MOT, the Ministry of Agriculture and Rural Affairs (MARA), and China Coast Guard (CCG), formulated the *Action Plan for Comprehensive Remediation of Key Sea Areas.* According to the *Action Plan*, comprehensive remediation will be carried out in the three key sea areas of Bohai Sea, Yangtze River Estuary-Hangzhou Bay and adjacent waters of the Pearl River Estuary, involving Tianjin, Shanghai and other "2 + 24" coastal cities. A head from MEE said that the *Action Plan* deploys eight special actions such as the investigation and rectification of sea outfalls.

In March 2022, the work reports of provincial governments in coastal areas were released one after another, and 11 provinces (municipalities and autonomous regions), made arrangements for sea-related work in 2022. Building of a strong maritime province, marine ecological restoration, development of coastal economic zones, and land-sea integration became key words in a number of *Report on the Work of the Government*. On January 20, 2022, Liaoning Province released its *Report on the Work of the Government*. The *Report on the Work of Government of Hebei Province* stresses that we will strengthen river basin remediation of water pollution, carry out a special campaign to improve the water quality of rivers entering the sea and offshore waters, and create beautiful rivers, lakes and bays.

In order to implement the comprehensive remediation of key sea areas, Guangdong Province has successively issued the *Plan for Marine Ecological and Environmental Protection of Guangdong Province during the 14th Five-Year Plan Period* and the *Implementation Plan for the Comprehensive Remediation of Sea Areas Adjacent to the Pearl River Estuary*. Focusing on reducing the total amount of nitrogen entering the sea, the province paid close attention to the management and control of total nitrogen in rivers entering the sea. To implement the *Opinions of the General Office of the Ministry of Ecology and Environment on the Management and Control of Total Nitrogen and Other Pollutants in Rivers Entering the Sea in Key Sea Areas*, Guangzhou, Zhuhai, Zhongshan and Jiangmen were organized to formulate and implement the "one-policy for one river" management and control plans. The Provincial Department of Ecology and Environment has organized the development of water pollutant discharge standards for the Sha River and Qijiang River Basins, and proposed total nitrogen discharge limits.

MEE has taken a number of measures to develop marine carbon sinks. On the one hand, the *Guiding Opinions on Coordinating and Strengthening the Work of Addressing Climate Change and Protecting the Ecological Environment* was issued and implemented. On the other hand, efforts to enhance marine response and adaptation to climate change were incorporated into the *National Plan for Marine Ecological and Environmental Protection during the 14th Five-Year Plan Period*, to systematically deploy relevant priority tasks. In 2023, China's provinces successively introduced various policies and plans for marine carbon sinks. In March 2023, Zhejiang

Province issued the *Guiding Opinions on Enhancing the Capacity of Marine Carbon Sinks in Zhejiang Province*, which proposes five major tasks.

4.5 CCICED Policy Recommendations

Policy recommendations submitted to the Chinese government from the CCICED 2022 AGM are listed below:

Reinforcing ecological conservation and restoration of river basins: It is suggested to strengthen the control of soil erosion and desertification in mountainous and hilly areas and implement the "returning space to rivers" campaign to restore river and lake basins; systematically manage hydropower projects to ensure that they undergo science-informed, robust and participatory EIA prior to project development, safeguard hydrological integrity and ecological water demand, and reduce ecological impacts through measures such as ecological scheduling and installation of fish passage facilities; promote 'grain-for-green' in ecologically sensitive areas and strengthen ecological restoration; pay attention to the melting of glaciers at the source of the Yangtze River and strengthen monitoring and early warning; and improve the safety and security mechanisms for vulnerable groups—especially women in disaster-prone areas such as villages, small towns, and flood storage areas.

Strengthening integrated water and land management of basin: Promoting the transition of the downstream industrial port shoreline into an ecological shoreline and a shoreline for residents. Concerted efforts should be made to formulate and supervise the implementation of the "three lines and one list" for shoreline protection and utilization; integrate green and low-carbon objectives in the basin-wide law, regulations, standards and guidelines; carry out the optimal utilization and vacating and replacing of the shoreline; reserve land on the shoreline to provide flexibility for future green development in compliance with spatial plans; explore the cultural and economic values of water; and promote shoreline renewal and public space construction.

Strengthening the protection and restoration of marine ecosystems: Harnessing the value of marine carbon sinks. Extensive measures should be taken to strictly enforce zoning management systems to avoid further destruction of marine habitats and coastal wetlands, and to restore degraded or damaged coastal wetlands and strictly protect critical marine habitats; invest in the creation of resilient, well-connected networks of marine protected areas covering national parks, nature reserves, and marine areas within the ecological redlines; align large marine protected areas and habitats of major importance with carbon storage. Referring to the Intergovernmental Panel on Climate Change (IPCC) guidelines to include oceanic carbon sinks in the national greenhouse gas inventory, scientifically assess blue carbon in marine and costal ecosystems under climate-smart integrated management for inclusion in China's updated NDC; anticipate the forthcoming global treaty on tackling plastic pollution, by taking early measures that encourage reduction, reuse, recycling and replacement, and support international cooperation; initiate pilot projects to tackle plastic pollution before the completion of the global plastics treaty.

5 Ecosystem and Biodiversity Conservation

5.1 Intensifying Integrated Ecosystem Management

Active efforts have been made to implement major projects to conserve biodiversity nationwide. In January 2022, the *Biodiversity Conservation and Utilization Plan in Hebei Province (2021–2030)* was released to promote major biodiversity protection projects and strengthen the building and management of the nature reserve system. In May 2022, the Yunnan Provincial Party Committee and Provincial People's Government issued the *Plan for Pioneering Ecological Civilization in Yunnan Province (2021–2025)*. In November 2022, the *Implementation Opinions of Further Strengthening Biodiversity Conservation in Shaanxi Province* was issued. In February 2022, the Department of Ecology and Environment of Jiangsu Province issued the *Plan for Biodiversity Observation Capacity Building (Phase I) in Jiangsu Province.*

1. Ecological redline system

MEE promulgated the *Measures for Ecological and Environmental Supervision of Ecological Redlines (Trial)* on December 27, 2022. In the meantime, national ecological and environmental supervision of ecological redlines will be carried out. At present, the delineation of ecological redlines has been basically completed. The area of ecological redlines in the terrestrial area accounts for about 30% or above, with 90% of key ecosystem types and 74% of wild animals and plants being protected.

Guided by the national policy, many regions have been pioneering. The *Provisions on the Management of Ecological Redlines in Hainan Province*, which came into effect at the end of May 2022, clarifies that the Hainan Provincial Department of Ecology and Environment, in conjunction with the relevant departments, will coordinate the building of a supervision platform for ecological redlines, and implement dynamic supervision of ecological redlines. Jiangsu Province has effectively put the ecological redline protection into practice. In May 2023, Shandong Province issued the *Measures for Ecological and Environmental Supervision of Ecological Redlines in Shandong Province (Trial)* to explore and innovate the supervision procedures.

2. Ecological compensation policy

On April 26, 2022, after deliberation and adoption by the Commission for Deepening Overall Reform of the CPC Central Committee, MEE, in conjunction with 10 relevant departments including the Supreme People's Court (SPC), the Supreme People's Procuratorate (SPP), MOST and the Ministry of Public Security (MPS), issued the *Provisions on the Administration of Compensation for Ecological and Environmental*

Damages (HFG [2022] No. 31). The *Provisions* clearly defines the division of tasks among departments and the responsibilities of local party committees and governments, makes clear and detailed provisions on key aspects, such as screening of case clues, case jurisdiction, claims initiation, investigation of damages, appraisal and assessment, claims negotiation, judicial confirmation, compensation litigation, and assessment of remediation effects, and proposes to develop sound safeguard mechanisms for development of appraisal and assessment organizations, appraisal and assessment technology methods, fund management, public participation and information disclosure, strengthen supervision and appraisal, and guides comprehensive and in-depth reform. On January 18, 2023, the Jiangsu Provincial Department of Ecology and Environment and 15 other departments issued the *Implementation Opinions on Implementing the Provisions on the Administration of Compensation for Ecological and Environmental Damages.*

5.2 Strengthening the Protection System of Mountains, Rivers, Forests, Farmlands, Lakes, Grasslands and Deserts

On June 7, 2022, MOF released the results of the competitive selection of second-batch projects for the Integrated Protection and Restoration Program of Mountains, Rivers, Forests, Farmlands, Lakes, Grasslands and Deserts (hereinafter referred to as the "China Restoration Program"), and identified 9 second-batch projects for the China Restoration Program.

On June 29, 2022, MOF, MNR and MEE held a meeting to promote the China Restoration Program. The meeting emphasized that the China Restoration Program is a concrete practice to implement the concept of "mountains, rivers, forests, farmlands, lakes, grasslands and deserts are part of the community of life".

During the 14th FYP period, on the basis of summarizing the experience of pilot projects, the central government has further advanced the China Restoration Program, and supported 19 provinces in systematic management, and the projects are being actively carried out in an orderly manner.

In December 2022, the United Nations announced at CBD COP 15 (Part 2) in Montreal, Canada that the China Restoration Program. The China Restoration Program is a landmark program practicing the concept of "mountains, rivers, forests, farmlands, lakes, grasslands and deserts are part of the community of life". Since the 13th FYP period, the China Restoration Program had deployed and implemented 44 projects in the important ecological barrier areas of "three zones and four belts", and completed ecological protection and restoration of more than 3.5 million ha, with the goal of restoring 10 million ha of natural ecology by 2030. The selection of the China Restoration Program as one of the "World Restoration Flagships" demonstrates that China is contributing solutions and wisdom to global biodiversity conservation.

5.3 *Further Exploring Ways to Realize the Value of Ecological Products*

In April 2021, the General Office of the CPC Central Committee and the General Office of the State Council issued the *Opinions on Establishing a Sound Mechanism for Realizing the Value of Ecological Products*. On June 7, 2022, the National Forestry and Grassland Administration (NFGA) and the National Bureau of Statistics (NBS) jointly issued a notice and decided to carry out pilot forest resource value accounting in five provinces.

Accounting for the value of ecological products is the key foundation for realizing the value of ecological products. In October 2022, NDRC and NBS commissioned the People's Publishing House to publish a single-volume version of the *Specification for Accounting for the Total Value of Ecological Products*, which specifies the indicator system, specific algorithms, data sources, and statistical calibers of the total value of ecological products. In line with the *Opinions on Establishing a Sound Mechanism for Realizing the Value of Ecological Products*, various localities have successively made practical explorations. In November 2022, Guangdong Province issued the *Implementation Plan for Establishing a Sound Mechanism for Realizing the Value of Ecological Products in Guangdong Province*, which mainly sets out objectives.

5.4 *Increasing Awareness of Wildlife Protection*

China is one of the countries boasting the most diversified wildlife species, with more than 7300 species of vertebrates. China has always been at the forefront of wildlife conservation in the world.

On December 30, 2022, the 38th session of the 13th NPC Standing Committee voted to adopt the amended *Wildlife Protection Law*, which came into effect on May 1, 2023. The newly amended *Wildlife Protection Law* strengthens the protection of wildlife habitats, making it clear that crucial wildlife habitats will be delineated into national parks, nature reserves and other nature protection areas for strict protection in accordance with the law. Terrestrial wildlife with important ecological, scientific and social values will be included in the scope of emergency rescue, the capacity of wildlife sheltering and rescue will be strengthened, and sheltering and rescue sites will be set up with appropriate professional and technical personnel, rescue tools, equipment and medicines.

5.5 Deepening the Building of a Management System for National Parks

In December 2022, NFGA, MOF, MNR, and MEE jointly issued the *Spatial Layout Program for National Parks*, which proposes that by 2035, China will basically build the world's largest national park system.

First, 49 national park candidate areas will be selected. In terms of concepts and objectives, the national park concepts of ecological protection priority, national representativeness and public welfare for all will be upheld, so as to protect the originality and integrity of natural ecosystems, safeguard national ecological security, and build a firm ecological foundation for building a beautiful China and modernization of harmony between humanity and nature.

Second, the total size of the protected area will rank first in the world. The *Spatial Layout Program for National Parks* is closely linked to the major projects for the protection and restoration of nationally important ecosystems centered on the "three zones and four belts", and covers the most critical areas of the ecological security barrier in the national territory. A cluster of national parks will be formed in the Qinghai-Tibetan Plateau, with a total area of about 770,000 km^2, to systematically and holistically protect the "Third Pole of the Earth"; and a number of national park candidate areas will be set up in the Yangtze River Basin and the Yellow River Basin, greatly contributing to protection of the Yangtze River, as well as ecological protection and high-quality development of the Yellow River Basin.

Third, the first batch of national parks will be built at a high standard and from a high starting point. The building of a national park system is a major institutional innovation in the reform of the ecological civilization system.

5.6 CCICED Policy Recommendations

CCICED proposed a number of valuable policy recommendations on ecosystem and biodiversity conservation in 2022. These are summarized below:

Reinforcing ecological conservation and restoration of river basins: Measure should be taken to strengthen the control of soil erosion and desertification in mountainous and hilly areas and implement the "returning space to rivers" campaign to restore the river and lake basins; systematically manage hydropower projects to ensure that they undergo science-informed, robust and participatory EIA prior to project development, safeguard hydrological integrity and ecological water demand, and reduce ecological impacts through measures such as ecological scheduling and installation of fish passage facilities; promote 'grain for green' in ecologically sensitive areas and strengthen ecological restoration; pay attention to the melting of glaciers at the source of the Yangtze River and strengthen monitoring and early warning; and improve the safety and security mechanisms for vulnerable groups—especially women in disaster-prone areas such as villages, small towns, and flood storage areas.

Strengthening the protection and restoration of marine ecosystems: Extensive measures should be taken to strictly enforce zoning management systems to avoid further destruction of marine habitats and coastal wetlands, and to restore degraded or damaged coastal wetlands and strictly protect critical marine habitats; invest in the creation of resilient, well-connected networks of marine protected areas covering national parks, nature reserves, and marine areas within the ecological redlines.

6 Regional and International Engagement

6.1 Leading the Global Biodiversity Conservation Process into a New Phase

On 15 November 2022, during the UNFCCC COP 27, the Chinese Presidency of CBD COP 15 and Canada, the host country of COP 15 Part 2, co-hosted the "Ministerial Event on Action for Biodiversity: the Road Leading to the Success of CBD COP 15", in Sharm el-Sheikh, Egypt. The event emphasized the importance of the *Post-2020 Global Biodiversity Framework* (GBF) in halting and reversing the loss of biodiversity and worked towards the Framework. Ministers of the Parties were invited to discuss the mission and resource mobilization of GBF, with a view to providing direction and building consensus on the key issues of GBF, and encouraging further political will and motivation to support and contribute to the achievement of a balanced, ambitious and pragmatic Framework at COP 15.

From December 7 to 19, 2022, COP 15 Part 2 was held in Montreal, where the CBD secretariat is based. China, as the Presidency of CBD COP 15, led the substantive and political affairs of the Conference, at which the Kunming-Montreal Global Biodiversity Framework (the "Framework") was successfully adopted, comprising four global long-term goals and 23 specific action targets to be achieved by 2030, depicting a new blueprint for global biodiversity governance up to 2030 and beyond. The agreement made it clear that developing countries must receive the support they need in terms of financial resources, technologies and capacity building. The Framework will guide global efforts on jointly halting and reversing biodiversity loss, promoting biodiversity recovery and realizing the vision of "living in harmony with nature" by 2050.

6.2 Actively Getting in Involved in International Response to Climate Change

As the world's largest developing country, China has implemented a series of strategies, measures and actions to address climate change, participated in global climate governance and achieved positive results in addressing climate change.

As the world's second-largest economy and the largest developing country, China has taken the initiative to shoulder its responsibilities as a major country in addressing climate change, participating in global climate governance and deeply engaging in global environmental governance. In this regard, China has established the China-EU High-Level Dialogue on Environment and Climate, actively carried out exchanges and dialogues such as the meetings of the environment ministers of the member states of the Shanghai Cooperation Organization (SCO) and the China-ASEAN Forum on Environmental Cooperation, strengthened South–South cooperation as well as cooperation with neighboring countries, and supported projects and actions in Africa, South–East Asia and South Asia in such areas as biodiversity conservation, green economy, chemical management, and implementation of international environmental conventions, which are now yielding good results.

From March 20 to 21, 2023, the Climate Ministerial Conference, co-convened by Denmark, the Egyptian Presidency of UNFCCC COP 27, and the UAE Presidency of COP 28, was held in Copenhagen. China expressed its readiness to fully support the UAE in successfully hosting COP 28, successfully completing the first Global Stocktake (GST) of the Paris Agreement, and promoting positive outcomes on key negotiation issues such as adaptation, finance, loss and damage, and mitigation. During the conference, China was invited to hold bilateral talks with COP 28 President-designate Sultan al-Jaber, UN Assistant Secretary-General Selwin Hart, UNFCCC Executive Secretary Simon Stiell, as well as ministerial representatives from Denmark, Germany, France, UK, Canada and Australia.

On July 4, 2023, Chinese Vice Premier Ding Xuexiang and Executive Vice President of the European Commission Frans Timmermans attended the 4th China-EU High-Level Dialogue on Environment and Climate in Beijing and reached a broad consensus. The two sides emphasized the need to make full use of the China-EU High-Level Dialogue on Environment and Climate, to hold regular high-level dialogue meetings, strengthen communication and coordination, and deepen cooperation in key areas.

From July 13 to 14, 2023, the 7th Ministerial Conference on Climate Action, co-organized by China, the EU and Canada and hosted by the EU, was held in Brussels, Belgium. China is ready to work with all parties to ensure the success of COP 28 and the building of a fair, equitable and win–win global climate governance system in accordance with the principles of openness, transparency, broad participation, driving by parties and consensus building.

6.3 Making Steady Progress in South–South Cooperation

On January 12, 2022, FAO officially launched the third phase of the FAO-China South–South Cooperation Program. On April 15, the China-WFP Digital Workshop on South–South Cooperation and the launching ceremony of the South–South Cooperation Knowledge Sharing Platform was successfully held in Beijing. China adheres to the pattern of international development cooperation, with North–South cooperation as the main channel and South–South cooperation as a complement, and encourages greater development assistance from developed countries to developing countries, in order to build a new, fair and balanced global development partnership, and create a favorable external environment for poverty reduction.

On June 24, 2022, the High-Level Dialogue on Global Development under the theme "Building A Global Partnership for Development in the New Era, and Joining Hands to Implement the 2030 Agenda for Sustainable Development", was held during the 14th BRICS Leaders' Meeting.

From August 27 to 28, 2022, the 6th Conference on South–South and Triangular Cooperation was held in New Delhi, India. Aiming to explore new models of development in the wake of global COVID-19 epidemic, international conflicts and climate crisis, and encourage countries to use globalization as an effective tool for cooperation on resources, knowledge and markets so as to trigger a shift in human lifestyles towards sustainable production and consumption.

In November 2022, three major regional and international meetings, including the ASEAN Summit, G20 Summit and APEC Summit, were held back-to-back in Phnom Penh, Bali and Bangkok, making Asia, especially East Asia, a focal spot of global economic governance. On the afternoon of November 10, then Premier Li Keqiang attended the 25th China-ASEAN (10 + 1) Leaders' Meeting. From November 14 to 19, President Xi Jinping visited Bali and Bangkok, and attended the G20 Leaders' Summit and the APEC Leaders' Informal Meeting in person.

6.4 Building Green and Low-Carbon "Belt and Road Initiative" In-Depth

To date, China has signed more than 200 cooperation documents on the co-building of the Belt and Road with 151 countries and 32 international organizations, including ESCAP and UNDP. As most of the BRI countries are still in the early stages of economic and social development, with heavy tasks of modernization, industrialization and urbanization, the trend of high carbon intensity will be maintained for periods of time, and the total amount of carbon emissions will continue to rise. Under the context of carbon neutrality, boosting green and low-carbon energy development is of outstanding strategic value and positive practical significance for the BRI countries to cope with the climate crisis and realize the UN 2030 SDGs.

On May 10, 2023, the Belt and Road Green Development Roundtable and the General Meeting of BRI International Green Development Coalition (BRIGC) were held in Beijing. China is ready to join hands with all parties to firmly support BRIGC in playing a more important role in the new historical stage, and making greater contributions to the high-quality building of the Belt and Road, Global Development Initiative, and green development and transition in the BRI countries.

6.5 Injecting New Elements into International Ocean Governance

The ocean is a strategic area for high-quality development. The *Outline of the 14th Five-Year Plan and Long-range Objectives Through the Year 2035* proposes the we will actively develop the blue partnership, deeply participate in the formulation and implementation of mechanisms and rules for international maritime governance, and promote the building of a just and equitable international maritime order and development of a marine community with a shared future.

In order to implement the UN 2030 Agenda for Sustainable Development, the 72nd and 75th sessions of the United Nations General Assembly adopted a resolution designating the period from 2021 to 2030 as the UN Decade of Ocean Science for Sustainable Development (hereinafter referred to as the "UN Ocean Decade"), and adopted a plan for its implementation, which was formally launched on 1 January, 2021.

In August 2022, with the approval of the State Council, MNR led and coordinated with relevant ministries and commissions to set up the China Committee for the UN Ocean Decade to plan, deploy and promote the relevant work. At present, China has successfully approved 1 UN Ocean Decade Collaborative Center, and 5 major science programs, including "Seamless Ocean and Climate Forecast System" and "Negative Global Ocean Emissions". The successful approval of the UN "Ocean Decade" Collaborating Center and the major science programs have laid a solid foundation and created new opportunities for China's participation in international cooperation on oceans under the UN framework.

Marine plastic debris is a widely concerned marine environmental problem. The global "Beat Plastic Pollution" campaign on World Environment Day 2023 called for global solutions to combat plastic pollution. China has taken multiple measures to promote the cleanup and control of marine plastic debris. At the regional level, under the framework of mechanisms such as China-Japan-ROK Environment Ministers' Meeting, China-Japan-ROK Leaders' Meeting, ASEAN-China Leaders' Meeting, G20 Summit and APEC, China has been actively cooperating with its neighboring countries in order to enhance the regional capacity to deal with marine plastic debris through joint scientific research, scientific and technological research and development, technical assistance and academic meetings. In addition, China has cooperated deeply with UNEP and other international organizations to create demonstration projects, share governance experience and promote practical experience.

6.6 CCICED Policy Recommendations

Policy recommendations from the CCICED 2022 AGM are as follows:

Strengthening international climate and biodiversity dialogues and exchanges and contributing to global environmental governance: It is advised to continuously promote bilateral and multilateral climate and biodiversity dialogues. In the COP 15 process, China should prepare for the implementation of GBF, including updating the National Biodiversity Strategy and Action Plan (NBSAP). Building on positive dialogue mechanisms such as EU–China High-Level Climate and Environment Dialogue, and the Ministerial Meeting on Climate Action, China and relevant parties should actively carry out Track 2 and Track 1.5 dialogues to control CO_2 and non-CO_2 greenhouse gases. Building on the China-EU, G20, UNEP and other initiatives, continued efforts should be made to identify the next steps in green financial mechanisms, including scaling-up of NbS.

Maintaining the momentum of linking nature and climate actions and promoting synergies: It is recommended that discussions on synergistic climate change governance be strengthened at COP 15 Part 2 and that further progress be made on synergies between biodiversity and climate change at the UNFCCC COP 27 to promote the integration of global climate and biodiversity governance into the Global Development Initiative.

Deepening international cooperation to support green and low-carbon development in BRI countries and exploring new paths of green and low-carbon cooperation under the reform of the global governance system: Relying on multilateral cooperation platforms such as the BRIGC and the Green Investment Principles (GIP) for the "Belt and Road Initiative", further efforts should be made to strengthen dialogues and exchanges among stakeholders, promote the establishment of green project development platforms under the framework of South–South cooperation, and deeply align with the green and low-carbon development needs of BRI countries. In conjunction with the Belt and Road South–South Cooperation Initiative on Climate Change and the Green Silk Road Envoys Program, China should help enhance the local capacity of BRI countries to address climate change and achieve an inclusive and resilient recovery; strengthen the South–North–South cooperation platform for low-carbon finance.

With regard to marine plastic pollution, it is proposed to anticipate the forthcoming global treaty on tackling plastic pollution, by taking early measures that encourage reduction, reuse, recycling and replacement, and strengthen international cooperation; and initiate pilot projects to tackle plastic pollution before the completion of the global plastics treaty.

6.7 Conclusions

CCICED Phase VII has paid more attention to major strategic issues related to China's long-term development, major issues affecting domestic and global sustainable development, and bringing into full play its role as a platform for two-way exchanges and sharing with the international community.

In Phase VII, the scope of CCICED's Chinese and international members has been expanded, including policymakers of the Chinese central government, heads of large enterprises, experts and scholars from renowned think tanks and universities, as well as representatives of multilateral financial institutions, international NGOs, major international organizations and professional institutions. It allows CCICED to hear different voices of the international community on environment and development issues, and facilitates the full exchange and sharing of experiences between China and the world on major environment and development issues.

Standing at a new historical starting point and with a more mature operational mechanism, CCICED Phase VII has given full play to its role as a high-end think tank in the field of environment and development. The systematic, strategic and forward-looking policy recommendations put forward over the past year reflect, to a certain extent, the superior capability of CCICED's Chinese and international members in Phase VII to anticipate domestic and international situations, as well as their insights into the international environment and development law.

Over the past year, the forward-looking proposals put forward by CCICED in the areas of the green and low-carbon transition, energy security, pollution control, green technology innovation and integrated ecosystem management, have been highly valued by the Chinese government, providing an important reference for future work on ecological civilization. China cannot achieve high-quality development and promote ecological progress without international cooperation. China has actively advanced South–South cooperation, responded to UN SDGs, and made new progress in issues of global concern, such as biodiversity conservation, addressing climate change, and marine ecological and environmental protection. China will continue to cooperate with the international community in a more open manner and contribute to the world's green prosperity.

Looking ahead, in terms of policy studies, CCICED, as an international high-end think tank, will keep abreast of the overall development trend of environment and development at home and abroad, attach importance to the "innovation" and "foresight" of research results, and put forward innovative and pioneering policy recommendations on issues of great concern at home and abroad, such as climate governance, low-carbon development, energy revolution, and just transition.

Chapter 14
Report on Gender Mainstreaming in SPS Research for the Period 2022–2023

1 Introduction

Since 2018, the CCICED Executive Committee has committed to mainstreaming gender perspectives in all aspects and processes of research. To support the integration of gender considerations as part of the Special Policy Studies (SPS) for the period 2022–2023, gender mainstreaming guidance was provided to each SPS at the beginning of the research process and continued at various touchpoints during the drafting process.

This report provides a detailed overview of the integration of gender perspectives as part of CCICED SPS final research papers. The report also identifies best practices and provides recommendations for further strengthening the mainstreaming of gender perspectives in the forthcoming phase of the research.

2 Gender Equality and the International Framework

The integration of gender issues and the participation of women in environmental policy research and development is a critical component of environmental, sustainable development, and climate change governance. The international community cannot afford to ignore gender equality if progress on the United Nations Sustainable Development Goals (SDGs) is to take place.

We know that advancements in gender equality can have profound positive impacts on social and environmental well-being. We know that working toward gender equality enhances environmental outcomes. We also know that when not managed

Drafted by the WPS Group, Kristine St Pierre and Jennifer Savidge and revised and edited by CCICED Secretariat and Secretariat International Support Office.

Green Empowerment and High Quality Development,
https://doi.org/10.1007/978-981-96-4218-2_14

properly, environmental measures can reinforce inequalities and lead to greater environmental degradation. However, while the rewards for addressing gender equality are high, how to do so remains a significant challenge.

Of even greater urgency is the recognition of the risks associated with treating gender equality and climate change mitigation goals and policies in isolation.[1] The ways in which societies respond to climate change will not only have environmental consequences, but social and economic consequences as well, directly impacting people's access to opportunities, resources, and living standards. It will be important to consider the potential consequences of climate change on gender relations and other forms of social equity.

3 CCICED's 2022–2023 Gender-Related Work Through Special Policy Studies: Key Observations

This section presents key observations on the importance of gender equality within each SPS research area and identifies opportunities to provide recommendations to further advance gender equality within environmental and climate change policy development and governance.

High-Quality Development of River Basins and Adaptation to Climate Change

The significance of gender equality and gender mainstreaming within sustainable water management is well-established globally. The integration of gender perspectives within river basin management in the context of climate change adaptation and mitigation is important, given that climate change impacts women and men differently. Ideally, women and men from a range of backgrounds can participate in and lead governance processes, representing diverse perspectives, including as they relate to climate change. This is particularly important given that women and men are impacted differently by climate change and often have different capacities to adapt to and mitigate climate change risks due to underlying gender discrimination; thus the perspectives and participation of women are ever more important.

The SPS on *High-Quality Development of River Basins and Adaptation to Climate Change* effectively integrates gender considerations through a focused section on gender equality and social inclusion in watershed governance, as well as a gender-focused policy recommendation. The focused section uses three case areas in the upper and middle reaches of the Yangtze River to illustrate the gaps in gender equality and social inclusion, including a look at gender dimensions of watershed management-related industries. The research finds that:

1. Women's participation in decision making of governance in the basin is still insufficient.

[1] IDRC, *Women's economic empowerment—the missing piece in low-carbon plans and actions*, 2022.

2. There is a relatively serious gender bias in the workplace in watershed management-related industries.
3. Most watershed management policies fail to effectively integrate a gender perspective.

The SPS provides entry points for addressing these gaps by (1) promoting women's participation and leadership in watershed governance through watershed collaboration, (2) implementing gender-specific statistics in multiple statistical indicators on watersheds and climate change, and (3) enhancing women's capacity to participate in watershed environmental protection and management through training and education. The SPS also points to the lack of gender-disaggregated statistics for disaster loss hazards and other climate change impacts as an important factor that contributes to the invisibility of women and their needs and priorities. It calls on government and educational institutions to work together to advocate for gender-disaggregated statistics on disaster loss hazards to provide an evidence base for informing the development of gender-responsive policies, programs, and budgets.

The SPS included one gender-focused recommendation on the importance of paying close attention to gender and social equity issues in watershed governance, including increasing the participation of women and other marginalized groups to ensure that their unique perspectives, needs, and capacities are considered and incorporated, contributing to sustainable development outcomes.

Collaborative Mechanism for Carbon Reduction, Pollution Reduction, Green Expansion and Growth

The *Collaborative Mechanism for Carbon Reduction, Pollution Reduction, Green Expansion and Growth* SPS includes a highly informative gender analysis section with a discussion of gender-related issues in the coordinated management of carbon reduction, pollution reduction, green expansion and growth, as well as key recommendations.

The issue of gender equality is often overlooked in the transition to a green economy. The SPS recognizes that the impacts of climate change and air pollution are not uniformly distributed worldwide and that certain social groups. The SPS is also clear about the risks of adopting a "gender-blind" approach to economic transformation on women's employment, participation, work environment, and opportunities for education and training. As such, the SPS sees introducing gender issues and actively practicing gender mainstreaming in the process of promoting green growth as key strategies for improving gender equality, as well as important foundations for promoting sustainable development.

The gender analysis included as part of the SPS discusses gender equality in the green and low-carbon transition in the power and the transportation sectors, providing a set of targeted recommendations for advancing gender equality in both sectors. This would involve policies that encourage more women to participate in the renewable energy industry, such as in solar and wind energy sectors. It would also require strengthening policy research and development from a gender perspective, including more gender-specific data to better understand the different impacts of the low-carbon transition on men and women and how to formulate more equitable policies. In the transportation sector, such policies would include those for enhancing women's skills and knowledge through education and vocational training, enabling them to find work in the emerging green transportation industry; enhancing women's representation and influence in transportation planning and decision making through public participation and social dialogue; and strengthening their leadership and influence in the green low-carbon transition by providing leadership training, promoting professional networks and mentoring systems, and increasing the proportion of women in critical decision-making positions.

The SPS makes it clear that gender mainstreaming is indispensable as China pushes forward to reduce carbon, decrease pollution, expand green development, and stimulate growth. As such, the SPS identifies three specific recommendations to promote gender equality in the green transition. They are:

(1) The development of a medium- to long-term research plan to track and thoroughly investigate the development of gender mainstreaming in China's green transition process, providing a scientific basis for policy-making and improvement.
(2) Address common gender issues at the same time as considering unique industry challenges.
(3) Learn from international practices on the integration of gender perspectives and promote international communications.

Promoting Digitalization and Green Technologies for Sustainable Development

To achieve gender equality in digitalization and unlock its potential to enable women's empowerment, it is important to understand the gendered dimensions and differentiated impacts of digitalization on women and men, and to identify and address gender-based barriers to ensure that women and men can equitably benefit from digitalization processes. This means ensuring women's access to digital tools and employment opportunities in information and communications technology and in the science, technology, engineering, and mathematics (STEM) fields, considering the structural and transformational changes that are needed as part of digital governance and within the digital ecosystem to ensure they are gender responsive and reduce the gender digital divide, and ensuring that the perspectives of a diversity of women and men are considered within digitalization processes.

The *Promoting Digitalization and Green Technologies for Sustainable Development* SPS effectively integrates relevant gender considerations throughout the research. One text box on artificial intelligence and biases. Another text box highlights, in practical terms, the gender dimensions of the avoid-shift-improve framework for gender-inclusive smart cities, and the need to consider gender and inclusion within urban design and decision making. Finally, the SPS includes a gender-focused section that explores the gender perspective within digitalization, examining the representation of women in STEM fields within China and globally, and considerations for ensuring their equitable employment, participation, and leadership within transition processes prompted by artificial intelligence and evolutions in digital infrastructure. It also incorporates an intersectional perspective in recognition that discrimination can compound based on the intersection of different identity factors, such as gender identity, disability, and socio-economic status, increasing the marginalization of women and other demographic groups and excluding them from the benefits of digitalization. It also recognizes the harmful consequences of excluding women from participating in the design and implementation of climate change policies and interventions on digitalization processes, and the importance of increasing their representation within STEM fields so that their full talents can be harnessed to find solutions within green transitions of the digital sector.

Such case studies could demonstrate how a gender analysis of the green transition of a digitalization process can be applied to identify and address gender-based barriers and opportunities in digitalization to enable equitable outcomes for women and men while simultaneously furthering sustainable development.

Climate Change

Gender equality and climate change intersect in important ways that have been recognized by the global community, within the SDGs and the UNFCCC and its associated Gender Action Plan. Traditional gender roles of women and men within society and as economic actors result in gender-differentiated impacts of climate change and a failure to include women's perspectives, participation, and leadership in efforts to address it. Entrenched gender stereotypes and discrimination often result in women having less access to and control of resources to enable them to mitigate and adapt to climate change, little input into decision making on climate action, and limited access to their rights.

International frameworks seek to promote women's participation and leadership across the spectrum of climate action and within all aspects of the transition to a green economy and clean energy investments.

The *Pathway to Carbon Neutrality and China's Role in Global Climate Governance* SPS effectively integrates gender into the research topic by including a dedicated chapter, *Analysis of Gender Mainstreaming*. This chapter is introduced with a brief examination of the reasons for climate change's disproportionate impact on women, including the reliance women have on natural resources, and the potential to maximize synergies in action on gender equality and climate change to achieve mutually reinforcing outcomes. The chapter also looks at how women can be better supported to adapt to climate change, given that adaptation measures currently tend not to incorporate gender considerations. Importantly, the SPS also considers how China can strengthen social accountability alongside environmental accountability—and promote gender equality specifically—with its overseas investments. Finally, the paper recommends that China strengthen international cooperation and promote knowledge sharing on gender equality amongst stakeholders at all levels.

Future research could focus on how gender considerations can be integrated into low-carbon development policies and could include case studies and practical examples from other countries to demonstrate good gender practices that have resulted in effective and equitable outcomes, with related recommendations for China.

Global Ocean Governance

As acknowledged within the SDGs, promoting gender equality is imperative to effectively protect and manage ocean and marine resources. Given the different gender roles of women and men, which tend to disadvantage women and restrict access to their rights, women and men are impacted differently by environmental issues, such as marine plastic pollution and climate change and have different needs to consider and capacities to contribute to solutions. Women are marginalized from fisheries governance despite their high concentration in onshore fish harvesting and processing activities, which results in their unique perspectives not being considered in decision making.

The *Sustainable Blue Economy Towards Carbon Neutrality* SPS effectively explores the role of women in small-scale fisheries in a gender-focused section. The research brings to light the disproportionate impact of climate change on women, specifically from climate disasters on women engaged in small-scale fisheries, and the need to implement targeted support measures tailored to the needs of different demographic groups and a gender-inclusive approach to climate change research and governance to increase the resilience of small-scale fishers. This section provides a gender analysis of the different roles of women and men within small-scale fisheries, where men are concentrated in fishing activities and women within pre-harvest and post-harvest activities.

Opportunities for future research include exploring the importance of incorporating a gender perspective into the marine plastics life cycle. Highlighting the greater vulnerability of women and other marginalized groups to the negative impacts of marine plastic pollution, while also emphasizing the potential for engaging women and other vulnerable groups as agents of change in implementing solutions to address the issue. This approach can better facilitate the resolution of marine plastic pollution issues and contribute to the advancement of the blue economy.

Green Finance

There is growing global recognition of the intersections between green finance and gender equality, with many sources establishing the mutually beneficial outcomes can be achieved from integrating gender equality within green finance. Gender-responsive approaches to green finance can be undertaken by public, private, and not-for-profit financial institutions in different types of financial vehicles.

The SPS on *Innovation Mechanisms for Sustainable Investments in Climate and Environment* incorporates gender considerations at key points throughout the research. Within the sustainable investing discussion, gender equality is reflected in the priorities of institutional investors (within diversity and inclusion) and is recognized as a critical social issue as part of the social component of responsible investing. It also acknowledges the concerns of NZAOA members around the social ramifications of organizational transitions, such as gender-differentiated impacts, and the importance of ensuring a fair transition. The research also referenced the importance to those exercising active ownership of ensuring that women are represented in sustainable investing and that the differentiated impacts on women are considered. This highlighted consideration of gender equality within investment decision making by using the example of the CPPIB's intent to vote against boards that do not consider gender diversity sufficiently, and the CPPIB's explicit commitment to gender-balanced senior management and diverse boards. Notably, within the recommendations, it is asserted that regulators should promote inclusive social development, with gender equality as a key component, alongside climate and environmental issues.

Future research might examine how the integration of gender equality considerations can support the scale-up of green finance, if effectively integrated, or hinder its scale-up if it is absent, and further explore the climate/environment-gender equality nexus within finance. Case studies of existing innovative finance instruments that simultaneously incorporate these priorities could be carried out, contributing to the evidence base for the efficacy of such instruments.

Green and Low-Carbon BRI

The numerous intersections between gender equality, renewable energy, and sustainable development are well-recognized globally and reflected within the SDGs and the UNFCCC. Entrenched gender roles within society, reinforced by stereotypes and discrimination, result in women's and men's differential energy access and gendered energy-related environmental and climate change impacts—differences often overlooked due to women's underrepresentation within the energy sector. As a result, women's unique needs and perspectives are often not considered within energy policy and financing, and risk being neglected in transitions to green and low-carbon energy, which can exacerbate gender inequalities and waste opportunities for women to act as agents of environmental change.

While the *Innovation Mechanism of BRI to Promote the Process of Sustainable Development* SPS references the social dimensions of green and low-carbon BRI, of which gender equality can play a part, it does not specifically integrate gender considerations. However, there are numerous opportunities for integrating gender into future iterations of the research on green and low-carbon BRI such as:

- Exploring practical measures to promote energy equity between women and men and among other marginalized groups within transitions to green and low-carbon energy.
- Examining how renewable energy financing can incorporate gender considerations to ensure that gender equality outcomes are promoted alongside climate change and environmental outcomes, in alignment with the Climate Change and Gender Outcomes resulting from COP 26.
- Including recommendations for key gender considerations within policy toolkits as gender good practices within the design and implementation phase of green BRI mechanisms and investments.

Future research could provide practical case studies and examples of how gender good practices institutionalized within development finance institutions.

Secure and Green Trade

At the time of writing the report on *Trade and Sustainable Supply Chains SPS*, the draft report did not include gender perspectives. However, gender has an enormous influence on sustainable consumption, largely due to the differing consumption patterns of women and men. While the green economy or low-carbon economy can bring new development opportunities to women. Their overall income level is lower than that of men, leaving them with comparatively less flexibility in adapting to either economic shifts or climate change. Because of China's global role in industrial production, as well as (increasingly) in consumption, and the disproportionate role of women as both producers and consumers, it is vital for policy-making to integrate the gender dimensions of sustainable production and consumption. Environmental concerns vary according to different manufacturing sectors, as do the roles women and men play in this area.

The literature on green industry likewise argues that promoting gender equality and integrating gender equality and women's empowerment principles in policy-making promotes more efficient and more profitable green industrial development.

Important research has been done to highlight the many gender dimensions of trade and global supply chains. As such, there remain important opportunities as part of this SPS to pay attention to gender equality and women's empowerment, to ensure that we dismantle gender stereotypes and other barriers so that women benefit equally from the opportunities offered by sustainable trade systems. It will also be important to ensure opportunities for the increased participation of women and gender experts in the development of policies and in decision-making processes. Conducting this research and looking at the gendered dimensions of supply chains will, in turn, enable us to build a more gender-responsive sustainable trade system that empowers women and men equally.

4 Gender Mainstreaming in SPS Research for the Period 2022–2023: Good Practices in the Integration of Gender Perspectives

The following good practices should continue to be incorporated into future SPS research and development:

(1) Conducting gender analysis designed to support the development of the research.
(2) Including gender in different areas of the SPS.
(3) Including case studies integrating gender considerations.

5 Recommendations to CCICED for the Forthcoming Research Phase

CCICED will continue to strengthen capacity for integrating gender perspectives within the framework of gender mainstreaming. The following recommendations are made to further strengthen the integration of gender within CCICED SPS:

- The requirement to include gender from the beginning of the research cycle should be considered.
- The inclusion of research and analysis on the impacts of environmental degradation and climate change on different social groups.
- The development of a gender-specific annotated bibliography to support and guide SPS teams in their research and analysis.

Annex: CICCED Phase VII Composition (As of December 2023)

Chinese Members

1	Mr. DING Xuexiang	**Chairperson of CCICED**
2	Mr. HUANG Runqiu	**Executive Vice Chairperson of CCICED** Minister, Ministry of Ecology and Environment
3	Mr. XIE Zhenhua	**Vice Chairperson of CCICED** China's Special Envoy on Climate Change
4	Mr. ZHOU Shengxian	**Vice Chairperson of CCICED** Former Minister of Environmental Protection
5	Mr. ZHAO Yingmin	**Secretary General of CCICED** Vice Minister, Ministry of Ecology and Environment
6	Mr. LIU Shijin	**Chinese Chief Advisor of CCICED** Former Vice President of the Development Research Center of The State Council
7	Mr. LIAO Min	Deputy Director of Central Financial and Economic Affairs Commission General Office; Vice Minister, Ministry of Finance
8	Mr. DOU Shuhua	Vice-chairperson, The Environmental Protection and Resources Conservation Committee
9	Mr. XIAO Yanshun	Member of the Leading Party Members Group, State Council Research Office
10	Mr. MA Zhaoxu	Vice Minister, Ministry of Foreign Affairs
11	Mr. HU Zucai	Vice Chairman, National Development and Reform Commission
12	Mr. XIN Guobin	Vice Minister, Ministry of Industry and Information Technology
13	Mr. ZHU Zhongming	Vice Minister, Ministry of Finance
14	Mr. WANG Hong	Vice Minister, Ministry of Natural Resources; Director of State Oceanic Administration

(continued)

Green Empowerment and High Quality Development,
https://doi.org/10.1007/978-981-96-4218-2

(continued)

15	Mr. DAI Dongchang	Vice Minister, Ministry of Transport
16	Mr. ZHANG Taolin	Vice Minister, Ministry of Agriculture and Rural Affairs
17	Mr. WANG Shouwen	China International Trade Representative (minister's level), and Vice Minister of Commerce, Ministry of Commerce
18	Mr. QIU Baoxing	Counsellor, the State Council
19	Mr. ZHANG Yaping	Vice President and Academician, the Chinese Academy of Sciences
20	Mr. CAI Fang	Former Vice President, Chinese Academy of Social Sciences
21	Mr. ZHANG Yuyan	Director, Researcher of the Institute of World Economics and Politics of Chinese Academy of Social Science (IWEP, CASS)
22	Mr. YU Yong	Deputy Administrator, China Meteorological Administration
23	Mr. DENG Xiuxin	Vice President and Academician, the Chinese Academy of Engineering
24	Ms. LUO Hui	Director General of Department of International Affairs (Hong Kong, Macao and Taiwan Exchange Office)
25	Mr. XUE Lan	Dean of Schwarzman College in Tsinghua University; Co-Chair of the Leadership Council of the UN Sustainable Development Solution Network (UNSDSN); Professor at School of Public Policy and Management at Tsinghua University
26	Mr. HE Kebin	Member of Chinese Academy of Engineering; Professor of the School of Environment and Dean of the Institute for Carbon Neutrality at Tsinghua University
27	Mr. ZHANG Yuanhang	Professor, College of Environment Sciences and Engineering, Pecking University; Academician, the Chinese Academy of Engineering
28	Mr. DAI Minhan	Chair Professor, Xiamen University; Academician of Chinese Academy of Sciences
29	Mr. FANG Jingyun	Professor, College of Urban and Environmental Sciences, Pecking University; Academician of Chinese Academy of Sciences
30	Mr. WANG Jinnan	President of Chinese Academy of Environmental Planning; Academician of Chinese Academy of Engineering
31	Mr. ZHANG Xiaoye	Research Fellow of Chinese Academy of Meteorological Sciences; Academician of Chinese Academy of Engineering
32	Mr. WANG Yi	Vice Chair, National Expert Committee on Climate Change; Professor of School of Public Administration, University of Chinese Academy of Sciences; Member, Standing Committee of the National People's Congress of China
33	Mr. SHU Yinbiao	Academician of Chinese Academy of Engineering, President of Chinese Society for Electrical Engineering; The 36th President of International Electrotechnical Commission
34	Mr. QIAN Zhimin	Chairman of the Board, State Power Investment Corporation Limited

(continued)

(continued)

35	Mr. HUANG Haiqing	Executive Director and the Chief Executive Officer, China Everbright International Limited
36	Mr. WANG Tianyi	Professor, Hong Kong University of Science and Technology (Guangzhou); International Consultant, Temasek
37	Ms. Marjorie YANG	Chairman, Esquel Group
38	Mr. XIN Bao'an	Executive Chairman, State Grid Corporation of China
39	Mr. LEI Mingshan	Executive Chairman, China Three Gorges Corporation

International Members

1	Mr. Steven Guilbeault	**Executive Vice Chairperson** Minister, Environment and Climate Change Canada
2	Mr. Achim Steiner	**Vice Chairperson** Administrator, The United Nations Development Programme
3	Ms. Inger Andersen	**Vice Chairperson** Executive Director, The United Nations Environment Programme
4	Ms. Kristin Halvorsen	**Vice Chairperson** Director, CICERO Center for International Climate Research; Former Minister of Finance of Norway; Former Deputy Prime Minister of Norway
5	Mr. Scott Vaughan	**International Chief Advisor of CCICED** Former President and CEO, International Institute for Sustainable Development
6	Mr. Danny Alexander	Vice President, Policy and Strategy, The Asian Infrastructure Investment Bank
7	Mr. Peter Bakker	President and CEO, World Business Council for Sustainable Development
8	Mr. Manish Bapna	President and Chief Executive Officer, the Natural Resources Defense Council
9	Mr. Børge Brende	President, World Economic Forum
10	Mr. Francesco La Camera	Director-General, the International Renewable Energy Agency
11	Mr. Tomas Anker Christensen	Climate Ambassador of Denmark
12	Mr. Srun Darith	Secretary of State, Ministry of Environment, Cambodia
13	Mr. Aniruddha (Ani) Dasgupta	President and CEO of World Resources Institute
14	Mr. John J. DeGioia	President, Georgetown University

(continued)

(continued)

15	Jos DELBEKE	The first EIB Chair on Climate Policy and International Carbon Markets
16	Mr. Jan Hendrik Dronkers	Secretary-General, the Ministry of Infrastructure and Water Management, the Netherlands
17	Ms. Kate Hampton	CEO, Children's Investment Fund Foundation
18	Mr. Arthur Hanson	Senior Advisor and Former President of International Institute for Sustainable Development
19	Mr. Hal Harvey	President, Climate Imperative Foundation; CEO, Energy Innovation
20	Mr. Stephen Heintz	President and CEO, the Rockefeller Brothers Fund
21	Ms. Naoko Ishii	Professor, Executive Vice President and Director of the Center for Global Commons, University of Tokyo; Former CEO and Chair, Global Environment Facility
22	Mr. Rodolfo Lacy	Former Director for Climate Action and Environment for Latin America, Special Envoy on Climate Matters to the United Nations, Organization for Economic Co-operation and Development
23	Mr. Marco Lambertini	International Special Envoy, World Wide Fund for Nature
24	Mr. Stanley Loh	Permanent Secretary, Ministry of Sustainability and the Environment, Singapore
25	Mr. Michael McElroy	Gilbert Butler Professor of Environmental Studies, Harvard University
26	Mr. Dirk Messner	President, German Federal Environment Agency
27	Mr. Hideki Minamikawa	President, Japan Environmental Sanitation Center
28	Ms. Jennifer Morris	Chief Executive Officer, The Nature Conservancy
29	Mr. Gerd Müller	Director General of the United Nations Industrial Development Organization
30	Mr. Bruno Oberle	Director General, the International Union for Conservation of Nature
31	Mr. Jonathan Pershing	Program Director of Environment, William and Flora Hewlett Foundation
32	Mr. Frank Rijsberman	Director General, Global Green Growth Institute
33	Mr. Carlos Manuel Rodriguez	CEO and Chairperson of the Global Environment Facility; Former Environment and Energy Minister, Costa Rica
34	Ms. Gwen Ruta	Executive Vice President, Environmental Defense Fund
35	Mr. Ahmed M. Saeed	Vice President (Operations 2), Asian Development Bank
36	Ms. Kirsten Schuijt	Director General, World Wide Fund for Nature

(continued)

(continued)

37	Mr. Erik Solheim	Senior Advisor, World Resources Institute
38	Mr. Andrew Steer	President and CEO, Bezos Earth Fund
39	Ms. Eva Svedling	Former State Secretary, Ministry of Environment, Sweden
40	Mr. Sukanto Tanoto	Founder and Chairman, Royal Golden Eagle
41	Mr. James Thornton	Founding CEO, ClientEarth
42	Ms. Nomfundo Tshabalala	Director-General of the Department of Forestry, Fisheries, and the Environment, Republic of South Africa
43	Ms. Laurence Tubiana	CEO, European Climate Foundation
44	Ms. Jo Tyndall	Director, Environment Directorate, Organization for Economic Co-operation and Development
45	Ms. Christie Ulman	President, Sequoia Climate Foundation
46	Mr. Kurt VANDENBERGHE	Director-General, Directorate-General Climate Action (DG CLIMA)
47	Mr. Juergen Voegele	Vice President for Sustainable Development, the World Bank
48	Mr. Jan-Gunnar Winther	Specialist Director, Norwegian Polar Institute
49	Mr. Zhang Hongjun	Board Chair, Energy Foundation China; Partner, Holland and Knight LLP

CCICED Phase VII Special Advisors

Chinese Special Advisors		
1	Mr. ZHANG Yong	Director-General, Bureau of General Affairs, the Office of the Central Financial and Economic Affairs Committee
2	Ms. CHEN Wenling	Chief Economist, China Center for International Economic Exchanges
3	Mr. ZHANG Yansheng	Chief Researcher of China Center for International Economic Exchanges
4	Mr. LI Haisheng	President and Party Secretary of China Research Academy of Environmental Sciences
5	Mr. GUO Jing	President of BRI Green Development Institute (BRIGDI)
6	Mr. ZHOU Heng	Former Senior Inspectorate Advisor and Director General, Department of International Cooperation, China Meteorological Administration
7	Mr. YE Yanfei	Senior Inspectorate Advisor, Policy Research Bureau of the China Banking and Insurance Regulatory Commission

(continued)

(continued)

8	Mr. HU Baolin	Honorary Dean of Research Institute of China Green Development of Tianjin University
9	Ms. ZHANG Chenghui	Member of Committee of Academics, TJD Research Institute
10	Mr. ZHAI Panmao	Co-chair, Intergovernmental Panel on Climate Change (IPCC) Working Group I; Chief Scientist and Principle Investigator of Chinese Academy of Meteorological Sciences
11	Mr. ZHANG Yongsheng	Director-General and Research Fellow, Research Institute for Eco-civilization, Chinese Academy of Social Sciences
12	Mr. MA Jun	Chairman, Green Finance Committee of China Society for Finance and Banking
13	Mr. LI Xiaojiang	Professor and former President, China Academy of Urban Planning and Design
14	Mr. YU Ping	Former President of China Council for the Promotion of International Trade
15	Mr. ZHAI Qi	Executive Secretary General, China Business Council for Sustainable Development
16	Mr. TANG Jie	Professor of Harbin Institute of Technology (Shenzhen), Director of Development Strategy Committee, Director of the Chinese University of Hong Kong (Shenzhen), and former vice mayor of Shenzhen
17	Mr. XU Lin	Chairman of China-US Green Fund, Committee Member of Demonstration Area in the Yangtze River Delta on Ecologically Friendly Development, Former Director General of the Department of Fiscal and Financial Affairs and the Department of Development Planning, NDRC
18	Mr. LI Zhenguo	Founder and President of LONGi
19	Mr. WANG Yusuo	Founder and Chairman of the Board of ENN Group
International Special Advisors		
1	Mr. Iskandar Abdullaev	Deputy Director, Central Asia Regional Economic Cooperation Institute; Former Executive Director, The Regional Environmental Center for Central Asia
2	Mr. Knut Halvor Alfsen	Former Head Research Director, Center for International Climate and Environmental Research Oslo
3	Mr. Howard Bamsey	Honorary Professor, School of Regulation and Global Governance of Australian National University; Former Chair, Global Water Partnership; Former Executive Director of GCF

(continued)

(continued)

4	Mr. Dimitri de Boer	Regional Director of Programmes for Asia and Chief Representative of China, ClientEarth
5	Mr. Guillermo Castilleja	Senior Advisor, Global Alliance for the Future of Food
6	Ms. Galit Cohen	Director, Program on Climate Change and National Security, Israel Institute for National Security Studies (INSS)
7	Mr. Stephan Contius	Special Advisor on SDGs at the Foundations Platform F20, former Commissioner for the 2030 Agenda for Sustainable Development and Director at the Federal Environment Ministry, Germany
8	Mr. Kevin P. Gallagher	Director, Global Development Policy Center, Boston University
9	Ms. Shenyu G. Belsky	China Program Director/China Chief Representative Rockefeller Brothers Fund Beijing, China
10	Mr. Mark Halle	Former European Representative and Director for Trade and Investment, International Institute for Sustainable Development
11	Ms. Jeanne-Marie Huddleston	Director General of Bilateral Affairs and Trade, the International Affairs Branch of Environment and Climate Change Canada
12	Ms. Bernice Lee	Research Director, Futures, Chatham House-Royal Institute of International Affairs
13	Mr. LEI Hongpeng	Global Director, Climate, Children's Investment Fund Foundation
14	Mr. LIU Jian	Director of the Science Division, United Nations Environment Programme
15	Mr. LO Sze Ping	Program Director, China and Southeast Asia, Sequoia Climate Foundation
16	Mr. Zafar Makhmudov	Executive Director, the Regional Environmental Centre for Central Asia
17	Mr. Hans Mommaas	Director-General, PBL Netherlands Environmental Assessment Agency
18	Ms. Neo Gim Huay	Managing Director, Centre for Nature and Climate, the World Economic Forum
19	Ms. Oyun Sanjaasuren	Director, External Affairs, Green Climate Fund
20	Mr. Ismo Tiainen	Director-general, Administration and International Affair, Ministry of the Environment, the Republic of Finland

(continued)

(continued)

21	Ms. Verónica Tomei	Head of the Division for Unit I I 4 United Nations, 2030-Agenda, Developing and Emerging Countries; Environmental Aspects of International Climate and Energy Policy of the German Federal Ministry for the Environment, Nature Conservation, Nuclear Safety and Consumer Protection (BMUV)
22	Mr. ZOU Ji	CEO and President of Energy Foundation China